Ecological Restoration in India: Perspective and Practices

Dheeraj Mittal • Dharmendra Kumar Gupta
Editors

Ecological Restoration in India: Perspective and Practices

Springer

Editors
Dheeraj Mittal
Indira Paryavaran Bhawan
Ministry of Environment, Forest and Climate Change
New Delhi, Delhi, India

Dharmendra Kumar Gupta
Indira Paryavaran Bhawan
Ministry of Environment, Forest and Climate Change
New Delhi, Delhi, India

ISBN 978-981-96-8942-2 ISBN 978-981-96-8943-9 (eBook)
https://doi.org/10.1007/978-981-96-8943-9

This Springer imprint is published by the registered company Springer Nature Singapore Pte Ltd.
The registered company address is: 152 Beach Road, #21-01/04 Gateway East, Singapore 189721, Singapore

Preface

Ecosystem restoration means assisting in the recuperation of ecosystems that have been degraded or destroyed, as well as conserving the ecosystems that are still intact. Unfortunately, in the last century, the world has lost not less than 30% of its forests and 35% of its wetlands. According to an estimate India with 2.4% of total global land area is home for about 8% of the global biological diversity. It is also true that in providing goods and services to 18% of world's population, mega diverse nations are losing their biological wealth. The degradation of terrestrial and aquatic ecosystems has been the key factor in accelerating this irreversible and unrepairable loss. Indian subcontinent ecosystems are subjected to brutal and enormous land degradation. The fragmentation of natural habitats and their disappearances due to anthropogenic drivers are playing key roles in loss of various species, associated ecosystem services and functions, and human well-being. The scientific evidence indicates that ecosystem degradation is a major force behind recent extreme climate events.

This book *Ecological Restoration in India: Perspective and Practices* presents the latest perspective for ecological restoration in India with few chapters on best practices especially on grassland restoration, wetland restoration, habitat restoration for wildlife, etc. The book has chapters on the policy framework in India, analyzing the legislative mandate as per relevant acts and rules in India, and provides international perspective. This book is offering a thorough examination of updates on the very important and emerging field of ecological restoration in India to authorities (Government Officers, Administrators and professionals dealing with the subject matters), foresters, ecologists, master's students, and PhD scholars of ecological sciences, environment, and climate change and will be a valuable addition to the libraries of various colleges and universities. Further, since it is a subject which is of international importance being debated and deliberated in various international forums, various NGOs, stakeholders, and international organizations such as UNFCCC, UNCBD, UNCCD, and international universities and colleges will be interested to see this work.

The editors (Dr. Dheeraj Mittal and Dr. Dharmendra K. Gupta) personally thank the authors for contributing with their valuable time, knowledge, and eagerness to bring this book in this shape. We also thank Sh. Raju Rawat for providing technical assistance.

New Delhi, India Dheeraj Mittal
New Delhi, India Dharmendra Kumar Gupta

Contents

1 Nurturing Nature: Roadmap to Ecological Restoration Through India's G20 Presidency 1
Dheeraj Mittal, Tania Banerjee, and Ashish Kulkarni

2 Eco-Restoration in India: A Policy Perspective 33
Charan Jeet Singh and Dheeraj Mittal

3 International Processes and Commitments to Land Restoration 55
Muralee Thummarukudy and Dheeraj Mittal

4 Ecological Restoration of Mined-Out Areas for Achieving Land Degradation Neutrality: Opportunities and Challenges 61
Vivek Kumar Choudhary, Sudeshna Mazumdar-Leighton, and Cherukuri Raghavendra Babu

5 Ecological Restoration in India for Socio-Economic Well-Being 89
Pankaj Satija

6 Mangroves of India: Diversity, Conservation and Restoration Prospects 111
Karthigeyan Kaliyamurthy and Ashiho Asosii Mao

7 Post-fire Restoration Plan for Sustainable Forest Management 137
K. Ravichandran

8 Brave New Wild: Restoring Biodiversity and Values Amid Invaded Landscapes of India 153
Ninad Avinash Mungi, Rajat Rastogi, Yadvendradev V. Jhala, Qamar Qureshi, and Jens-Christian Svenning

9 Yamuna Biodiversity Park: Restoring Natural History of Yamuna Floodplain in Delhi Stretch 177
Faiyaz A. Khudsar, Mohan Singh, Umendra Sirohi, Harmeek Singh, Ekta Khurana, M. Faisal, and A. K. Singh

10 Ecological Restoration in Delhi: Challenges and Their Mitigation . . . 193
Rita Rath, Sadhna Gupta, Neeraja Sood, Versha Goel, and Roopa Rani Samal

11 Ecological Restoration of Grasslands in Saurashtra . . . 235
Sunil Kumar Berwal

12 Ecosystem Restoration Through Umbrella Species: Tracking Ecological Response of Tiger Recovery in Panna Tiger Reserve, Central India . . . 249
Ramesh Krishnamurthy, Supratim Dutta, and Sankarshan Chaudhuri

13 An Experiment in Quasi-Eco-Restoration of a Creek Slope: A Collaborative Approach on a University Campus in India . . . 279
Padamnabhi S. Nagar, Shishir R. Raval, Haribhai R. Kataria, Ankur Rajwadi, Snehal Chavda, Jaydeep J. Sharma, Gulzar Malek, Roshan D. Virola, and Tanmay Rohit

Correction to: An Experiment in Quasi-Eco-Restoration of a Creek Slope: A Collaborative Approach on a University Campus in India . . . C1
Padamnabhi S. Nagar, Shishir R. Raval, Haribhai R. Kataria, Ankur Rajwadi, Snehal Chavda, Jaydeep J. Sharma, Gulzar Malek, Roshan D. Virola, and Tanmay Rohit

Editors and Contributors

About the Editors

Dheeraj Mittal worked as Assistant Inspector General of Forests in the Forest Conservation Division in the Ministry of Environment, Forest and Climate Change, Government of India, New Delhi. He graduated from University of Delhi and was awarded PhD in Plant Molecular Biology in 2013. He joined Indian Forest Service in the year 2012 and worked as Forest Officer in the State of Gujarat, India. He has worked in Gir National Park and Wildlife Sanctuary as protected area manager. He joined the Ministry of Environment Forest and Climate Change, Government of India, as Assistant Inspector general of Forest and is working for Project Lion, Project Elephant, and Forest Conservation. He has published research articles, review articles/book chapters in international peer-reviewed journals.

Dharmendra Kumar Gupta is working as Director(S)/Scientist-F at Ministry of Environment, Forest and Climate Change, Government of India, New Delhi. He has been awarded several international awards such as the Japan Society for the Promotion of Science (JSPS), the Belgian Science Policy Office (BELSPO), JAE-Doc. (Spain), TWAS-CNPq (Italy), Royal Society India Fellow (UK), MASAV (Israel), and Jawahar Lal Nehru Scholarship (India). He has published more than 130 international peer-reviewed original research/review articles/book chapters with total SCI factor of 454.80, h-Index 45, and i10 Index 100. The total citation of his work by other researchers is 8755 (Google Scholar). He has successfully completed 10 multidisciplinary research projects from international/national bodies. Besides, he also published 26 books from leading publishers. Till date, his books have been downloaded more than 300,000 times. He also served as a Book Series editor for two book series.

Contributors

Cherukuri Raghavendra Babu Centre for Environmental Management of Degraded Ecosystems (CEMDE), University of Delhi, New Delhi, India

Tania Banerjee Boston Consulting Group, Gurgaon, India

Sunil Kumar Berwal Forest Department, Government of Gujarat, Morbi, Gujarat, India

Sankarshan Chaudhuri Wildlife Institute of India, Dehradun, Uttarakhand, India

Snehal Chavda Department of Botany, Faculty of Science, The M.S. University of Baroda, Vadodara, India

Vivek Kumar Choudhary Plant-Biotic Interactions Group, Department of Botany, University of Delhi, New Delhi, India

Centre for Environmental Management of Degraded Ecosystems (CEMDE), University of Delhi, New Delhi, India

Supratim Dutta Wildlife Institute of India, Dehradun, Uttarakhand, India

M. Faisal Biodiversity Parks Programme, Centre for Environmental Management of Degraded Ecosystems, University of Delhi, Delhi, India

Versha Goel Department of Zoology, Maitreyi College, University of Delhi, New Delhi, India

Sadhna Gupta Department of Zoology, Dyal Singh College, University of Delhi, New Delhi, India

Yadvendradev V. Jhala Wildlife Institute of India, Dehradun, India

Indian National Science Academy, National Centre for Biological Sciences, Bangalore, India

Karthigeyan Kaliyamurthy Headquarters, Botanical Survey of India, Salt Lake City, Kolkata, West Bengal, India

Haribhai R. Kataria Department of Mathematics, Faculty of Science, The M.S. University of Baroda, Vadodara, India

Faiyaz A. Khudsar Biodiversity Parks Programme, Centre for Environmental Management of Degraded Ecosystems, University of Delhi, Delhi, India

Ekta Khurana Biodiversity Parks Programme, Centre for Environmental Management of Degraded Ecosystems, University of Delhi, Delhi, India

Ramesh Krishnamurthy Wildlife Institute of India, Dehradun, Uttarakhand, India

Faculty of Forestry, University of British Columbia, Vancouver, BC, Canada

Ashish Kulkarni Boston Consulting Group, Gurgaon, India

Gulzar Malek Department of Botany, Faculty of Science, The M.S. University of Baroda, Vadodara, India

Ashiho Asosii Mao Headquarters, Botanical Survey of India, Salt Lake City, Kolkata, West Bengal, India

Sudeshna Mazumdar-Leighton Plant-Biotic Interactions Group, Department of Botany, University of Delhi, New Delhi, India

Dheeraj Mittal Indira Paryavaran Bhawan, Ministry of Environment, Forest and Climate Change, New Delhi, Delhi, India

Ninad Avinash Mungi Center for Ecological Dynamics in a Novel Biosphere Section of Ecoinformatics (ECONOVO); Department of Biology, Aarhus University, Aarhus, Denmark

Wildlife Institute of India, Dehradun, India

Padamnabhi S. Nagar Department of Botany, Faculty of Science, The M.S. University of Baroda, Vadodara, India

Qamar Qureshi Wildlife Institute of India, Dehradun, India

Ankur Rajwadi Department of Botany, Faculty of Science, The M.S. University of Baroda, Vadodara, India

Rajat Rastogi Wildlife Institute of India, Dehradun, India

German Center for Integrative Biodiversity Research (iDiv) Halle-Jena-Leipzig, Leipzig, Germany

Rita Rath Department of Zoology, Dyal Singh College, University of Delhi, New Delhi, India

K. Ravichandran Indian Institute of Forest Management (IIFM), Bhopal, Madhya Pradesh, India

Shishir R. Raval The M.S. University of Baroda, Vadodara, India

Tanmay Rohit Department of Botany, Faculty of Science, The M.S. University of Baroda, Vadodara, India

Roopa Rani Samal Department of Zoology, Dyal Singh College, University of Delhi, New Delhi, India

Pankaj Satija Ferro Alloys & Minerals Division, Tata Steel Limited, Bhubaneswar, India

Jaydeep J. Sharma Department of Botany, Faculty of Science, The M.S. University of Baroda, Vadodara, India

A. K. Singh Biodiversity Parks Programme, Centre for Environmental Management of Degraded Ecosystems, University of Delhi, Delhi, India

Charan Jeet Singh Indira Paryavaran Bhawan, Ministry of Environment, Forests and Climate Change, New Delhi, Delhi, India

Harmeek Singh Biodiversity Parks Programme, Centre for Environmental Management of Degraded Ecosystems, University of Delhi, Delhi, India

Mohan Singh Biodiversity Parks Programme, Centre for Environmental Management of Degraded Ecosystems, University of Delhi, Delhi, India

Umendra Sirohi Biodiversity Parks Programme, Centre for Environmental Management of Degraded Ecosystems, University of Delhi, Delhi, India

Neeraja Sood Department of Zoology, Dyal Singh College, University of Delhi, New Delhi, India

Jens-Christian Svenning Center for Ecological Dynamics in a Novel Biosphere Section of Ecoinformatics (ECONOVO); Department of Biology, Aarhus University, Aarhus, Denmark

Muralee Thummarukudy G20 Global Land Initiative, United Nations Convention to Combat Desertification, Berlin, Germany

Roshan D. Virola Department of Botany, Faculty of Science, The M.S. University of Baroda, Vadodara, India

Nurturing Nature: Roadmap to Ecological Restoration Through India's G20 Presidency

1

Dheeraj Mittal, Tania Banerjee, and Ashish Kulkarni

Abstract

This chapter establishes the urgency of land restoration, its socioeconomic imperatives, and approaches of different countries toward addressing this global challenge. This chapter further discusses the concerted efforts through calls for global action on land degradation at various platforms and the Sustainable Development Goals (SDGs) and emphasizes the role of international cooperation. This chapter goes on to describe the journey of land restoration efforts across previous G20 presidencies, brings about the successes unlocked during India's G20 presidency, and paves a path forward with clear recommendations, such as integrating restoration in developmental policies, innovating financing, ensuring enforcement, scaling up ecosystem-based approaches, and involving local communities with Indigenous knowledge toward comprehensive land restoration efforts in India.

Keywords

Land degradation · Ecological restoration · Ecosystem services · Policy implications · Biodiversity loss · Finance for restoration · Community participation

D. Mittal (✉)
Indira Paryavaran Bhawan, Ministry of Environment, Forest and Climate Change, New Delhi, Delhi, India
e-mail: dmittal.ifs@gov.in

T. Banerjee · A. Kulkarni
Boston Consulting Group, Gurgaon, India
e-mail: banerjee.tania@bcg.com; kulkarni.ashish@bcg.com

D. Mittal, D. K. Gupta (eds.), *Ecological Restoration in India: Perspective and Practices*, https://doi.org/10.1007/978-981-96-8943-9_1

1.1 Ecological Restoration: The Global Urgency

As the world confronts unprecedented environmental crises, it is evident that the health of Earth's ecosystems underpins the well-being of human civilization. Ecosystem degradation—driven by climate change, biodiversity loss, pollution, desertification, deforestation, and ocean degradation—has left natural systems in a precarious state. Human activities, including industrialization, unsustainable agriculture, and urban sprawl, have pushed these ecosystems to the brink, threatening both environmental stability and human welfare. The loss of ecosystem services is worth more than 10% of our global economic output (UNEP 2021a, b). The urgency to restore degraded ecosystems has never been more pressing, and ecological restoration has emerged as a critical intervention to reverse damage, rebuild resilience, and ensure the sustainability of both human and natural systems.

Recognizing these interconnected challenges, the G20[1] has, across multiple presidencies, underscored the necessity of accelerating restoration efforts to safeguard ecosystems that are rapidly deteriorating. Given the nature and scale of impact the G20 nations are well-positioned to have, their commitment reflects an understanding that environmental health is inextricably linked to economic stability and global prosperity. As India took up the mantle of leading G20 Presidency in 2023, the fact that key environmental crises must be addressed, such as **climate change; biodiversity loss; pollution; desertification; deforestation; water quality, availability, and accessibility; and land and ocean degradation, emerged as a collective view**. Each of these crises poses a threat to ecosystems and the communities that rely on them, making restoration not just an environmental necessity but a socioeconomic imperative.

1.1.1 Degradation and the Imperative for Ecosystem Restoration

Ecosystems form the foundation of life on Earth, offering critical services such as carbon sequestration, water filtration, flood control, and biodiversity support. However, when ecosystems degrade, their capacity to perform these functions diminishes, with far-reaching consequences for human society. The destruction of forests, wetlands, coral reefs, and grasslands disrupts natural processes, increasing vulnerability to disasters and diminishing resources vital for human survival. The scale of this degradation is staggering: **Over 991 million hectares of land globally have been impacted by degradation** (Crossman, 2017), with **80% of arable land**

[1] The Group of 20 (G20) is an intergovernmental forum comprising 19 sovereign countries, the European Union (EU), and the African Union (AU). The sovereign countries include Argentina, Australia, Brazil, Canada, China, France, Germany, India, Indonesia, Italy, Japan, the Republic of Korea, Mexico, Russia, Saudi Arabia, South Africa, Turkiye, the United Kingdom, and the United States.

affected by poor land management practices (United Nations Environment Programme 2021a, b).

Land degradation is not just an environmental issue—it directly impacts food security, livelihoods, and public health. **Every minute, 23 hectares of arable land are lost due to desertification and drought** (United Nations Environment Programme 2018). Without urgent action, this crisis will continue to displace communities, deepen poverty, and exacerbate social inequality, particularly in developing countries. For instance, degraded agricultural land leads to lower yields, pushing small-scale farmers into poverty and driving migration to urban areas in search of work. This rural exodus can destabilize economies and place additional pressures on cities, further compounding the socioeconomic challenges arising from environmental degradation.

Forests are not the only ecosystems under threat. Wetlands, which provide natural flood control, habitat for diverse species, and water filtration, have been reduced by 35% since 1970 (Ramsar Convention Secretariat, 2018). Coastal ecosystems, such as mangroves and coral reefs, are also deteriorating due to human activity and climate change. These ecosystems play critical roles in protecting shorelines from storms, supporting fisheries, and storing carbon. Without concerted restoration efforts, the loss of these vital coastal ecosystems will accelerate, leaving the planet far less capable of sustaining life.

Marine ecosystems are equally under siege. **Plastic pollution**, ocean acidification, and overfishing are devastating ocean health, disrupting food chains, and threatening marine biodiversity. **Globally, 85% of marine litter is plastic** (United Nations Environment Programme 2021a, b), with between **4.8 and 12.7 million metric tons of plastic entering the oceans annually** (National Geographic, 2017). As plastic waste accumulates, it entangles marine life and infiltrates the food chain through microplastics, threatening both biodiversity and human health. By **2040, up to 50 kg of plastic waste could accumulate for every meter of coastline plastic** (United Nations Environment Programme 2021a, b), significantly worsening the already precarious state of marine ecosystems.

In addition to pollution, **ocean acidification and warming**—driven by excess atmospheric carbon dioxide (CO_2)—are expanding marine "dead zones" where oxygen levels plummet, decimating marine life. Overfishing, which affects over a third of the world's fish stocks, compounds these problems, further threatening biodiversity. As a result, by **2100**, half of the world's marine species could be at risk (Ibid). **Mangroves, seagrass beds, and salt marshes**—essential coastal ecosystems that provide protection from storms, filtration of pollutants, and carbon sequestration—are increasingly at risk from both human activity and the changing climate.

1.1.2 Human Drivers of Ecosystem Degradation

The scale of ecological degradation is deeply tied to human activities, with deforestation, mining, pollution, and unsustainable agricultural practices being among the primary drivers. Over the past three decades, **420 million hectares of forest have**

been lost (Food and Agriculture Organization 2022), primarily due to agricultural expansion and logging. This deforestation not only accelerates biodiversity loss but also heightens the effects of climate change by releasing stored carbon into the atmosphere. Forests, which act as critical carbon sinks, lose their capacity to absorb CO_2 when cleared, further contributing to global warming.

Mining operations, especially in developing regions, are another significant contributor to land degradation. The extraction of minerals significantly impacts its ecological integrity, leaving behind scarred landscapes that take decades to recover, if at all. Forest fires have significantly increased worldwide; in 2015 alone, approximately 98 million hectares of forests were affected by fires (Food and Agriculture Organization 2022). Climate change and land use shifts induced by increasing human activity are making forest fires more frequent and intense. These forest fires are also hugely attributed to illegal and unattended campfires, illegal expansion for agriculture, stubble burning in close proximity to the forests, and illegal logging. These forest fires emerge as one of the biggest causes for anthropogenic emissions, leading to global warming and enhancing the availability of dry forest waste, fueling fresh fires.

Human activities such as drainage of coastal ecosystems for agriculture, dredging and channelization for navigation and flood protection, and gray infrastructure construction, like dykes, dams, and seawalls, have been causing environmental degradation across estuaries, intertidal wetlands, and the ocean (Sam A., 1981).

However, the degradation of ecosystems is not solely driven by the above causes. Poverty and lack of access to sustainable livelihoods force many rural communities to rely on fragile ecosystems for subsistence. In regions where deforestation or overgrazing occurs, local communities may have few alternatives, leading to a cycle of environmental degradation and poverty. This highlights the need for an integrated approach to ecological restoration that balances economic growth with environmental sustainability. The G20 nations during India's Presidency have acknowledged that without addressing these underlying socioeconomic issues, efforts to restore all these ecosystems will remain incomplete.

1.1.3 Climate Change and Ecosystem Vulnerability

The effects of **climate change** have intensified the degradation of ecosystems across the globe. Rising temperatures, erratic weather patterns, and extreme events, such as wildfires, droughts, and floods, exacerbate existing environmental pressures. Forests, wetlands, and peatlands, which play critical roles in storing carbon and regulating water cycles, are losing their resilience in the face of these changes. As ecosystems degrade, they release vast amounts of stored carbon, contributing further to the problem.

Damaged peatlands alone are responsible for nearly 5% of global anthropogenic CO_2 emissions, despite covering just **0.3% of the Earth's surface** (International Union for Conservation of Nature 2021). Similarly, the **loss of forest cover** reduces the ability of natural systems to mitigate climate change while simultaneously increasing the risk of wildfires, which, in turn, degrade ecosystems further. The feedback loop created by ecosystem degradation and climate change requires urgent intervention.

The G20, in recognizing this challenge, has emphasized the importance of implementing **nature-based solutions (NbS)**, ecosystem-based approaches (EbA), and other management practices to restore degraded landscapes. These strategies provide critical co-benefits beyond ecosystem health, including disaster risk reduction, improved food security, and enhanced resilience for vulnerable populations.

1.1.4 Biodiversity Loss: A Global Emergency

The world is experiencing an alarming rate of **biodiversity loss**, driven by habitat destruction, pollution, and climate change. **Approximately 85% of the world's biodiversity is currently under threat from habitat loss** (World Wildlife Fund 2024), and the extinction rate is now **100–1000 times higher than prehuman levels** (World Wildlife Fund 2024). This loss has profound implications for the stability of ecosystems and the human societies that depend on them. Pollinators, for instance, play a crucial role in the global food system, yet their populations are plummeting due to habitat destruction and pesticide use. The economic consequences of this decline are staggering, with a potential loss of **$235–$577 billion annually** in crop output (Begosh et al. 2022).

The oceans, too, are witnessing widespread biodiversity loss, as **overfishing**, pollution, and climate change devastate marine ecosystems. Coral reefs, home to a quarter of all marine species, are in sharp decline, threatening not only biodiversity but also the livelihoods of the millions of people who rely on them. As marine biodiversity continues to erode, the stability of coastal communities becomes increasingly precarious, and the broader ecological services provided by the oceans are at risk.

1.1.5 The Economic and Social Drivers for Restoration

The degradation of ecosystems comes with significant economic costs. Restoring ecosystems, by contrast, offers substantial economic benefits. By restoring 350 million hectares of degraded land, the world could generate $9 trillion in ecosystem services by 2030 (United Nations Decade on Ecosystem Restoration 2021–2030, 2019). These services—ranging from carbon sequestration to flood control and biodiversity conservation—are critical for both environmental sustainability and economic resilience.

Restoration efforts also create employment opportunities. For example, the restoration sector in the United States alone generates 126,000 jobs directly and contributes an additional 95,000 jobs indirectly (World Business Council for Sustainable Development, 2015). Scaling up restoration projects globally could provide similar benefits, particularly in developing countries, where unemployment and poverty are closely linked to ecosystem degradation.

Nature-based solutions (NbS) *are actions to protect, sustainably manage, and restore natural and modified ecosystems that address societal challenges effectively and adaptively, simultaneously benefiting people and nature. They have the potential to make an essential contribution to reaching net-zero* CO_2 *globally by around 2050, alongside other decarbonization strategies. Nature-based solution projects of high quality and integrity could have the mitigation of up to 10 gigatonnes of carbon dioxide (*$GtCO_2$*) per year, corresponding to approximately 27% of the current global annual emissions* (UNEP FI, 2023). ***Ecosystem-based approaches (EbA)*** *seek to balance ocean, coastal, and freshwater resource use with conservation, acknowledging ecosystem connectivity. It involves coordinated land and sea activity management to minimize cumulative impacts on marine and coastal ecosystems. This approach overcomes the limitations of single-sector strategies, offering holistic human–ecosystem interaction management* (UNEP 2024).

Moreover, ecosystem degradation often exacerbates social inequalities, disproportionately affecting the poorest and most vulnerable communities. These populations are often the most reliant on natural resources for their livelihoods and the least able to adapt to environmental changes. Restoration efforts, therefore, not only protect biodiversity and mitigate climate change but also offer a pathway to poverty alleviation and sustainable development.

1.1.6 The Call for Global Action

The urgency of this moment is undeniable. As ecosystems collapse and biodiversity is lost at an unprecedented pace, the global community must act swiftly. The declaration of the UN Decade on Ecosystem Restoration from 2021 to 2030 was a direct response to the global recognition that we are running out of time. One hundred and fifteen governments have committed to restoring nearly 1 billion hectares of land (Sewell et al. 2020). Restoration is not just a response to damage already done—it is a proactive strategy to build resilience and ensure a sustainable future for generations to come.

The G20, accounting for approximately 85% of global GDP and 80% of global emissions (UNFCCC 2023), has a unique role to play in driving ecosystem restoration efforts. G20 initiatives such as the **Global Land Initiative**[2] **(GLI)** to prevent,

[2] In November 2020, at the Riyadh (Virtual) Summit of the G20 leaders, the G20 launched the Global Initiative on Reducing Land Degradation and Enhancing Conservation of Terrestrial Habitats. The ambition of this Global Initiative is to prevent, halt, and reverse land degradation and reduce degraded land by 50% by 2040.

halt and reverse land degradation and the **Osaka Blue Ocean Vision**[3] have already set the stage in the past, but the work is far from over. India's G20 Presidency made major strides to continue and propel this momentum; however, the path forward requires collaboration, innovation, and the collective will of nations to recognize that the health of ecosystems is intrinsically linked to the health of the economy and society as a whole.

Effective restoration requires more than top-down interventions—it demands the active involvement of local communities. Indigenous and traditional knowledge systems, which have long sustained ecosystems, are essential for ensuring that restoration efforts are sustainable and context-specific. In many regions, local communities possess a deep understanding of the landscapes they inhabit, making them critical stakeholders in restoration projects.

The G20 recognizes the importance of inclusivity in ecological restoration efforts, stressing that marginalized groups, including **women, youth, Indigenous Peoples, and rural communities**, must be engaged in decision-making processes. Restoration projects that involve these communities not only benefit from their expertise but also create opportunities for economic development and social cohesion.

Local initiatives, such as community-based reforestation projects or sustainable agricultural practices, are vital for maintaining the integrity of ecosystems and preventing further degradation. These efforts can be scaled up through partnerships between governments, NGOs, and the private sector, ensuring that restoration aligns with broader goals of environmental sustainability and social justice.

1.2 Global Dialogue on Ecological Restoration

Ecological restoration has steadily emerged as a central pillar of global environmental policy. As the world grapples with increasingly severe environmental crises, the recognition that ecosystem restoration is critical to mitigating biodiversity loss, climate change, and land degradation has grown stronger. This global consensus is driven by an evolving framework of international commitments, shaped over decades and recently consolidated in initiatives such as the **UN Decade on Ecosystem Restoration** and the **Kunming-Montreal Global Biodiversity Framework (KMGBF)**. These efforts are further advanced by the concerted actions through the G20 platform, which has positioned ecological restoration as a key strategy for sustainable development and climate resilience over the past few Presidencies.

[3] At the G20 Ministerial Meeting on Energy Transitions and Global Environment for Sustainable Growth, Karuizawa, held in June 2019, the "G20 Implementation Framework for Actions on Marine Plastic Litter" was established, and it was endorsed by the G20 leaders at the subsequent G20 Osaka Summit. As a common global vision, the "Osaka Blue Ocean Vision" was shared by the leaders at the Summit, aiming to reduce additional pollution by marine plastic litter to zero by 2050.

The progression from early environmental concerns to today's comprehensive frameworks underscores the urgency and importance of coordinated restoration efforts. The G20, with its economic and political clout, has played a pivotal role in shaping this global agenda, using its presidencies to drive action on restoration and conservation. These efforts are not just aimed at reversing ecological damage but also at ensuring that restoration becomes a core component of economic recovery, poverty alleviation, and resilience building.

1.2.1 The Foundations of Global Restoration Efforts

The global discourse on environmental degradation and restoration can be traced back to the **1972 United Nations Conference on the Human Environment** in Stockholm. This was a foundational moment in international environmental governance, where the relationship between economic development and environmental degradation was formally recognized. It led to the creation of the **United Nations Environment Programme (UNEP)**, an institution that has since been at the forefront of advocating for sustainable development and environmental protection.

However, it was not until the **1992 Rio Earth Summit** that ecological restoration was brought into sharper focus. The summit gave rise to three significant conventions, collectively known as the **Rio Conventions**: the **Convention on Biological Diversity (CBD)**, the **United Nations Framework Convention on Climate Change (UNFCCC)**, and the **United Nations Convention to Combat Desertification (UNCCD)**. These conventions laid the groundwork for global environmental policy by framing biodiversity conservation, climate mitigation, and land restoration as interconnected challenges. Restoration, for the first time, was recognized not only as a reactive measure but also as a proactive strategy to safeguard biodiversity and ecosystem services.

1.2.2 The 1992 Rio Earth Summit: A New Milestone for Restoration

The Rio Earth Summit in 1992 marked a watershed moment in global environmental policy. Central to this summit was the creation of the **Convention on Biological Diversity (CBD),** which remains one of the most comprehensive global treaties focused on biodiversity conservation. For the first time, restoration was formally recognized as essential for addressing the growing loss of biodiversity (SPREP 2012). The **CBD** introduced key principles, such as the **sustainable use of biodiversity** and the fair and equitable sharing of benefits arising from the use of genetic resources. It pushed the global conversation efforts beyond preservation to **active restoration**, signaling that conservation alone was not sufficient to reverse the damage done by human activities.

In the years that followed, the understanding of restoration deepened. Scientific research highlighted the scale of the degradation problem, showing how ecosystems worldwide were deteriorating at unprecedented rates. With this realization, new international initiatives began to take shape, aimed specifically at scaling up restoration efforts.

The G20 nations that make up close to half of the Earth's total land area (G20 Global Land Initiative, 2024) recognize that restoration is not just a matter of protecting ecosystems but also actively repairing those that have been degraded, underscoring the critical role restoration plays in reversing biodiversity loss.

1.2.3 From Aichi to the SDGs: Setting Global Targets for Restoration

In 2010, the **CBD adopted the Aichi Biodiversity Targets**, a set of global goals aimed at halting biodiversity loss. Central to these targets was the acknowledgment that **restoration is essential to conserving biodiversity**. This marked a pivotal shift in the global restoration dialogue, embedding the idea that ecological restoration is not merely an afterthought but a necessary tool for ensuring the health of ecosystems.

Building on the momentum of the Aichi Targets, the **United Nations launched the Sustainable Development Goals (SDGs) in 2015**, further integrating ecological restoration into global development strategies. Several SDGs directly address the need for restoration. For instance, **SDG 15 (Life on Land)** focuses on protecting, restoring, and promoting the sustainable use of terrestrial ecosystems, while **SDG 14 (Life Below Water)** highlights the importance of restoring marine and coastal ecosystems. These goals stress that without healthy ecosystems, achieving sustainable development is impossible. The G20 recognizes these linkages and **reaffirms its commitment to ensuring that the restoration of ecosystems remains integral to global sustainability efforts**.

The Paris Agreement, adopted in 2015, added urgency to the restoration agenda by connecting ecosystem restoration with climate action. Restored ecosystems, from forests to wetlands, serve as critical carbon sinks, absorbing CO_2 and mitigating the impacts of global warming. The G20 has consistently highlighted that ecological restoration is a crucial strategy for both climate mitigation and adaptation, recognizing the profound role ecosystems play in stabilizing global temperatures, and it witnessed further recognition through the Indian Presidency.

> ***The Paris Agreement*** *is a legally binding international treaty on climate change. It was adopted by 196 parties at the UN Climate Change Conference (COP21) in Paris, France, on 12 December 2015. It entered into force on 4 November 2016. Its overarching goal is to hold "the increase in the global average temperature to well below 2 °C above pre-industrial levels" and pursue efforts "to limit the temperature increase to 1.5 °C above pre-industrial levels."*

1.2.4 The UN Decade on Ecosystem Restoration: A Global Call to Action

In response to the escalating environmental crisis, the **United Nations** launched the **UN Decade on Ecosystem Restoration (2021–2030)**, an ambitious global initiative aimed at preventing, halting, and reversing the degradation of ecosystems worldwide. The **UN Decade** represents the most significant collective effort to restore ecosystems across land and sea, and it serves as a rallying call for nations, businesses, and civil society to invest in restoration efforts that address climate change, food insecurity, biodiversity loss, and poverty.

The **UN Decade** targets the restoration of **at least 1 billion hectares of degraded land** by 2030 (Sewell et al. 2020), in alignment with existing initiatives like the **Bonn Challenge**, which set a goal of restoring **350 million hectares of degraded land** by 2030 (UNEP; FAO 2022). It has significantly raised the profile of ecosystem restoration on the global agenda, generating political momentum, as evidenced by more than 70 countries supporting the initiative (UNECE 2021). But the scope of the **UN Decade** extends far beyond land—it encompasses oceans, coastal ecosystems, urban environments, and farmlands, recognizing the interconnectedness of these ecosystems and their collective impact on global sustainability.

The initiative is grounded in a holistic approach to restoration that goes beyond the recovery of ecosystems; it aims to build social and economic resilience through nature-based solutions. Ecosystems such as **mangroves, wetlands, forests, and coral reefs**, once restored, provide critical services that buffer communities from the impacts of climate change, improve food security, and enhance biodiversity. By restoring degraded ecosystems, the **UN Decade** seeks to catalyze significant socio-economic benefits, particularly for communities that depend on natural resources for their livelihoods.

Importantly, the **UN Decade** emphasizes that restoration is not merely about healing the planet—it is also a development strategy. By integrating restoration into national policies and global frameworks, countries can align ecological goals with broader sustainable development objectives, such as poverty reduction, gender equality, and economic growth. The G20, as a critical player in this initiative, has contributed by pushing forward discussions on land degradation, marine restoration, and the role of **nature-based solutions** in achieving global restoration targets.

1.2.5 Kunming-Montreal Global Biodiversity Framework: Charting a Path Forward

The **Kunming-Montreal Global Biodiversity Framework (KMGBF)**, adopted in 2022 during the **UN Biodiversity Conference (COP15)**, marks a turning point in global biodiversity conservation. It sets out a roadmap to halt biodiversity loss by **2030** and ensure that humanity lives in harmony with nature by **2050**. Signed by **196 countries**, the KMGBF introduces a set of ambitious targets, including conserving **30% of the world's land and sea areas**, restoring **30% of degraded ecosystems**, and reducing harmful subsidies that contribute to biodiversity loss (UNCTAD 2022).

The **KMGBF** builds on the lessons of its predecessor, the **Aichi Biodiversity Targets**, and addresses gaps in international biodiversity efforts by setting clear, measurable targets for restoration and conservation. The G20, through its leadership on climate and environmental sustainability, has committed to aligning with the **KMGBF** by integrating biodiversity restoration into broader strategies for economic recovery and development. This is particularly important for **G20 countries** like **India**, **Brazil**, and **Indonesia**, which are rich in biodiversity and hold significant responsibility for global conservation efforts.

The framework also calls for increased financial support, particularly for developing countries, to implement restoration projects. The G20 has recognized this need, advocating for the mobilization of both public and private financing to support large-scale restoration efforts. The commitment to the **KMGBF** aligns with the G20's broader goals of tackling **land degradation**, protecting **marine ecosystems**, and achieving **climate resilience**.

1.2.6 The G20 Journey of Advancing Restoration

The **G20** has emerged as a crucial player in driving global restoration efforts. The nations that are members of G20 wield considerable influence in shaping environmental policies and directing financial resources toward restoration. Over the past decade, successive G20 presidencies have advanced ambitious restoration agendas that have culminated in landmark initiatives such as the **Global Land Initiative (GLI)** and the **Osaka Blue Ocean Vision**. Saudi Arabia's Presidency in 2020 launched the GLI, the Global Initiative on Reducing Land Degradation and Enhancing Conservation of Terrestrial Habitats, which aims to halve the world's degraded land by 2040 (Group of Twenty (G20)—Saudi Arabia Presidency, 2020). This initiative marked a significant step in global restoration efforts, aligning with broader goals such as the Bonn Challenge and the UN Decade on Ecosystem Restoration. During Italy's G20 Presidency, an integrated approach was reflected in the G20's commitment to Land Degradation Neutrality (SDG 15.3) and the use of restoration to support carbon sequestration and enhance resilience, highlighted through the adoption of the Naples Charter. Indonesia's 2022 G20 Presidency brought to the forefront the topic of Ocean Economy and stressed the restoration of mangroves, seagrass beds, and other coastal ecosystems. These initiatives have not only reinforced the G20's role in global marine conservation but also demonstrated that the restoration of marine ecosystems is integral to both environmental and economic resilience (Fig. 1.1).

India's Presidency in 2023 set out with an ambitious aim of securing consensus on the need to execute restoration of mining- and forest fire–affected areas. The nations noted the need for remediation of mining-degraded lands and acknowledged that high-intensity forest fires/wildfires may affect the conservation and regeneration of ecosystems and their functions and services and can exacerbate climate change, land degradation, and biodiversity loss. Moreover, the G20 collectively adopted the "Chennai High-Level Principles for a Sustainable and Resilient Blue/Ocean-Based Economy" that directly calls upon all G20 nations to halt and reverse

Fig. 1.1 G20 countries' flags at the fourth Environment and Climate Sustainability Meeting under India's Presidency in July 2023. (Source: MoEFCC, India)

biodiversity loss and conserve coastal and marine ecosystems, among other elements, prior to handing the baton over to Brazil for Presidency. The Fifth Report on Actions against Marine Plastic Litter led by the Indian Presidency also tackled the matter of marine plastic litter, continuing the legacy set by Japan's Presidency in 2019.

1.2.7 Challenges in Scaling Up Global Restoration Efforts: The Role of International Cooperation

Despite significant progress in advancing the global restoration agenda, challenges remain. The financial resources needed to achieve large-scale restoration are immense, and the gap between current investments and the required funding is vast. Many developing countries, which house the world's most vulnerable ecosystems, lack the capacity and resources to implement comprehensive restoration projects. Through the Indian Presidency in 2023, G20 nations acknowledged the need to strengthen **policies and mobilize financing** from all sources in a predictable and adequate manner to address **climate change, biodiversity loss, and environmental degradation,** marking a significant step in the right direction.

In addition to financial constraints, the global community must overcome governance challenges that hinder effective restoration. Weak legal frameworks, poor enforcement of environmental regulations, and a lack of coordination between governmental and nongovernmental actors often impede restoration efforts. The **KMGBF** addresses these challenges by calling for stronger monitoring and reporting mechanisms to track the restoration progress and ensure accountability.

Finally, the ongoing degradation of ecosystems, driven by **deforestation**, **overfishing**, and **pollution**, continues to outpace restoration efforts in many parts of the world. The global community must not only ramp up restoration efforts but also address the root causes of degradation by promoting **sustainable land management**, **circular economies**, and **conservation practices**.

1.3 India's G20 Presidency: Championing Ecological Restoration

India's G20 Presidency in 2023 marked a pivotal moment in the global movement toward ecological restoration. As environmental crises intensify globally, India's leadership came at a time when urgent, coordinated action was needed to reverse ecosystem degradation and chart a sustainable path forward. Building on the progress made by previous G20 presidencies, India amplified its commitment to addressing environmental challenges such as **climate change**, **biodiversity loss**, **pollution**, **desertification**, **deforestation**, and **ocean degradation**. These crises are interconnected and require comprehensive, integrated responses. India's G20 Presidency not only acknowledged this complexity but also set forth an ambitious direction to drive restoration efforts at both the national and global levels.

India's Presidency was characterized by a forward-looking agenda that placed ecological restoration at the core of sustainable development and climate resilience. The Presidency built on the work of previous presidencies, such as that of **Saudi Arabia**, **Italy**, **Japan**, and **Indonesia**, while introducing new themes, strategies, and frameworks that solidified restoration as a priority for the G20. By advocating for **nature-based solutions**, **ecosystem-based approaches**, and **sustainable development**, India's leadership underscored the importance of restoring degraded ecosystems to safeguard the well-being of both people and the planet (Fig. 1.2).

1.3.1 G20's Shared Vision for the Future of Restoration

The G20's commitment to ecological restoration is clear. Through its leadership on land degradation, marine conservation, and biodiversity and coastal protection, the G20 has positioned restoration as a key pillar of its climate and environmental

Fig. 1.2 G20 Environment and Climate Sustainability Working Group (ECSWG) ministerial meeting, July 2023. (Source: MoEFCC)

sustainability agenda. The **UN Decade on Ecosystem Restoration** and the **Kunming-Montreal Global Biodiversity Framework** provide the necessary global frameworks for action, but a major driver for success is the G20's collective responsibility to ensure that restoration efforts are scaled up, adequately financed, and aligned with broader goals of **climate resilience** and **economic recovery**.

As the world continues to confront the challenges of **climate change**, **biodiversity loss**, and **ecosystem degradation**, the G20 must lead by example. By integrating restoration into their national policies and committing to international frameworks, G20 countries can help secure a future where both **nature** and **humanity** can thrive.

1.3.2 Building on Previous G20 Presidencies: A Progressive Legacy

India's approach to ecological restoration did not emerge in isolation. It was built on the robust foundations laid by previous G20 presidencies, which had established significant initiatives and frameworks that guided global environmental actions.

- **Saudi Arabia's Presidency, 2020**: Launched the GLI, the Global Initiative on Reducing Land Degradation and Enhancing Conservation of Terrestrial Habitats, which aims to halve the world's degraded land by 2040 (Group of Twenty (G20)—Saudi Arabia Presidency, 2020). This initiative marked a significant step in global restoration efforts, aligning with broader goals such as the Bonn Challenge and the UN Decade on Ecosystem Restoration. The Saudi Presidency not only put land degradation at the center of the G20's environmental agenda but also emphasized the importance of integrating biodiversity conservation into global recovery efforts.
- **Italy's Presidency, 2021**: Further solidified the connection between climate action, biodiversity conservation, and restoration. During Italy's G20 Presidency, an integrated approach was reflected in the G20's commitment to Land Degradation Neutrality (SDG 15.3) and the use of restoration to support carbon sequestration and enhance resilience, highlighted through the adoption of the Naples Charter. Italy's presidency underscored the need to address climate change and biodiversity loss through nature-based solutions, emphasizing that restoring ecosystems is essential for mitigating the impacts of climate change.
- **Indonesia's Presidency, 2022**: The Bali Joint Environment and Climate Ministers' Meeting underlined the importance of protecting marine habitats, which are critical for both biodiversity and the livelihoods of coastal communities. Indonesia's leadership in marine restoration has advanced the Osaka Blue Ocean Vision, which aims to eliminate additional pollution from marine plastic litter by 2050 (G20 Research Group (University of Toronto), 2020). Indonesia's 2022 G20 Presidency brought to the forefront the topic of Ocean Economy and

stressed the restoration of mangroves, seagrass beds, and other coastal ecosystems. These initiatives have not only reinforced the G20's role in global marine conservation but also demonstrated that the restoration of marine ecosystems is integral to both environmental and economic resilience.

1.3.3 Prioritized Themes: A Comprehensive Approach to Restoration

During its G20 Presidency, India prioritized several critical themes that reflected the country's commitment to comprehensive ecological restoration. These themes spanned a wide range of ecosystems, including **forests**, **degraded lands**, **coastal and marine ecosystems**, and **biodiversity hotspots**. By focusing on these areas, India aimed to address the interconnected environmental crises that threaten both natural systems and human well-being.

1.3.3.1 Forest and Land Restoration

India's leadership in forest restoration was a key focus of its G20 Presidency. Forest ecosystems are vital for maintaining **biodiversity**, sequestering carbon, and providing ecosystem services that benefit millions of people, particularly in **rural areas**. India's **National Mission for a Green India**, part of the **National Action Plan on Climate Change**, seeks to increase forest cover, restore degraded forest landscapes, and enhance carbon sequestration. This aligns closely with the G20's broader goals of addressing **deforestation** and **land degradation**, as outlined in the G20 ECSWG Outcome Document and Chair's Summary (G20—India Presidency 2023).

G20 ECSWG Outcome Document and Chair's Summary 2023 *was the final outcome of the ministerial meeting of the Indian Presidency's ECSWG, unanimously agreed to by all G20 delegations, except for paragraphs 63–66, which pertain to the Chair's Summary. It covered the thematic priorities of Accelerating Climate Action; Preventing, Reducing and Reversing Land Degradation; Accelerating Ecosystem Restoration and Halting and Reversing Biodiversity Loss; Sustainable and Integrated Water Resource Management; Protecting and Conserving the Ocean, Promoting a Sustainable and Resilient Blue/Ocean-Based Economy; and Promoting Resource Efficiency and Circular Economy, Preventing Pollution and Minimizing Its Risks.*

The Working Group also delivered ten Presidency Documents, as presented below[4].

[4] The ECSWG Outcome Document and Chair's Summary, as well as the Presidency Documents, can be accessed via this URL: https://www.g20.in/es/media-resources/documents/doc-outcomes.html

G20 ECSWG Outcome Document & Chair's Summary 2023 *was the final outcome of the Ministerial Meeting of the India Presidency's ECSWG, unanimously agreed to by all G20 delegations except for paragraphs 63-66, which pertains to the Chair's Summary. It covered the thematic priorities of Accelerating Climate Action, Preventing, Reducing and Reversing Land Degradation; Accelerating Ecosystem Restoration and Halting and Reversing Biodiversity Loss, Sustainable and Integrated Water Resource Management, Protecting and Conserving the Ocean, Promoting a Sustainable and Resilient Blue/Ocean-Based Economy, and Promoting Resource Efficiency and Circular Economy, Preventing Pollution and Minimizing Its Risks.*

The Working Group also delivered 10 Presidency Documents, as presented below.

10 Presidency Documents

Land & Biodiversity

1. Presidency's **Gandhinagar Implementation Roadmap (GIR)** and **Gandhinagar Information Platform (GIP)** for strengthening the G20 Global Land Initiative
2. Compendium of **Best Practices** on **Restoration of Mining Affected Areas**
3. Compendium of **Best Practices** on **Restoration of Forest Fire Affected Areas**
4. Compendium of **Best Practices** for **Water Management**

Blue Economy

5. Technical Study on **Accelerating the Transition to a Sustainable and Resilient Blue Economy**
6. G20 Report on **Actions Against Marine Plastic Litter** – Fifth Information Sharing Based on G20 Implementation Framework

Circular Economy

7. Knowledge Exchange on **Circular Economy in Steel Sector**
8. Knowledge Exchange on **EPR for Circular Economy**
9. Knowledge Exchange on **Circular Bioeconomy**
10. Resource Efficiency and **Circular Economy Industry Coalition (RECEIC)**

India also placed a strong emphasis on restoring **degraded lands** affected by **forest fires** and **mining** activities. Forest fire–impacted land restoration became a central theme, reflecting the increasing threat posed by wildfires, which are exacerbated by **climate change**. India called for global commitments to restore fire-affected landscapes through **reforestation** and **natural regeneration** techniques, underscoring the need for both **prevention** and **restoration** in combating the long-term effects of forest degradation.

Mining-induced land degradation was another critical focus area. India's G20 Presidency highlighted the importance of **sustainable mining practices** and the need for **ecological restoration** of abandoned mining sites. By advocating for the **remediation of degraded mining lands**, India emphasized the urgency of restoring ecological integrity to lands scarred by industrial activities. Sustainable restoration practices, including **soil remediation** and **reforestation**, were promoted as key strategies to reverse the environmental damage caused by mining.

1.3.3.2 Coastal and Marine Ecosystem Restoration

Coastal ecosystems, such as **mangroves**, **coral reefs**, and **seagrass beds**, are crucial for protecting coastal communities from extreme weather events, supporting fisheries, and storing carbon. India's G20 Presidency prioritized the restoration of these ecosystems, recognizing their vulnerability to both **human activities** and **climate change**. Coastal ecosystems play a vital role in mitigating the impacts of **sea level**

rise and **storm surges**, making their restoration a critical element of climate adaptation strategies.

The **Osaka Blue Ocean Vision**, which aims to reduce marine plastic pollution to zero by **2050**, provided a framework for India to expand its focus on marine restoration. The Indian Presidency developed the 5th Report on Actions against Marine Plastic Litter, marking the highest participation of countries and international organizations thus far. The Presidency advanced initiatives to combat **marine plastic pollution**, emphasizing the need for **nature-based solutions** and enhanced **waste management** systems to reduce the flow of plastics into the oceans. In addition, India advocated for the protection and restoration of **marine biodiversity**, highlighting the need to restore coral reefs and seagrass meadows, which are essential for maintaining healthy marine ecosystems and supporting **coastal livelihoods**.

A key achievement unlocked by this Presidency was the adoption of the "**Chennai High-Level Principles for a Sustainable and Resilient Blue/Ocean-Based Economy**" that directly calls upon all G20 nations to halt and reverse biodiversity loss and conserve coastal and marine ecosystems as one of the nine principles negotiated among the G20 countries. It marked the formal recognition of blue/ocean-based economy as a priority, which has a massive role to play in the protection of marine and coastal ecosystems.

1.3.3.3 Biodiversity Conservation

Biodiversity conservation was a key priority for India's G20 Presidency, aligning with the goals of the **Kunming-Montreal Global Biodiversity Framework (KMGBF)**, which calls for the restoration of **30% of degraded ecosystems** by **2030**. India advocated for the restoration of **wetlands, forests**, and other biodiversity-rich ecosystems, recognizing that biodiversity loss threatens the stability of ecosystems and the services they provide.

The restoration of **wetlands** emerged as a critical focus, particularly in the context of **water management** and **climate resilience**. Wetlands provide vital ecosystem services such as **water filtration, flood control**, and habitat for diverse species, yet they are among the most degraded ecosystems globally. India's Presidency supported the **Ramsar Convention on Wetlands** and called for greater international efforts to restore these ecosystems, recognizing that wetlands play an essential role in supporting both biodiversity and human livelihoods.

1.3.4 Key Strategies for Advancing Ecological Restoration

India's G20 Presidency emphasized the importance of comprehensive and integrated strategies for advancing ecological restoration. Several key strategies emerged during the Presidency, focusing on **financing, science–policy integration**, and **community engagement**.

1.3.4.1 Financing Mechanisms

One of the critical challenges in scaling up ecological restoration is securing adequate financing. India's G20 Presidency called for innovative financial mechanisms that could mobilize resources for restoration projects. **Payment for Ecosystem Services (PES)**[5] was highlighted as a key tool for incentivizing landowners and communities to engage in conservation and restoration activities. PES schemes allow landowners to be compensated for maintaining ecosystems that provide services such as **carbon sequestration**, **water purification**, and **biodiversity support**.

In addition to PES, India advocated for the expansion of **green bonds**[6] and **environmental impact bonds**, which can provide upfront capital for restoration projects. These financing mechanisms were promoted as essential tools for aligning financial incentives with environmental goals, ensuring that restoration projects are adequately funded and sustainable in the long term.

1.3.4.2 Science–Policy Integration

India recognized the critical role of **science–policy integration** in ensuring the success of restoration efforts. During its G20 Presidency, India pushed for greater collaboration between scientific institutions and policymakers to ensure that restoration initiatives were informed by the latest ecological research and technological innovations. Digital tools such as **remote sensing**, **GIS mapping**, and **data analytics** were promoted as key enablers for monitoring restoration progress and improving resource allocation.

India's focus on **evidence-based decision-making** highlighted the need for real-time data and scientific research to inform restoration efforts, ensuring that projects are not only effective but also scalable. The Presidency emphasized that successful restoration requires a strong foundation of scientific knowledge to address the complexities of **ecosystem degradation** and develop targeted restoration strategies.

1.3.4.3 Community Participation and Local Knowledge

India's G20 Presidency also prioritized **community engagement** in restoration projects. Local communities, particularly **Indigenous Peoples** and **rural populations**, play a critical role in maintaining the health of ecosystems, and their involvement in restoration efforts is essential for long-term success. India advocated for the inclusion of **Indigenous knowledge systems** and **traditional practices** in restoration initiatives, recognizing the value of local expertise in managing ecosystems sustainably.

[5] Payment for ecosystem services (PES) as defined by the Intergovernmental Science-Policy Platform on Biodiversity and Ecosystem Services (IPBES) is a type of market-based instrument that is increasingly used to finance nature conservation. Payment of ecosystem service programs allow for the translation of the ecosystem services that ecosystems provide for free into financial incentives for their conservation, targeted at the local actors who own or manage the natural resources.

[6] Green bonds are financial instruments that finance green projects and provide investors with regular or fixed-income payments. In 2013, the International Finance Corporation issued the market's first global US dollar benchmark-sized green bonds, with two $1 billion issuances in that year; this set a precedent as the largest green bonds at the time of issuance and helped solidify the market (Source: The World Bank).

By empowering local communities to take ownership of restoration efforts, India's Presidency underscored the need for socially inclusive restoration projects that benefit both **people** and **nature**. Community-driven projects, such as **community-based reforestation** and **sustainable agriculture**, were promoted as models for scaling up restoration efforts in a way that supports **livelihoods** while enhancing **ecosystem resilience.**

1.3.5 Successes Unlocked During India's Presidency

India's G20 Presidency achieved a series of critical successes, which collectively strengthened the global agenda for ecological restoration. These accomplishments reflected India's leadership in navigating the complex intersection of **environmental, economic, and social challenges**. By focusing on **nature-based solutions**, **ecosystem-based approaches**, and **other management and conservation approaches**, India ensured that the G20's environmental strategy was comprehensive, resilient, and adaptable to diverse global ecosystems. The following are some of the key successes that emerged under India's G20 Presidency:

1.3.5.1 Gandhinagar Implementation Roadmap (GIR) and Gandhinagar Information Platform (GIP)

A key milestone of India's G20 Presidency was the acknowledgment of the Gandhinagar Implementation Roadmap (GIR) and the Gandhinagar Information Platform (GIP), centered around the restoration of landscapes of forest fire–impacted areas and mining-affected areas, in order to reinforce the G20 Global Land Initiative (GLI) implementation. The initiative has two key objectives:

1. **Enhance focus on ecological/ecosystem restoration of land** by reinforcing the capacity of habitats to **sustain the diversity of their species, in particular the keystone/indicator/umbrella species**, in accordance with the respective priorities as determined by scientific evidence
2. **Accelerate action** on the identified landscapes by operationalizing the Roadmap's tangible pillars of work to help achieve the objectives under the Kunming-Montreal Global Biodiversity Framework and other relevant frameworks and initiatives, including the G20 voluntary target of restoration of 50% of degraded land by 2040

The G20 nations **recognized the initiative by** noting the release of the Gandhinagar Implementation Roadmap (GIR) and Gandhinagar Information Platform (GIP) paper during the Indian Presidency's Environment and Climate Sustainability Working Group's (ECSWG) fourth ministerial meeting.

The India G20 ECSWG Outcome Document reflected this achievement by explicitly stating that the G20 nations will **promote integrated action and commit to enhancing ecosystem resilience through protection, conservation, sustainable use, and restoration measures**, which include **nature-based solutions, ecosystem-based approaches**, and other management approaches. The acceptance of this statement marked a critical moment in international environmental policy, reflecting a unified global strategy that balances diverse ecosystem needs and socio-economic considerations.

This success established a new global standard for restoration, emphasizing that these approaches should be pursued not in isolation but as interconnected strategies that can deliver **climate change mitigation**, **biodiversity conservation**, and **socio-economic benefits** simultaneously. India's leadership was pivotal in embedding this comprehensive approach into the G20's environmental policy, ensuring that future restoration efforts remain multidimensional and resilient.

1.3.5.2 Balance Between Conservation and Trade: A Win for Developing Nations

India's G20 Presidency also achieved a significant diplomatic win by addressing the concerns raised by developing nations regarding the balance between **conservation** and **economic development**. It was highlighted that while many nations, particularly in the **Global South**, are pressured to conserve their forests, they face restrictions on the trade of forest products, which are often critical to their economies.

India's Presidency facilitated crucial discussions that led to alignment, ensuring that **green economy policies**, including unilateral ones, **should not constitute a means of arbitrary or unjustifiable discrimination or act as disguised restrictions on international trade.**

This was a major victory for developing countries, reflecting a commitment to **equity** in environmental policies. India's Presidency played a key role in securing this agreement, demonstrating its ability to mediate complex international issues and deliver balanced solutions that promote both **sustainability** and **economic sovereignty**.

1.3.5.3 Strengthening the Blue Economy Agenda

Another notable success of India's G20 Presidency was its leadership in advancing the **Blue Economy** agenda, particularly in promoting the sustainable management of **marine ecosystems** and **coastal restoration**. India strengthened the **Osaka Blue Ocean Vision**, which set the ambitious goal of eliminating additional marine plastic pollution by 2050. By building on this vision, India pushed for a comprehensive approach to **marine conservation**, emphasizing the need to restore **mangroves**, **coral reefs**, and **seagrass beds**, all of which are critical for **carbon sequestration**, **coastal protection**, and **biodiversity conservation**.

India also championed efforts to reduce marine plastic pollution by integrating **circular economy principles**. This involved promoting **resource efficiency**, reducing single-use plastics, and enhancing **waste management systems** to prevent plastics from entering marine ecosystems. These efforts were vital in aligning environmental goals with economic sustainability, particularly for coastal communities that depend on marine resources.

The **Chennai High-Level Principles for a Sustainable and Resilient Blue/Ocean-Based Economy**, adopted during India's Presidency, further underscored the importance of **marine conservation** as a central element of **economic**

resilience. By advancing this agenda, India ensured that **marine ecosystem restoration** would remain a priority in future G20 actions, contributing to the broader goals of climate adaptation and biodiversity protection.

1.3.5.4 Advancing Global Commitments on Land Degradation

India's G20 Presidency also achieved significant progress in advancing global commitments to tackle **land degradation**. India played a key role in rallying G20 countries around the goal of achieving **Land Degradation Neutrality (SDG 15.3)** by 2030, with a particular focus on restoring **agricultural lands**, **forests**, and **mining-degraded areas**.

Through the **Global Initiative on Reducing Land Degradation**, India reinforced the need for **sustainable land management** practices, such as **agroforestry**, **reforestation**, and the remediation of lands affected by industrial activities. A central part of this effort was securing commitments to mobilize **financial resources** for restoration, particularly through **innovative financing mechanisms** like **green bonds**, **environmental impact bonds**, and **Payment for Ecosystem Services (PES)** programs.

India also emphasized the importance of **monitoring and reporting systems** to track progress on land restoration targets. By promoting **real-time monitoring technologies** and **data-driven decision-making**, India ensured that restoration efforts would be scalable, measurable, and transparent.

1.3.5.5 Mobilizing Financial Resources for Restoration

A crucial aspect of India's leadership was the emphasis on mobilizing the financial resources necessary to support **large-scale restoration projects**. Recognizing the significant costs associated with restoring degraded ecosystems, India pushed for both **public** and **private sector** engagement in financing restoration efforts.

India advocated for the expansion of **blended finance models**, which leverage public investment to attract private capital, thereby reducing financial risks and increasing overall investment in restoration. India's leadership also highlighted the role of **multilateral development banks (MDBs)** and **international financial institutions (IFIs)** in providing the necessary financial flows to **developing countries**, many of which face the greatest challenges in addressing land and marine degradation.

The promotion of **green bonds**, **impact funds**, and other **innovative financial instruments** ensured that restoration efforts could be sustained over the long term, providing critical support to communities and ecosystems in need of urgent action.

1.3.6 India's Legacy in Global Ecological Restoration

India's G20 Presidency marked a turning point in the global effort to restore degraded ecosystems. Through its ambitious agenda, India demonstrated its commitment to addressing **climate change**, **biodiversity loss**, and **environmental degradation** while promoting **inclusive, sustainable development**. By building on the achievements of previous presidencies and introducing new strategies for advancing restoration, India solidified its role as a global leader in ecological restoration. India's G20 Presidency not only advanced the restoration agenda at the global level but also provided a model for how countries can integrate **restoration** into their **national development strategies**. As the world continues to confront the challenges of **ecosystem degradation** and **climate change**, India's leadership in promoting restoration offers a path forward to build a more **resilient** and **sustainable future** for both nature and humanity.

1.4 How India's G20 Presidency Progress Can Drive Ecological Restoration Implementation in India

India's leadership during the 2023 G20 Presidency has set the stage for transformative changes in global ecological restoration. While India has been at the forefront of advancing global commitments, the lessons and strategies developed on the world stage now need to be implemented at home. With some of the most biodiverse ecosystems in the world, India faces urgent environmental challenges, from widespread biodiversity loss and escalating climate change impacts to critical issues around land degradation and pollution. Yet the potential for restoration holds tremendous promise for improving not just the health of ecosystems but also the livelihoods of millions who depend on them.

The pressing need for ecological restoration in India is not just a matter of environmental concern—it directly ties to the country's economic stability, food security, and social equity. The complex interplay between environmental degradation and socioeconomic issues, as outlined in the G20 Outcome Document, reflects the broader reality that environmental health is inseparable from human well-being. Restoration, therefore, is not an isolated initiative but a critical pathway toward building resilience against climate risks, restoring biodiversity, and achieving sustainable growth.

1.4.1 India's Environmental Challenges: Urgency and Complexity

The environmental landscape in India presents a myriad of challenges that require urgent and comprehensive solutions. **Biodiversity loss** has been accelerating, driven by rapid **deforestation**, **habitat destruction**, and **urban expansion**. Coastal ecosystems, once thriving, are increasingly vulnerable to the impacts of climate change and human activity. The country is also grappling with **air and water pollution**, with the growing menace of plastic waste and industries contributing to hazardous emissions. Meanwhile, vast areas of agricultural land are deteriorating, threatened by **soil erosion** and unsustainable practices that contribute to land degradation and **desertification**.

These issues are intricately connected, as highlighted in the Indian Presidency ECSWG Outcome Document. Ecosystem degradation exacerbates the effects of climate change, increases poverty, and threatens food security, particularly in rural areas. For a nation whose development is intertwined with its natural resources, restoring degraded ecosystems is not just a matter of environmental protection but a strategic imperative for sustainable growth.

Biodiversity loss in India poses one of the greatest challenges. Home to several biodiversity hotspots, India is witnessing alarming declines in species populations due to habitat fragmentation and the overexploitation of resources. The collapse of these ecosystems could have far-reaching consequences—not just for the environment but for the many communities that rely on these ecosystems for agriculture, clean water, and other resources.

At the same time, **climate change** is intensifying, manifesting in more frequent and severe natural disasters. From deadly floods to prolonged droughts, the changing climate is disrupting food production, displacing communities, and undermining economic stability. Coastal regions, in particular, face existential threats from rising sea levels and extreme weather events, jeopardizing the lives and livelihoods of millions.

India's **pollution crisis** compounds these challenges. Air quality in urban centers is among the worst globally, with pollutants contributing to both public health crises and environmental degradation. Rivers such as the Ganges are severely polluted by industrial waste and plastic, damaging aquatic ecosystems and threatening water security.

Finally, **land degradation** is eating away the country's arable land, with **desertification** expanding in some of the most drought-prone areas. Without significant intervention, the loss of fertile soil could undermine agricultural productivity, further stressing food security in a country with a rapidly growing population.

1.4.2 Overcoming Policy Gaps for Restoration Success

Despite having comprehensive environmental frameworks, India faces substantial gaps when it comes to enforcement and execution. Policies that address **deforestation, waste management**, and **pollution** often exist in silos, with weak coordination between ministries. Additionally, while laws such as the **Forest Conservation Act** and the **Water Prevention Act** are in place, they are hampered by inconsistent enforcement and the challenge of bureaucratic inefficiency.

To translate the global momentum from the G20 into meaningful national action, India must create stronger linkages between its environmental and development policies. Restoration needs to be embedded into broader national development frameworks, ensuring that initiatives in **agriculture**, **infrastructure**, and **urban planning** align with ecological sustainability goals. This would reflect the G20's emphasis on inclusive economic growth that integrates environmental health as a foundational pillar.

The **Kunming-Montreal Global Biodiversity Framework** offers a roadmap for India to bolster its biodiversity efforts by aiming for the restoration of **30% of degraded ecosystems by 2030**. Additionally, adopting the goals of the **Osaka Blue Ocean Vision**, which targets the reduction of **marine plastic pollution** to zero by 2050, would help India tackle its growing plastic waste problem, particularly in its coastal regions. There is a clear need for integrated policy frameworks that can drive coordinated action across government departments.

1.4.3 Strategies for Ecological Restoration: Scaling Impact Across India

India's G20 Presidency demonstrated the power of **nature-based solutions** and **ecosystem-based approaches** in addressing environmental crises. These solutions, which include the restoration of **forests**, **wetlands**, and **mangroves**, are integral to building resilience against climate change, improving biodiversity, and enhancing ecosystem services. Scaling these approaches domestically will require a strategic push across sectors, engaging local communities, leveraging innovative financing models, and advancing science-driven restoration practices.

The restoration of **forests and degraded lands** is central to India's ecological goals. Reforestation and **agroforestry systems**, which integrate trees with crops, can provide both environmental benefits—such as carbon sequestration—and economic opportunities for rural communities. Initiatives such as the **National Mission for a Green India** aim to increase forest cover, but scaling up these efforts will require a stronger alignment with local stakeholders, particularly in forest-dependent regions.

Coastal restoration is another critical area, particularly in light of rising sea levels and the vulnerability of India's extensive coastline. The restoration of **mangroves**, which act as natural barriers against storm surges and provide habitat for marine life, will be essential in protecting coastal communities and promoting **climate resilience**. Expanding projects that restore mangroves and **seagrass beds** can also enhance the productivity of coastal fisheries, creating sustainable livelihoods while restoring ecosystems.

Similarly, addressing the degradation of **wetlands** is vital for improving **water security** and regulating floods. India's wetland systems are among the most endangered ecosystems, yet they provide essential services such as **water purification** and **biodiversity conservation**. Strengthening efforts to restore wetlands, particularly through international frameworks like the **Ramsar Convention on Wetlands**, would position India as a leader in water management and biodiversity protection.

Community involvement will be critical to the success of these restoration efforts. Engaging **Indigenous Peoples**, **local communities**, and **women**—as emphasized in the **G20 Outcome Document**—is key to creating sustainable projects that are grounded in local knowledge and practices. Community-based projects, such as **community-driven reforestation** and **sustainable agriculture**, can drive both ecological and socioeconomic benefits, ensuring that restoration efforts support livelihoods while enhancing ecosystem resilience.

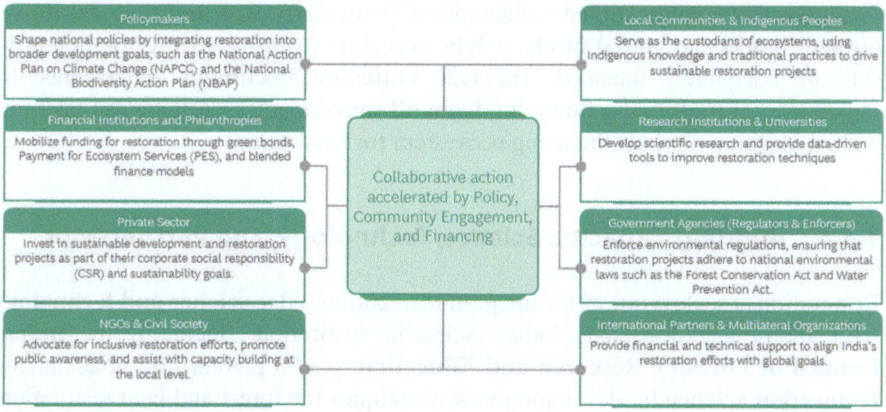

Fig. 1.3 Key stakeholders driving ecological restoration in India

1.4.4 Scaling Impact Across India: A Collaborative Effort

Achieving large-scale ecological restoration across India will require the coordinated efforts of various stakeholders—each playing a critical role in driving the restoration agenda forward. From **policymakers** establishing supportive regulations to **local communities** implementing restoration projects on the ground, a multi-stakeholder approach is essential to ensuring both the success and sustainability of these efforts.

The integration of **financial institutions**, **research organizations**, and **government agencies** will be necessary to mobilize the financial, technical, and scientific resources required for long-term impact. At the same time, the active engagement of **NGOs**, **civil society**, and **international partners** will ensure that restoration efforts are inclusive and aligned with global goals (Fig. 1.3).

1.4.5 Financing Ecological Restoration: Unlocking Investment

Large-scale restoration requires significant financial resources, and one of the key challenges India faces is mobilizing adequate funding. The success of India's restoration ambitions will depend on leveraging both **public and private sector investment**, creating a framework where environmental projects are economically viable and attractive to investors. India's G20 Presidency made substantial progress in advocating for innovative **financing mechanisms**, such as **green bonds** and **Payment for Ecosystem Services (PES)**, which will be critical in driving long-term restoration.

India can build on these innovations by promoting **blended finance models**, which combine public funding with private capital to de-risk investments in **restoration projects**. By aligning restoration with **economic benefits**—such as **carbon credits**, **sustainable forestry**, and **ecotourism**—India can tap into new financial streams that make large-scale restoration both feasible and profitable.

Furthermore, international collaboration, particularly with **multilateral development banks** and **global funds**, will be necessary to ensure that restoration projects are adequately financed. The **G20 Outcome Document** underscores the importance of **mobilizing financing from all sources**, a lesson India can take forward to develop a robust financing ecosystem for restoration.

1.4.6 Building Capacity: Science, Technology, and Partnerships

Restoration at scale requires the integration of cutting-edge **science and technology** with traditional knowledge. India's scientific institutions, including the **Indian Council of Forestry Research and Education**, play a pivotal role in advancing **restoration science** by developing new techniques for forest and land restoration. These innovations, coupled with **remote sensing technologies** and **data-driven decision-making tools**, can dramatically improve the precision and efficiency of restoration projects.

Public–private partnerships (PPPs) are also crucial in scaling up restoration efforts. By fostering collaboration between the government, private sector, and civil society, India can unlock resources, technical expertise, and community support to implement projects that are scalable and sustainable. Partnerships with international organizations, such as **UNEP** and the **World Bank**, can provide India with the technical expertise and financial support needed to drive large-scale restoration.

1.4.7 A Path Forward for Restoration in India

India's leadership during its **G20 Presidency** has laid the foundation for ambitious **ecological restoration efforts** at both the global and national levels. By leveraging the progress made during the G20 Presidency, India can now scale up its **restoration efforts** domestically, addressing the urgent environmental challenges it faces, from **biodiversity loss** to **climate change impacts** and **land degradation**. Through a combination of **science-based solutions, inclusive community engagement**, and **innovative financing mechanisms**, India has the opportunity to become a global leader in ecological restoration, setting an example for the world on how to restore ecosystems while promoting **sustainable development** and **social equity**.

1.5 A Recommended Five-Point Action Agenda for Mainstream Ecological Restoration in India

India's G20 Presidency has established a powerful global platform for ecological restoration, offering a unique opportunity to carry this momentum into national action. With India facing urgent environmental challenges—including **biodiversity loss**, **climate change**, **land degradation**, and **pollution**—it is crucial to develop a clear and actionable strategy to mainstream restoration and integrate it into national policy and practice. This five-point action agenda sets forth specific recommendations that will help India integrate restoration into its broader development framework while aligning with global goals and ensuring sustainable economic growth.

1.5.1 Integrating Restoration into National Development Policies

Ecological restoration should not be seen as a stand-alone environmental effort; rather, it must be woven into India's broader **development strategy**. This includes integrating restoration goals into key sectors such as **agriculture**, **urban planning**, **infrastructure development**, and **industry**. Doing so will ensure that ecosystem health becomes a foundational pillar of India's economic and social progress.

To achieve this, India must prioritize **mainstreaming restoration** in its **national development plans**, aligning initiatives like the **National Action Plan on Climate Change (NAPCC)** and the **National Biodiversity Action Plan (NBAP)** with restoration goals. Policies that govern **agriculture**, **forestry**, and **urbanization** must be revisited to include **nature-based solutions**, **ecosystem-based approaches**, and **other management and conservation practices** that support long-term ecological stability.

Aligning restoration efforts with broader development policies will also require creating linkages between restoration and critical socioeconomic goals such as **poverty alleviation**, **rural development**, and **disaster risk reduction**. Restoring ecosystems is not just about the environment; it is about ensuring **food security**, **water availability**, and **community resilience**. Integrating restoration into India's development frameworks will create a multiplier effect, where environmental health fuels economic growth.

1.5.2 Scaling Up Nature-Based Solutions and Ecosystem-Based Approaches

India has already recognized the importance of **nature-based solutions (NbS)** and **ecosystem-based approaches** in addressing climate change and environmental degradation, but these efforts must be scaled up to make a tangible impact. The **G20 Outcome Document** highlights the need for **promoting integrated action** through **nature-based solutions**, **ecosystem-based approaches**, and other management strategies to restore and enhance ecosystem resilience. This multidimensional approach can be expanded across India's landscapes to address **deforestation**, **land degradation**, **wetland loss**, and **coastal erosion**.

A large-scale expansion of **reforestation** and **agroforestry** efforts can provide **carbon sequestration**, support **biodiversity**, and enhance **livelihoods** for rural communities. Similarly, restoring **mangrove ecosystems** and **wetlands** is essential for buffering coastal populations from climate impacts while simultaneously supporting **fisheries** and improving **water quality**.

India's action agenda must prioritize the restoration of **degraded lands**, particularly through the scaling up of **natural regeneration** and **reforestation** initiatives. These efforts should also extend to the **restoration of coastal ecosystems**, including **mangroves**, **seagrass meadows**, and **coral reefs**, to protect against **rising sea levels** and extreme weather events. By embedding these strategies within national programs, India can maximize the environmental, economic, and social benefits of its restoration projects.

1.5.3 Mobilizing Innovative Financing for Large-Scale Restoration

Restoration at the scale India requires cannot be achieved without robust financial mechanisms. To mobilize the necessary investment, India needs to develop a strong framework that attracts both **public and private sector finance**. India's G20 Presidency laid the groundwork by advocating for innovative financial tools, such as **green bonds**, **Payment for Ecosystem Services (PES)**, and **blended finance models**, which will be key to unlocking the capital needed for large-scale restoration projects.

A targeted financial strategy must include the following:

- **Green bonds** that provide upfront capital for restoration projects and enable investors to contribute to sustainable initiatives
- **Payment for Ecosystem Services (PES)**, allowing landowners and communities to be compensated for maintaining and restoring ecosystems that provide critical services, such as **carbon sequestration** and **water filtration**
- **Blended finance models** that combine **public investment** with **private capital**, reducing risks and incentivizing private sector participation in restoration efforts

Additionally, India should leverage **international funds** such as the **Green Climate Fund** (**GCF**) and **Global Environment Facility** (**GEF**), as well as engage **multilateral development banks (MDBs),** to secure the financial resources required to scale up restoration initiatives. The private sector, including industries dependent on natural resources, should be encouraged to invest in **restoration projects** as part of their **corporate social responsibility** (**CSR**) efforts and **sustainability goals**.

1.5.4 Strengthening Policy Coherence and Enforcement

India's existing environmental policies need to be strengthened and better coordinated to ensure that restoration efforts are effective and sustained. The **G20 Outcome Document** emphasizes the importance of **policy coherence** and **effective enforcement**. India must take a similar approach by aligning its environmental laws with global frameworks, such as the **Kunming-Montreal Global Biodiversity**

Framework, which sets clear targets for restoring degraded ecosystems and conserving biodiversity.

Policy coherence must also extend across sectors. For instance, forest conservation policies need to be integrated with land use planning, agricultural policies, and water management strategies. Furthermore, stricter **enforcement mechanisms** are needed to curb illegal activities that harm ecosystems, such as **illegal logging**, **mining**, and **pollution**. A unified approach that involves stronger enforcement from **local governments**, along with **capacity-building** programs for environmental agencies, will ensure that **restoration projects** are implemented effectively and achieve long-term success.

Improved **monitoring systems** and the use of **digital tools**—such as **GIS mapping**, **remote sensing**, and **data analytics**—can help track the restoration progress and ensure compliance with environmental regulations.

1.5.5 Engaging Local Communities and Indigenous Knowledge

The success of ecological restoration hinges on the active participation of local communities, including **Indigenous Peoples**, **women**, and other **marginalized groups**. India must prioritize **community-driven restoration projects** that empower local stakeholders and incorporate **Indigenous knowledge systems**, which have long been central to ecosystem management.

Community-based reforestation, **sustainable agriculture**, and **wetland restoration** projects that involve local populations can not only enhance **ecosystem resilience** but also support **livelihoods** and improve **social equity**. The G20 Outcome Document underscores the importance of engaging **local communities** in decision-making processes and ensuring that restoration projects are socially inclusive.

Building partnerships with **local NGOs**, **civil society organizations**, and **private sector entities** will be critical in ensuring that restoration projects are tailored to the specific needs of communities and ecosystems. Community engagement should also extend to the **monitoring** and **management** of restored ecosystems, ensuring that they are maintained over the long term and continue to deliver **socio-economic** and **environmental** benefits.

1.5.6 Conclusion: A Roadmap for India's Ecological Future

India stands at a pivotal moment in its journey toward ecological restoration. The progress made during its G20 Presidency provides a strong foundation for transformative national action. By implementing this **five-point action agenda**, India can mainstream restoration into its national development framework, scaling up efforts to restore ecosystems, improve biodiversity, and enhance climate resilience.

India has established global leadership through the execution of a successful G20 Presidency in 2023 and ensuring continuity of progress in Brazil's Presidency as well, made evident by the consideration of topics such as Payment for Ecosystem

Services and Oceans as thematic priorities. India's presence at COP29 and UNCCD COP16 is significantly prominent, reiterating the nation's commitment to combat land degradation and foster global drought resilience. India has hosted the 14th session of the Conference of the Parties to the United Nations Convention to Combat Desertification (UNCCD) and at COP28 in 2023, India's prime minister, Mr. Narendra Modi, proposed hosting the 33rd Conference of the Parties of the UNFCCC (COP 33) in 2028. This showcases **India's readiness to lead international fora on a massive scale,** and the initiatives, propositions, and experiences garnered from the G20 Presidency and leading the UNCCD COP and others would offer a template to be presented to the world toward COP 33.

Through **integrated policies, community engagement**, and **innovative financing**, India has the opportunity to lead the global effort in ecological restoration, ensuring a sustainable future for both nature and humanity.

References

Begosh A, Smith LM, McMurry ST (2022) Major land use and vegetation influences on potential pollinator communities in the High Plains of Texas. J Insect Conserv *26*(2):231–241. https://link.springer.com/article/10.1007/s10841-022-00380-9

Climate Scorecard (2023) The Saudi Green Initiative. https://www.climatescorecard.org/2023/07/the-saudi-green-initiative/

El-Kogali SET (2024) Fostering a sustainable future: the role of land restoration in Saudi Arabia. World Bank Blogs. https://blogs.worldbank.org/en/arabvoices/fostering-a-sustainable-future%2D%2Dthe-role-of-land-restoration-in-saudi-arabia

Food and Agriculture Organization of the United Nations (2002) Chapter 8: The role of agriculture in the development of least-developed countries and their integration into the world economy. https://www.fao.org/4/v4360e/V4360E08.htm

Food and Agriculture Organization of the United Nations (2022) Kunming-Montreal global biodiversity framework. https://www.fao.org/biodiversity/kunming-montreal-global-biodiversity-framework/en

G20 Global Land Initiative (2024) Reducing land degradation and enhancing conservation of terrestrial habitats. https://g20land.org/

G20 India (2023) G20 background brief. https://www.g20.in/en/docs/2022/G20_Background_Brief.pdf

G20 Indonesia (2022) G20 Chair's summary: joint environment and climate ministers' meeting, August 31, 2022

G20 Research Group (2020) Road to the G20 Riyadh Summit: G20 environment ministers meetings. University of Toronto

Haryanto R, Affandi D, Tanaya S (2022) 6 ways the G20 can maximize the role of forests in climate action. World Resources Institute. https://www.wri.org/insights/6-ways-g20-can-maximize-role-forests-climate-action

International Union for Conservation of Nature (2021) Peatlands and climate change. https://iucn.org/sites/default/files/2022-04/iucn_issues_brief_peatlands_and_climate_change_final_nov21.pdf

Jambeck JR, Geyer R, Wilcox C, Siegler TR, Perryman M, Andrady A, Narayan R, Law KL (2015) Plastic waste inputs from land into the ocean. Science 347(6223):768–771. https://doi.org/10.1126/science.1260352

Meakin S (1992) The Rio Earth Summit: summary of the United Nations Conference on Environment and Development (BP-317E). Government of Canada Publications. https://publications.gc.ca/Collection-R/LoPBdP/BP/bp317-e.htm

Natural History Museum (2024) What is mass extinction and are we facing a sixth one? https://www.nhm.ac.uk/discover/what-is-mass-extinction-and-are-we-facing-a-sixth-one.html
PlasticsEurope (2020) Plastics – the facts 2020. https://www.plasticseurope.org/en/resources/publications/4312-plastics-facts-2020
Secretariat of the Pacific Regional Environment Programme (2012) MEA series: factsheet 2. https://www.sprep.org/attachments/Publications/FactSheet/MEA_Factsheet_2_Web.pdf
United Nations Conference on Trade and Development (2022) Kunming-Montreal global biodiversity framework. https://unctad.org/topic/trade-and-environment/biotrade/kunming-montreal-global-biodiversity-framework
United Nations Convention to Combat Desertification (2017) Ecological restoration: a global priority. https://www.unccd.int/sites/default/files/2018-06/16.%20Ecological%2BRestoration__N_D_Crossmann.pdf
United Nations Convention to Combat Desertification (2019) Data and facts about recent assessments of land degradation. https://www.unccd.int/sites/default/files/relevant-links/2019-06/Summary_Assessement_Land_Degradation.pdf
United Nations Convention to Combat Desertification (2022) Global land outlook second edition: land restoration for recovery and resilience
United Nations Decade on Ecosystem Restoration (2021) UN Decade on Ecosystem Restoration (2021–2030). https://www.decadeonrestoration.org/
United Nations Economic Commission for Europe (2021) UN decade on ecosystem restoration. https://unece.org/un-decade-ecosystem-restoration
United Nations Educational, Scientific and Cultural Organization (2018) Worsening land degradation impacts 3.2 billion people worldwide. https://www.unesco.org/en/articles/worsening-land-degradation-impacts-32-billion-people-worldwide
United Nations Environment Programme (2018) #FridayFact: Every minute, we lose 23 hectares of arable land worldwide to drought. https://www.unep.org/news-and-stories/story/fridayfact-every-minute-we-lose-23-hectares-arable-land-worldwide-drought
United Nations Environment Programme (2021a) Ecosystem restoration playbook. https://wedocs.unep.org/bitstream/handle/20.500.11822/36251/ERPNC.pdf
United Nations Environment Programme (2021b) Comprehensive assessment of marine litter and plastic pollution. https://www.unep.org/news-and-stories/press-release/comprehensive-assessment-marine-litter-and-plastic-pollution
United Nations Environment Programme (2022a) Global peatlands assessment 2022. https://www.unep.org/resources/global-peatlands-assessment-2022
United Nations Environment Programme (2022b) Kunming-Montreal global biodiversity framework. https://www.unep.org/resources/kunming-montreal-global-biodiversity-framework
United Nations Environment Programme (2024) What is ecosystem restoration? https://www.decadeonrestoration.org/what-ecosystem-restoration
United Nations Framework Convention on Climate Change (2023) Simon Stiell & Dr. Sultan Al Jaber: leadership by the G20 is indispensable. https://unfccc.int/news/simon-stiell-dr-sultan-al-jaber-leadership-by-the-g20-is-indispensable
United Nations Office for Disaster Risk Reduction (2019) Terminology: high impact practices. https://www.undrr.org/understanding-disaster-risk/terminology/hips/en0016
United Nations Regional Information Centre (2022) Oceans in danger: the threats they face. https://unric.org/en/oceans-in-danger-the-threats-they-face/
United States Environmental Protection Agency (2001) Threats to wetlands. https://www.epa.gov/sites/default/files/2021-01/documents/threats_to_wetlands.pdf
World Resources Institute (2024) Forest loss. https://research.wri.org/gfr/forest-extent-indicators/forest-loss
World Wide Fund for Nature (2024) Habitat loss and degradation. https://wwf.panda.org/discover/our_focus/wildlife_practice/problems/habitat_loss_degradation/

Eco-Restoration in India: A Policy Perspective

2

Charan Jeet Singh and Dheeraj Mittal

Abstract

India, the seventh-largest country globally, occupies 2.4% of the world's geographical area yet supports 16% of its population, reflecting immense pressure on its land resources. With per capita arable land declining from 0.15 ha to an anticipated 0.09 ha by 2075, land degradation poses a severe threat to sustainability. The country is vulnerable to soil erosion, biodiversity loss, and habitat destruction due to anthropogenic activities like urbanization, industrialization, and deforestation. Restoring degraded ecosystems is crucial for mitigating climate impacts, preserving biodiversity, and sustaining ecosystem services essential for human well-being.

Aligned with the National Forest Policy (NFP) of 1988, India has undertaken significant afforestation initiatives, including a pledge under the Bonn Challenge to restore 26 million hectares of degraded land by 2030. However, this represents only 1.5% of India's degraded land, indicating the need for enhanced efforts. Current interventions focus on Land Degradation Neutrality (LDN) targets across 15 states, with innovative approaches like agroforestry and Joint Forest Management Committees (JFMCs) driving restoration efforts. Agroforestry, in particular, offers a promising solution by converting wastelands into productive landscapes, enhancing biodiversity, and supporting rural economies.

Despite progress, challenges persist, including fragmented scientific knowledge and limited private sector and nongovernmental organization (NGO) involvement. Comprehensive assessments and cross-sectoral collaboration are critical to addressing land degradation effectively. Integrating scientific expertise, local knowledge, and practical technology can ensure sustainable land man-

C. J. Singh (✉) · D. Mittal
Indira Paryavaran Bhawan, Ministry of Environment, Forest and Climate Change,
New Delhi, Delhi, India
e-mail: c.jsingh1@gov.in; dmittal.ifs@gov.in

D. Mittal, D. K. Gupta (eds.), *Ecological Restoration in India: Perspective and Practices*, https://doi.org/10.1007/978-981-96-8943-9_2

agement and restoration. India's commitment to ecological restoration is a cornerstone for achieving global environmental goals and securing a healthier, more resilient future.

Keywords

Degradation · Restoration · Compensatory afforestation · Ecosystem services

2.1 Introduction

India is the seventh-largest country in the world, with a total geographical area of 329 million ha, which is about 2.4% of the world's total geographical area, and supports approximately 16% of the world's population (NAP 2023). Per capita arable land in India is around 0.15 ha, which is expected to decrease to a meager 0.09 ha by 2075, implying that land is highly vulnerable to degradation. Given the natural process of reclamation, which generally takes a long period of up to 1000 years to form an inch of the topsoil, it is necessary to promote eco-restoration of degraded forest lands of the country (LRUMG 2016). Increasing GDP growth is expanding urbanization and industrialization, and, therefore, more and more land is being utilized for non-various developmental purposes (Maji et al. 2010).

India is one of the most biodiverse countries in the world, hosting a myriad of ecosystems, from forests and wetlands to grasslands and coastal areas. In consonance with the National Forest Policy (NFP), 1988, which envisages the national goal to have a minimum of one-third of the total land area under forest or tree cover, the Central Government has been taking several initiatives through various afforestation-related schemes implemented by the Ministry of Environment, Forest and Climate Change and other ministries of the Government of India, which aim at increasing and improving forest and tree cover, thereby combating desertification. However, the pressures of population growth, economic development, and environmental degradation have necessitated a focused approach toward ecological restoration (Singh et al. 2021).

Ecological restoration is a critical approach for reviving ecosystems that have been degraded, damaged, or destroyed. In India, rapid urbanization, industrialization, and agricultural intensification have severely impacted ecosystems, leading to biodiversity loss, habitat destruction, and climate instability. The fragmentation of natural habitats and their disappearance due to anthropogenic drivers are playing key roles in the loss of various species, associated ecosystem services and functions, and human well-being (Bawa et al. 2020).

The scientific evidence indicates that ecosystem degradation is a major force behind recent extreme climate events, such as heat waves, floods, droughts, and landslides, in ecologically sensitive regions of the country (Mishra et al. 2022; Singh et al. 2022). This degradation of the life-supporting environment is a serious challenge for the much-discussed sustainable development of India. The policy momentum to restore degraded ecosystems is well reflected in India's various initiatives and missions (Bawa et al. 2020). It is the need of the hour to formulate sustainable eco-restoration strategies to ensure lesser soil loss and land degradation (Thakur et al. 2022).

Restoring our planet's imperiled ecosystems intrinsically connects us with a chance at a healthier future. The Bonn Challenge is a global effort to bring 150 million hectares of deforested and degraded land into restoration by 2020 and 350 million hectares by 2030. The Government of India made a Bonn Challenge pledge to bring under restoration 13 million hectares of degraded land by 2020 and an additional eight million hectares by 2030 (Borah et al. 2018). India joined the Bonn Challenge in 2015 with a pledge to restore 13 million hectares of degraded and deforested land by 2020 and 26 million hectares by 2030 to be able to create a committed carbon sink of 2.5–3 billion tons of carbon dioxide equivalent; there are ecologists also who suggest that whatever has been committed is just 1.5% of the total degraded land in India.

2.2 Causes of Land Degradation in India

Lack of adequate information on soil resources, coupled with improper land use planning, many of the present-day land degradation problems in our country have resulted, including salinity / alkalinity and waterlogging in command areas, severe erosion in catchments leading to siltation of reservoirs, and a decrease in productivity of crops.

During 2015–16, the statistics of land degradation revealed that water erosion is the major cause of land degradation in India, accounting for 509.96 lakh ha, which is 55.91% of the total land degradation and 15.53% of the total geographical area (TGA) (UNCCD 2022). Wind erosion accounted for 142.8 lakh ha, contributing to 15.66% of total land degradation in India and 4.35% of the total geographical area. Salinization/alkalization was spread over 64.65 lakh ha or 7.09% of total land degradation, and acidification over 30.41 lakh ha (3.33% of total land degradation) (Souvik 2024). Paddy lands, dense forest areas, and plantations have been excluded from this spatial extent. Anthropogenic processes include industrial effluent areas, mining and dumps, and brick kilns. These are spread over an area of 6.55 lakh ha. Barren rocky areas, mining and industrial areas and snow covered areas/snow caps, categorized under "others" category account for 12.13% of total land degradation. Compared to 2005–06, there has been a substantial increase in the area under the anthropogenic category of land degradation, while all the other categories saw a decline, though marginal (Table 2.1).

2.3 Status of Degraded Land in India

There is no periodic conduct of scientific survey/study specifically to identify the extent and nature of soil erosion and land degradation across the country. However, assessments have been made by different agencies/organizations using different methodologies/criteria from time to time. Details of assessments of degraded forest lands in the country, as assessed by the various organizations, are given in Table 2.2.

Table 2.1 Types of land degradation in India

Degradation type	Area (in M ha)	Land use
Water erosion		
Loss of topsoil	13.25	Mainly agricultural areas
Gully formation	8.31	Mainly wastelands and partly agricultural
Ravines	2.06	Mainly wastelands
Wind erosion		
Loss of topsoil	3.76	Mainly agricultural areas
Overblowing	1.89	Partly agricultural and partly wastelands
Terrain deformation	3.24	Mainly wastelands
Chemical degradation		
Salt-affected soils	6.73	Partly agricultural and partly wastelands
Land degradation due to acidity	16.03	Mainly agricultural areas and partly wastelands
Physical degradation		
Waterlogging		
Surface ponding	1.66	Mainly wastelands
Subsurface waterlogging	4.75	Mainly agricultural areas
Vegetal degradation with water erosion	40.16	Mainly wastelands
Others		
Mining and industrial waste	0.13	Wastelands
Barren rocky/stony waste	6.46	Wastelands
Snow-covered area/ice caps	5.58	Wastelands
Total area	**114.01**	

Source: Degraded and Wastelands of India Status and Spatial Distribution, June 2010

Table 2.2 Assessment of land degradation in India by different agencies/organizations

Agencies/organizations	Year	Area (million hectares)
National Commission on Agriculture	1976	148.09
Ministry of Agriculture (Soil and Water Conservation Division)	1978	175.00
National Remote Sensing Agency (NRSA)	1985	53.28
Ministry of Agriculture	1985	173.64
Ministry of Agriculture	1994	107.43
ICAR-National Bureau of Soil Survey and Land Use Planning (NBSS&LUP)	1994	187.70
ICAR-NBSS&LUP (Revised)	2004	146.82
Indian Council of Agricultural Research, National Academy of Agricultural Sciences (NAAS) based on harmonized database	2010	120.70
Space Applications Centre (SAC), Indian Space Research Organization (ISRO), Ahmedabad (based on Indian Remote Sensing Satellite (IRS) Advanced Wide Field Sensor (AWiFS) data)	2016	96.40
Department of Land Resources in collaboration with the National Remote Sensing Centre (Wastelands Atlas of India)	2019	55.77

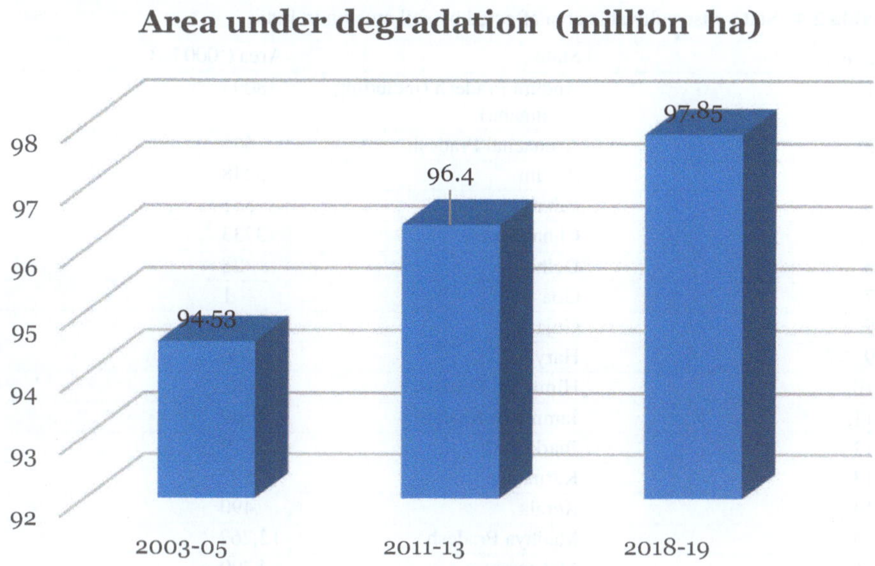

Fig. 2.1 Land area under degradation. (Source: SAC 2021)

Table 2.3 Factors responsible for land degradation

S. no.	Agents of land degradation	2003–05	2011–13	2018–19
1.	Water erosion	10.83	10.98	11.01
2.	Vegetation degradation	8.60	8.91	9.15
3.	Waterlogging	0.18	0.20	0.25
4.	Frost shattering	0.95	1.02	1.05
5.	Mass movement	0.26	0.28	0.29
6.	Man made	0.11	0.12	0.19
7.	Settlement	0.47	0.57	0.69
8.	Wind erosion	5.58	5.55	5.46
9.	Salinity	1.22	1.12	1.11
10.	Barren/rocky	0.57	0.57	0.57

Source: SAC (2021)

The most recent mapping (2018–2019) of degraded lands in India by the Space Applications Centre (SAC) reveals that an area of 97.85 mha (29.77% of the total geographical area (TGA) is under degradation (Fig. 2.1). From 2003–05 to 2018–19, there is an increase of about 3.32 mha of land under degradation (Fig. 2.1). The major drivers identified for this increase are water erosion (affecting 11.01% of TGA), vegetation degradation (affecting 9.15% of TGA), and wind erosion (affecting 5.46% of TGA) (NAP 2023) (Table 2.3).

The extent of soil erosion (soil loss of more than 10 tonnes/ha/yr) in cultivable land of the country was 92.4 million hectares, as assessed by the ICAR using the harmonized database. The state-wise details are provided in Table 2.4.

Table 2.4 State-wise cultivable area affected by soil erosion in India

S. no.	State	Area ('000 ha)
1.	Andhra Pradesh (including Telangana)	8093
2.	Arunachal Pradesh	666
3.	Assam	3248
4.	Bihar	851
5.	Chhattisgarh	3733
6.	Delhi	28
7.	Goa	1
8.	Gujarat	984
9.	Haryana	306
10.	Himachal Pradesh	982
11.	Jammu & Kashmir	1369
12.	Jharkhand	3219
13.	Karnataka	7522
14.	Kerala	490
15.	Madhya Pradesh	12,262
16.	Maharashtra	8799
17.	Manipur	122
18.	Meghalaya	302
19.	Mizoram	–
20.	Nagaland	46
21.	Orissa	2227
22.	Punjab	229
23.	Rajasthan	19,029
24.	Sikkim	45
25.	Tamil Nadu	2308
26.	Tripura	109
27.	Uttar Pradesh	13,075
28.	Uttarakhand	1018
29..	West Bengal	1332
	Total	**92,400**

Source: Degraded and Wastelands of India—Status and Spatial Distribution, NAAS, New Delhi Publication, June 2010

2.4 Assessment of Degraded Forest Land by the Forest Survey of India

As per the India State of Forest Report, 2021, the total degraded forest land in the country is approximately 3.54 lakh sq. km, which is almost 11% of the total geographical area of the country. The states with the most degraded forest land include Madhya Pradesh, Arunachal Pradesh, Assam, Andhra Pradesh, Chhattisgarh, and Gujarat. The state-wise details of degraded forest land are presented in Table 2.5.

Table 2.5 Degraded forest lands in India

State/UT	Geographical area (GA)	2021 assessment (in sq km)		
		Open forest (canopy density up to 0.4)	Scrub (canopy density up to 0.10)	Total
Andhra Pradesh	1,62,968	13,861	8276	22,137
Arunachal Pradesh	83,743	15,197	797	15,994
Assam	78,438	15,304	228	15,532
Bihar	94,163	3762	236	3998
Chhattisgarh	1,35,192	16,370	615	16,985
Delhi	1483	131.68	0.38	132
Goa	3702	1130	0	1130
Gujarat	1,96,244	9516	2828	12,344
Haryana	44,212	1130	159	1289
Himachal Pradesh	55,673	5180	322	5502
Jharkhand	79,716	11,431	584	12,015
Karnataka	1,91,791	13,212	4611	17,823
Kerala	38,852	9837	30	9867
Madhya Pradesh	3,08,252	36,619	5457	42,076
Maharashtra	3,07,713	21,475	4247	25,722
Manipur	22,327	9465	1215	10,680
Meghalaya	22,429	7326	663	7989
Mizoram	21,081	11,948	1	11,949
Nagaland	16,579	6530	824	7354
Odisha	1,55,707	23,948	4924	28,872
Punjab	50,362	1043	34	1077
Rajasthan	3,42,239	12,208	4809	17,017
Sikkim	7096	688	296	984
Tamil Nadu	1,30,060	11,792	758	12,550
Telangana	1,12,077	10,471	2911	13,382
Tripura	10,486	1863	33	1896
Uttar Pradesh	2,40,928	8162	563	8725
Uttarakhand	53,483	6482	392	6874
West Bengal	88,752	9587	156	9743
A & N Islands	8249	383	1	384
Chandigarh	114	8.01	0.38	8
Dadra & Nagar Haveli and Daman & Diu	602	140.79	4.85	145.64
Jammu & Kashmir Shapefile Area* (54,624)	2,22,236	9115	284	9399
Ladakh Shapefile Area* (1,68,055)		1758	279	2037
Lakshadweep	30	11.01	0.00	11.01
Puducherry	490	35.77	0.00	35.77
Total	**32,87,469**	**3,07,120**	**46,539**	**3,53,659**

Source: India State of Forest Report (2021)

* indicates that area for individual UTs of Jammu & Kashmir and Ladakh have not been received from SoI. The geographical area reported for the unified J&K in Census 2011 is 2,22,236 km^2

2.5 Need for Ecological Restoration in India

2.5.1 Biodiversity Loss

India is home to nearly 8% of the world's biodiversity, yet it faces significant threats from habitat destruction and fragmentation. Ecological restoration helps in recovering native species and preserving genetic diversity. A study by the Indian Ministry of Environment, Forest and Climate Change reported that about 21% of the total geographical area of India is degraded, impacting almost 300 million people dependent on these ecosystems. The Global Environment Facility (GEF) estimates that over 30% of India's biodiversity is at risk due to land degradation. According to the Indian Network on Climate Change Assessment (INCCA), India has lost nearly 39% of its mangroves over the last 25 years, profoundly impacting coastal biodiversity. Land degradation is a critical factor driving biodiversity loss in India. Forest degradation has put about 1215 species in India at risk of extinction, according to the *IUCN Red List*. Addressing this issue requires integrated conservation strategies, better land management practices, reforestation efforts, and community engagement to restore degraded ecosystems and protect India's rich biodiversity. Promoting sustainable agricultural practices and enforcing laws against deforestation are vital steps toward mitigating these impacts.

2.5.2 Land Degradation

Land degradation is a pressing environmental challenge in India, significantly affecting both forest and agricultural lands. An estimated 120 million hectares in India are degraded due to deforestation, overgrazing, and unsustainable agricultural practices. Restoring these lands enhances productivity and supports food security. This degradation stems from various anthropogenic activities, leading to adverse effects on the environment, biodiversity, and livelihoods. India's forest cover has been substantially impacted by the use of forest land for various non-forestry purposes such as commercial activities, agriculture, urban expansion, and logging. As per the Forest Survey of India (2019), the total forest cover in India is approximately 21.67% of the geographical area, with a reported loss of 2.89 million hectares between 2017 and 2019. Key factors contributing to this degradation include encroachment of habitats for agriculture and development, soil erosion, and biodiversity loss.

2.5.3 Climate Change Mitigation

Climate change represents one of the most significant environmental challenges facing India today, with profound implications for land degradation across various ecosystems, including forests, agricultural lands, and other critical landforms. Rising temperatures, shifting rainfall patterns, and extreme weather events collectively threaten the integrity of India's landscapes, impacting biodiversity, food security, and rural livelihoods.

Increased temperatures, altered rainfall patterns, increased pest and disease outbreaks, soil erosion and salinity, reduced crop yields, water scarcity, etc., are directly linked to climate change and have been impacting the various landforms, leading to their degradation. Besides this, climate change also contributes to the degradation of wetlands through altered hydrology and increased evaporation rates. Wetlands are crucial for biodiversity and serve as natural flood buffers, and their loss intensifies land degradation. Climate change is also adding to the desertification, as approximately 30% of India's geographical area is prone to desertification, worsened by climate change factors such as increased temperatures and erratic rainfall. This results in the loss of arable land and diminished productivity.

Climate change is a significant driver of land degradation in India, influencing forests, agricultural lands, and various other ecosystems. Addressing this interconnected challenge requires an integrated approach that combines sustainable land management, community engagement, and policy interventions. By proactively implementing measures to mitigate climate change and restore degraded lands, India can enhance its resilience, protect its rich biodiversity, and safeguard the livelihoods of millions dependent on these vital resources. The time for action is now, as the health of India's land ecosystems is crucial for its sustainable future.

2.6 Policy Initiatives to Restore the Degradation of Lands

With a view to check land degradation, various initiatives have been taken both at the international and national levels. These measures include policy interventions addressing land degradation with collaborative efforts across countries, global organizations, and multilateral frameworks. These interventions focus on promoting sustainable land management, reversing land degradation, and achieving environmental, social, and economic sustainability.

2.6.1 Policy Interventions Scenario: International Level

2.6.1.1 United Nations Convention to Combat Desertification (UNCCD)

The United Nations has three Rio Conventions, namely, the United Nations Framework Convention on Climate Change (UNFCCC), the Convention on Biological Diversity (CBD), and the United Nations Convention to Combat Desertification (UNCCD). Established in 1994, the United Nations Convention to Combat Desertification (UNCCD) is the only legally binding international agreement linking environmental and development issues to the land agenda. In 1994, the United Nations General Assembly declared 17 June the "World Day to Combat Desertification and Drought" to promote public awareness and the implementation of the UNCCD in the desertification-affected countries.

The UNCCD has taken milestone initiatives such as the Global Framework for countries to develop national action programs (NAPs) to combat land degradation and promote sustainable land management, the Land Degradation Neutrality (LDN) target, aiming to ensure that global land resources are managed in such a way that land degradation is minimized and restoration efforts counterbalance degradation, and Global Monitoring and Reporting to monitor progress on land degradation and coordinate with countries to ensure they meet their commitments to combating desertification.

2.6.1.2 Sustainable Development Goals (SDGs)

The SDGs are a bold universal set of 17 goals and 169 targets to help organize and streamline development actions for greater achievement of human well-being while leaving no one behind—by 2030 (NITI Ayog 2024). Goal 15 of the United Nations Sustainable Development Goals specifically addresses land degradation, aiming to protect, restore, and promote the sustainable use of terrestrial ecosystems. Targets under SDG 15 include combating desertification, restoring degraded land, and ensuring sustainable land management. Similarly, SDG 2 and SDG 13 also provide for land restoration by advocating sustainable agricultural practices and climate resilience, respectively.

2.6.1.3 United Nations Framework Convention on Climate Change (UNFCCC)

UNFCC, through its various policy interventions such as land-based climate solutions, nationally determined contributions (NDCs), and carbon sequestration, recognizes and promotes land-based solutions, such as afforestation and reforestation (A/R) and reduced emissions from deforestation and forest degradation (REDD+), which help to mitigate climate change, prevent land degradation, restore land, protect forests, etc.

2.6.1.4 Convention on Biological Diversity (CBD)

The CBD emphasizes the role of biodiversity in land management and promotes better management of biodiversity for healthy ecosystems, which are vital for soil fertility, water retention, and reducing the impacts of land degradation. The CBD also encourages the integration of biodiversity conservation into land-use policies, including those aimed at addressing land degradation, to ensure long-term sustainability.

2.6.1.5 The Global Environment Facility (GEF)

GEF encourages an integrated landscape approach, supporting projects that work at the nexus of biodiversity, climate, and land degradation. It also extends financial support for projects aimed at combating land degradation, land restoration, sustainable agriculture, reforestation, and enhancing ecosystem services.

In addition to the above, international organizations like the International Food Policy Research Institute (IFPRI) and Global Partnerships, the World Bank and Regional Development Banks, the International Union for Conservation of Nature

(IUCN), Global Land Outlook (GLO), and International Frameworks on Financing Land Restoration, Regional Agreements and Cooperation also promote policies for sustainable development that focus on land restoration and combating land degradation.

International policy interventions to combat land degradation are multifaceted, involving legal frameworks, financial mechanisms, technical support, and global cooperation. Through international conventions, financing programs, and collaborative efforts, countries are working together to address the global challenge of land degradation, aiming for land restoration, sustainable management, and resilience to environmental changes. These efforts are essential to ensure that land resources are preserved for future generations and that ecosystems continue to provide vital services to humanity.

2.6.2 Policies and Legislations for Achieving Ecological Restoration in India

The Government of India has recognized the importance of ecological restoration and has instituted several policies and programs aimed at achieving this goal. Key policies, missions, and programs launched by the Government of India are given below:

2.6.2.1 National Forest Policy (1988)

The National Forest Policy of 1988 was promulgated to check the further depletion of forests in the country, which was attributable to relentless pressures arising from the ever-increasing demand for fuelwood, fodder, and timber; inadequacy of protection measures; etc. The policy, through its various objectives, emphasizes checking soil erosion and denudation in the catchment areas of rivers, lakes, and reservoirs in the interest of soil and water conservation, checking the extension of sand dunes in the desert areas and along the coastal tracts, and substantially increasing the forest/tree cover in the country through massive afforestation and social forestry programs, especially on all denuded, degraded, and unproductive lands. It has been inter alia envisaged in the policy that to attain the objectives of the policy, the national goal is to have a minimum of one-third of the total land area of the country under forest or tree cover, including that two-thirds of the area in hills and mountainous regions under forest and tree cover. The policy further provides that the goal specified in the policy should be attained in order to prevent erosion and land degradation and to ensure the stability of the fragile ecosystem.

The Hon'ble Supreme Court of India, in its judgment dated 6.07.2011 passed in Writ Petition No. 202/1995, also directed that the National Forest Policy, 1988, which lays down far-reaching principles, must necessarily govern the grant of permissions under Section 2 of the Forest (Conservation) Act, 1980, as the same provides the road map to ecological protection and improvement under the Environment (Protection) Act, 1986. The principles/guidelines mentioned in the National Forest Policy, 1988, should be read as part of the provisions of the Environment (Protection) Act, 1986, read together with the Forest (Conservation) Act, 1980.

Table 2.6 Progressive change in the Forest Cover of India compiled from the India State of Forest Reports

S. no.	Year of assessment	Total forest cover (in Sq. Km.)	% of the country's geographical area
1.	1991	639,182	19.44
2.	1993	640,107	19.47
3.	1995	638,879	19.43
4.	1997	633,397	19.27
5.	1999	637,293	19.39
6.	2001	675,538	20.55
7.	2003	678,333	20.64
8.	2005	677,088	20.60
9.	2011	692,027	21.05
10.	2013	697,898	21.23
11.	2015	701,673	21.34
12.	2017	708,273	21.54
13.	2019	712,249	21.67
14.	2021	713,789	21.71
15.	2023	715,342	21.76

Compiled from biennial ISFR, FSI, Dehradun

Table 2.7 Progressive change in the tree cover (trees outside forests). Forest cover of India compiled from the India State of Forest Reports

S. no.	Year of assessment	Total forest cover (in Sq. Km.)	% of the country's geographical area
1.	2005	91,663	2.79
2.	2009	92,769	2.82
3.	2011	90,884	2.76
4.	2013	91,266	2.78
5.	2015	92,572	2.82
6.	2017	93,815	2.85
7.	2019	95,027	2.89
8.	2021	95,748	2.91
9.	2023	1,12,014	3.41

Compiled from biennial ISFR, reports of FSI, Dehradun

The directives laid down in the National Forest Policy paved the way for unfurling the various centrally sponsored schemes launched for the conservation, development, and better management of our natural resources. As per the data compiled from the India State of Forest Reports, during the period 1991 to 2025, the forest cover of the country has increased from 19.44% to 21.76%. Similarly, the tree cover outside the forests during the period 2005 to 2023 has increased from 2.79% to 3.41%. Together, the forest and tree cover constitute 25.17% of the total geographical area of the country (ISFR, 2023), representing a progressive change in the restoration of degraded lands of the country. Progressive changes in the forest cover and tree cover of the country are represented in Tables 2.6 and 2.7.

2.6.2.2 National Afforestation and Eco-Development Board (NAEB)

The National Afforestation and Eco-Development Board, under the Ministry of Environment, Forests and Climate Change, was set up in August 1992 by the Central Government. The board is responsible for promoting afforestation, tree planting, ecological restoration, and eco-development activities in the country, with special attention to the degraded forest areas and lands adjoining the forest areas, national parks, sanctuaries, and other protected areas, as well as the ecologically fragile areas like the Western Himalayas, Aravallis, and the Western Ghats. To accomplish the various objectives entrusted to NAEB, various schemes and programs have been launched to promote afforestation and land restoration.

National Afforestation Programme and National Green Mission

The National Afforestation Programme/Plan (NAP) was launched in 2000 for the afforestation of degraded forest lands. NAP was merged with Green India Mission (GIM) in 2015–16, and the committed liabilities are to be given under Green India Mission (GIM).

The National Mission for Green India (GIM) is one of the eight missions outlined under the National Action Plan on Climate Change (NAPCC). It aims at protecting, restoring, and enhancing India's diminishing forest cover and responding to climate change by a combination of adaptation and mitigation measures (Fig. 2.2). It envisages a holistic view of greening and focuses on multiple ecosystem services, especially biodiversity, water, biomass, preserving mangroves, wetlands, critical habitats, etc., along with carbon sequestration as a co-benefit. This mission has adopted an integrated cross-sectoral approach, as it will be implemented on both public and private lands with a key role for the local communities in planning, decision-making, implementation, and monitoring. The goals set up under the mission include the following:

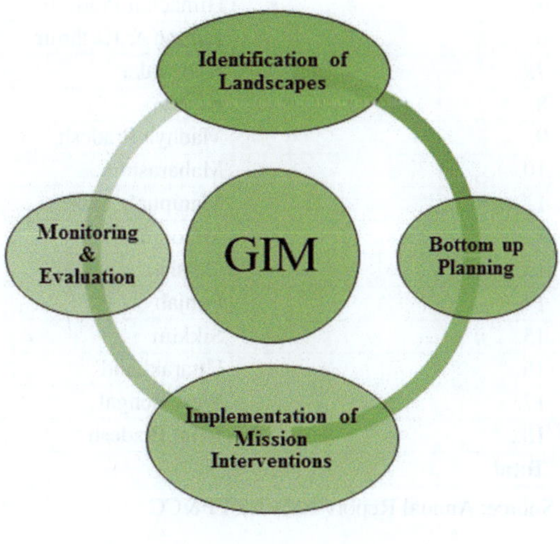

Fig. 2.2 Components of Green India Mission

- To increase forest/tree cover to the extent of five million hectares (mha) and improve the quality of forest/tree cover on another 5 mha of forest/non-forest lands.
- To improve/enhance ecosystem services like carbon sequestration and storage (in forests and other ecosystems), hydrological services, and biodiversity, along with provisioning services like fuel, fodder, and timber and non-timber forest products (NTFPs).
- To increase forest-based livelihood income of about three million households.

The GIM has also been converged with the Mahatma Gandhi National Rural Employment Guarantee Scheme (MNREGS) and statutory schemes envisaged under the Compensatory Afforestation Fund Act, 2016, to ensure a synergized approach. Further, efforts are underway to finalize convergence guidelines with other complementary schemes to set out the approach for coordination at the field level and to address the challenges being faced in the environment, forest, and wildlife sector, thereby contributing to ecological security in the context of climate change.

Under the GIM, so far, plantation over 1,55,130 ha has been carried out over forest and non-forest areas along with various livelihood improvement activities in 17 states and one union territory, as per the details provided in Table 2.8.

Table 2.8 Plantations raised under Green India Mission

S. no.	Name of the state	Creation of plantation (area in ha)
1.	Andhra Pradesh	1433
2.	Arunachal Pradesh	–
3.	Chhattisgarh	19,128
4.	Haryana	1301
5.	Himachal Pradesh	–
6.	Jammu & Kashmir	1066
7.	Karnataka	2722
8.	Kerala	12,297
9.	Madhya Pradesh	26,597
10.	Maharashtra	5223
11.	Manipur	14,432
12.	Mizoram	19,643
13.	Odisha	20,711
14.	Punjab	6568
15.	Sikkim	6567
16.	Uttarakhand	14,836
17.	West Bengal	2606
18..	Uttar Pradesh	–
Total		**155,130**

Source: Annual Report 2024 MoEF&CC

Nagar Van Yojana

The Nagar Van Yojana was launched on June 5, 2020, by revamping the Nagar Van Udyan Scheme of 2015. The scheme envisages creating forest/green spaces in urban areas by developing Nagar Van/Vatikas to protect forest land within or at the fringes of cities/towns from degradation and encroachment. The scheme proposes to create 1000 Nagar Van/Vatikas in the country during the period of 2020–21 to 2026–27 by utilizing CAMPA funds, managed by the National Authority of CAMPA. So far, more than 421 proposals for setting up Nagar Van/Vatikas have been sanctioned by the Central Government, involving a financial outlay of Rs. 286.04 crores.

School Nursery Yojana

The School Nursery Yojana, being implemented under the aegis of MoEF&CC, aims to provide an environment for the students to understand and appreciate the significance of plants in maintaining and sustaining the natural ecosystem. Around 5000 school nurseries are proposed to be developed in five years, from 2020–21 to 2024–25. The scheme is being funded from statutory schemes envisaged under the Compensatory Afforestation Fund Act, 2016, and an amount of Rs. 49.50 crore has been allocated for the plan period for implementation of the scheme in the country. So far, nearly 743 schools across the 21 states/UTs of the country have set up school nurseries under the scheme.

2.6.2.3 Desert Development Programme

It was launched in 1995 to minimize the adverse effects of drought and to rejuvenate the natural resource base of the identified desert areas. The Programme was launched for hot desert areas of Rajasthan, Gujarat, and Haryana and cold desert areas of Jammu & Kashmir and Himachal Pradesh.

2.6.2.4 National Action Programme to Combat Desertification

It was prepared in 2001 to address issues of increasing desertification and to take appropriate actions.

In addition to the above, the Ministry of Agriculture and Farmers' Welfare in India has been working on several schemes and programs aimed at the restoration of degraded lands. These initiatives focus on improving the productivity, fertility, and sustainability of degraded agricultural land and preventing further degradation. Some key schemes and programs under the ministry, as well as related ministries, include the National Mission for Sustainable Agriculture (NMSA), Watershed Development Programs, Soil Health Management and Soil Health Card Scheme, Conservation and Management of Soil and Water Resources, and Land Degradation Neutrality (LDN) Program. These schemes and programs aim to address both the environmental and agricultural concerns related to land degradation. They encourage the adoption of sustainable farming practices, soil health restoration, and water resource management to bring back productivity to degraded lands. Additionally, these efforts are supported by various state-level and local initiatives to create a more sustainable agricultural environment across India.

2.6.3 Legislations

2.6.3.1 The Van (Sanrakshan Evam Samvardhan) Adhiniyam, 1980, and Compensatory Afforestation

The Van (Sanrakshan Evam Samvardhan) Adhiniyam, 1980, is an important central statute for the conservation of forests in the country. It provides that the de-reservation of reserved forests, use of forest land for non-forest purposes, assigning by way of lease or otherwise of forest land to any private person or to any authority, corporation, agency, or any other organization not owned, managed, or controlled by the government, and clearing of naturally grown trees for the purpose of reafforestation require prior permission of the Central Government. During the last four decades, after the enactment of adhiniyam, the average annual rate of diversion of forest land has been reduced to approximately 23,700 ha. During the last 44 years, approximately 10.4 lakh ha of forest land was diverted for non-forest purposes with adequate mitigating measures. Details are provided in Table 2.9.

2.6.3.2 Compensatory Afforestation in Lieu of Diversion of Forest Land

The forest land diverted for various non-forest purposes is compensated by rejuvenating the ecosystem goods and services by undertaking compensatory afforestation (CA) and enrichment of biodiversity in the surrounding forests. Compensatory afforestation (CA) is one of the important requirements for obtaining the prior approval of the Central Government for non-forest use of forest land. The provisions of CA are levied to compensate for the loss of land by land and trees by trees (Table 2.10). Following the principle of land for land and tree for tree, the provisions of compensatory afforestation are stipulated in all approvals granted by the Central Government for the diversion of forest land for non-forestry purposes. Any

Table 2.9 Forest area diverted for non-forestry use

S. no.	Categories of proposals	Forest area diverted for various non-forest purposes (in Ha)
1.	Encroachments and village conversion	253498.24
2.	Mining and quarrying	161345.80
3.	Others	159368.77
4.	Irrigation and canals	135709.30
5.	Hydel	93692.35
6.	Road and approach access	73785.90
7.	Defense	66665.09
8.	Power transmission line	54251.16
9.	Rehabilitation	20078.66
10.	Railway	16202.67
11.	Thermal	7778.30
12.	Wind power	3916.51
Grand total		**1046292.75**

Table 2.10 Detail of compensatory afforestation undertaken in lieu of diversion of forest land

S. no.	States/UTs	Area stipulated for compensatory afforestation (ha)
1.	Andaman and Nicobar	2380.56
2.	Andhra Pradesh	41544.51
3.	Arunachal Pradesh	55484.04
4.	Assam	11022.22
5.	Bihar	12898.89
6.	Chandigarh	123.14
7.	Chhattisgarh	126370.24
8.	Dadra & Nagar Haveli	87.53
9.	Delhi	291.12
10.	Goa	2924.52
11.	Gujarat	61568.58
12.	Haryana	18301.01
13.	Himachal Pradesh	47674.89
14.	Jammu and Kashmir	19994.90
15.	Jharkhand	90352.47
16.	Karnataka	33643.99
17.	Kerala	59947.68
18.	Madhya Pradesh	272290.55
19.	Maharashtra	74108.89
20.	Manipur	7630.77
21.	Meghalaya	4966.00
22.	Mizoram	10853.28
23.	Odisha	110092.04
24.	Punjab	25017.80
25.	Rajasthan	61967.63
26.	Sikkim	6149.64
27.	Tamil Nadu	4397.70
28.	Telangana	30242.62
29.	Tripura	10787.93
30.	Uttar Pradesh	31370.95
31.	Uttarakhand	74820.23
32.	West Bengal	4900.92
	Total	**13,14,055.39**

proposal submitted by the State/UT Government seeking prior approval of the Central Government under the Van (Sanrakshan Evam Samvardhan) Adhiniyam, 1980, must be accompanied by the comprehensive scheme for compensatory afforestation, duly approved by the competent authority of the concerned state/UT administration.

Compensatory afforestation (CA) is a critical environmental strategy aimed at counterbalancing the loss of forest cover due to developmental activities such as urbanization, infrastructure projects, and agricultural expansion. This practice plays a pivotal role in ecological restoration and seeks to mitigate the adverse impacts of diverting forest land for non-forestry uses. With increasing pressure on forests, understanding the importance of compensatory afforestation is more crucial than ever.

Importance of Compensatory Afforestation

1. *Biodiversity Conservation*: Forests are biodiversity hotspots, home to a vast array of flora and fauna. Compensatory afforestation helps in preserving these species by restoring their habitats. By planting native tree species, biodiversity can be maintained and even enhanced, supporting various ecosystems.
2. *Carbon Sequestration*: Trees are significant carbon sinks, absorbing carbon dioxide from the atmosphere and thus playing an essential role in combating climate change. Compensatory afforestation initiatives help offset the carbon emissions resulting from deforestation and land-use changes, contributing to global climate mitigation efforts.
3. *Soil Conservation*: Deforestation leads to soil erosion and degradation. By promoting afforestation, the soil structure is improved, preventing erosion and maintaining soil health. This not only supports local agriculture but also enhances water retention, reducing the risk of floods and droughts.
4. *Ecosystem Services*: Forests provide numerous ecosystem services, including clean air and water, climate regulation, and support for wildlife. Compensatory afforestation restores these essential services, contributing to overall environmental health and human well-being.
5. *Social and Economic Benefits*: Healthy forests provide livelihoods to millions. Through compensatory afforestation programs, local communities can benefit from sustainable forest management, agroforestry, and ecotourism, creating a balance between development and conservation.

2.6.3.3 The Compensatory Afforestation Fund (CAF) Act, 2016

The Compensatory Afforestation Fund (CAF) Act, 2016, provides for the establishment of funds under the public accounts of India and the states. The purpose of this act is to ensure the effective utilization of funds collected for compensatory afforestation, additional compensatory afforestation, penal compensatory afforestation, net present value of forest (NPV), catchment area treatment plan, or any money required to be spent for the protection of wildlife, forests, and biodiversity. The CAF Act, 2016, establishes the National Compensatory Afforestation Fund under the Public

Account of India and a State Compensatory Afforestation Fund under the Public Account of each state. These funds receive payments for: (i) compensatory afforestation, (ii) net present value of forest (NPV), and (iii) other project-specific payments. The national fund receives 10% of these funds, and the state funds receive the remaining 90%. These funds are primarily spent on afforestation to compensate for the loss of forest cover, regeneration of forest ecosystems, wildlife protection, and infrastructure development.

1. *National Biodiversity Action Plan (2008)*: This framework encourages the conservation and sustainable use of biodiversity, advocating for the restoration of degraded ecosystems as a means to achieve biodiversity goals.
2. *Wetlands (Conservation and Management) Rules (2017)*: These rules aim to protect and restore wetlands in India, which are crucial for biodiversity, flood regulation, and water quality. The framework promotes community involvement in wetland management.
3. *Pradhan Mantri Fasal Bima Yojana (PMFBY)*: While primarily a crop insurance scheme, PMFBY encourages sustainable agricultural practices that contribute to soil health, thereby linking agricultural recovery efforts with ecological restoration.
4. *National Afforestation Programme (NAP)*: This initiative focuses on reforesting degraded forest lands through collaborative efforts involving local communities, emphasizing the restoration of ecosystem functions and services.

2.6.3.4 Accredited Compensatory Afforestation

Accredited Compensatory Afforestation (ACA) is a proactive system of raising afforestation over non-forestry land by any agency, i.e., government or private, which can be swapped with the CA in lieu of diversion of forest land by the interested user agencies. ACA can be undertaken by the government as well as private entities on the lands on which the provisions of the Van (Sanrakshan Evam Samvardhan) Adhiniyam, 1980, are not applicable. Lands with a minimum of 10 ha of area bearing plantations of a minimum of 5 years of age with a density of 0.4 can be considered for ACA. Existing plantations meeting the preconditions can also be considered for ACA. Plantations raised by the agencies and found satisfactory by the local forest department can be accredited as ACA afforestation, which can be swapped for raising CA in lieu of diversion of forest land by the interested user agencies. The ACA can also be swapped across the states/UTs, subject to the consent of the respective states/UTs.

The ACA has been developed in line with the spirit of "Digital India," and capturing the essence of minimum government and maximum governance to ensure online monitoring and tracking of various programs, the Central Government has notified the guidelines on Accredited Compensatory Afforestation. It will further add to the ease of doing business initiative of the Government of India.

Raising such afforestation in degraded lands will also provide for the accrual of carbon credits and associated benefits thereof, which can not only help the agency but also increase the total carbon sequestered, thereby helping to achieve the NDCs targets and restoration of degraded lands. The proposed scheme will also encourage afforestation over fallow lands, thereby helping in achieving the national targets as envisaged in the National Forest Policy, 1988.

Way Forward

Soil and land degradation is a complex and pressing issue that results from both natural factors and human activities. Natural factors include climate conditions, such as droughts or excessive rainfall, as well as geological processes like erosion. Human activities, such as deforestation, overgrazing, urbanization, and agricultural practices, play a significant role in accelerating land degradation. The need for immediate action to combat and reverse these processes is critical, given the consequences of land degradation. These include reduced agricultural productivity, loss of biodiversity, desertification, and increased vulnerability to natural disasters such as floods and droughts. While scientific knowledge and technology are available to address these issues, they are often fragmented, inconsistent, and lack widespread consensus on the most effective strategies for land restoration. To effectively address land degradation, comprehensive assessments are essential.

Based on ecological and geographical conditions and extent of degradation, 15 states of India, including Andhra Pradesh, Chhattisgarh, Gujarat, Jharkhand, Karnataka, Madhya Pradesh, Odisha, Rajasthan, Telangana, Uttar Pradesh, Uttarakhand, Tamil Nadu, Himachal Pradesh, West Bengal, and Maharashtra, have been identified for interventions to achieve the Land Degradation Neutrality (LDN) targets (NAP 2023). These states account for more than 25% of the country's geographical area, with a total area of 83.78 mha under desertification. In the Open Forest Area, only 30% of open forest areas are suitable for afforestation and can be treated to achieve LDN.

Based on a survey of secondary data on interventions by government agencies, nongovernmental organizations (NGOs), and the private sector, the report titled "Bonn Challenge and India: Progress on Restoration Efforts across States and Landscapes" states that India restored approximately 9.8 million hectares of degraded land between 2011 and 2017, with nearly 95% of the restoration activities reported being led by government agencies. Despite the limited data on the private sector and NGO restoration efforts, the report emphasizes the vital role that all stakeholders play in the planning and implementation of restoration programs, highlighting NGOs' technical expertise and knowledge of local conditions and private companies' financing capacity (Borah et al. 2018). Among "transformational changes" brought about by public restoration programs, the Joint Forest Management Committee (JFMC) approach has increased biodiversity and forest productivity of forests by making forest restoration socially more inclusive. Some of the benefits listed include the creation of livelihood opportunities in remote areas by linking villages to markets for the sale of non-timber forest products. By providing villagers with opportunities to participate in micro-plans, the JFMC model has successfully involved the communities in the protection and sustainable harvesting of natural resources and synergically integrated expertise and aligned interests of multiple stakeholders in forest restoration.

Currently, various schemes and programs for land and ecosystem restoration are being implemented with the Central as well as State Government assistance. Convergences of these various schemes, catalyzed with public participation, hold the key to the restoration of degraded landscapes at a faster pace.

Agroforestry is also an ideal intervention for transforming wastelands to productive use that can play an instrumental role in achieving national goals and global commitments. At present, the area under agroforestry in India is about 24.43 million ha, while the area under wastelands is 55.76 million ha. This shows that the area under agroforestry can be raised to three times the existing area by bringing wastelands under agroforestry. The conversion of culturable wastelands into productive use through agroforestry will surely provide multiple benefits to the environment, expansion of tree cover, rural communities, and a wood-based economy (Neelam Patel et al. 2024).

By conducting regular and comprehensive assessments, policymakers, land managers, and scientists can develop targeted strategies to halt and reverse land degradation. This requires collaboration across disciplines, regions, and sectors to develop cohesive, long-term solutions that integrate scientific knowledge, local knowledge, and practical technology. Only through such coordinated efforts can we ensure the sustainable use and restoration of our land and soil resources.

References

Annual Report (2023–2024) Government of India, Ministry of Environment, Forest and Climate Change, New Delhi

Bawa SK, Sengupta A, Chavan V, Chellam R, Ganesan R, Krishnaswamy J (2020) Securing biodiversity, securing our future: a national mission on biodiversity and human well-being for India. TIDEE 19(4):513–514

Borah B, Bhattacharjee A, Ishwar NM (2018) Bonn challenge and India: progress on restoration efforts across states and landscapes IUCN India. Ministry of Environment, Forest and Climate Change

Indian Land Degradation Data Set Gridded Fractional Area of Land Degradation Compatible to Mesoscale Models Over Indian Region; Technical Document prepared by Soil & Land Resources Assessment Division, Land Resources & Land use Mapping & Monitoring Group (LRUMG), Remote Sensing Application Area (RSA), National Remote Sensing Centre (NRSC) ISRO, Department of Space, Govt. of India, February 2016. NRSC-RSAA-LRUMG-SLRAD-Feb 2016-TR-811

ISFR (2021) India State of Forest Report. Forest Survey of India, Ministry of Environment, Forest and Climate Change, Government of India, Dehradun

ISFR (2023) India State of Forest Report. Forest Survey of India, Ministry of Environment, Forest and Climate Change, Government of India, Dehradun

Maji AK, Reddy Obi GP, Sarkar D (2010) Degraded and Wastelands of India status and spatial distribution. A book published by Dr T P Trivedi, Project Director, Directorate of Information and Publications of Agriculture. Indian Council of Agricultural Research, Krishi Anusandhan Bhavan I. Pusa, New Delhi

NAP (2023) National Action Plan (NAP) to combat desertification and land degradation through forestry interventions. Ministry of Environment Forest and Climate Change, Government of India

Neelam Patel (2024) NITI Aayog, Technical Report on Greening and Restorations of Wastelands with Agroforestry (GROW), ISBN No. 978-81-956821-3-3

SAC (2021)Desertification and land degradation atlas of India (assessment and analysis of changes over 15 years based on remote sensing). Desertification and Land Degradation: 282

Singh K, Singh RP, Tewari SK (2021) Ecosystem restoration: challenges and opportunities for India. Restor Ecol 29(3):e13341

Singh K, Byun C, Bux F (2022) Ecological restoration of degraded ecosystems in India: science and practices. Ecol Eng 182:106708

Souvik C (2024) Soil degradation: a global climate issue in 2024. Published in Climate-Europe, Global Climate Care, online edition dated 15.01.2024. https://globalclimatecare.in/climate-europe/f/soil-degradation-a-global-climate-issue-in2024#:~:text=Wind%20erosion%20accounted%20for%20142.8,%25%20of%20total%20land%20degradation

Thakur TK, Dutta J, Upadhyay P, Patel DK, Thakur A, Kumar M, Kumar A (2022) Assessment of land degradation and restoration in coal mines of central India: a time series analysis. Ecol Eng 175:106493

UNCCD (2022) United Nations Convention to Combat Desertification, Data Dash Board, Accessed on 3.03.2025. https://data.unccd.int/land-degradation?grouping=SDG&country=IND

International Processes and Commitments to Land Restoration

3

Muralee Thummarukudy and Dheeraj Mittal

Abstract

Land degradation is a compelling issue driven by human activity and climate change; however, growing recognition and initiatives aim to address and restore degraded lands globally and nationally. Agricultural land is vital for food production but is at risk of degradation from unsustainable farming practices, overgrazing, and deforestation, leading to soil erosion and nutrient depletion. Forest lands are degraded through deforestation and illegal logging, resulting in a loss of biodiversity and ecosystem services. Urban and industrial areas often face degradation due to soil sealing from construction, which impacts water management and may lead to contamination from industrial waste. Wetlands are also threatened by drainage and pollution, losing their ecological functions. Rangelands, used for grazing, suffer from overgrazing and climate change effects. On a global scale, initiatives are being launched to restore degraded lands. This chapter provides an update on the various international initiatives and commitments.

Keywords

Land degradation · Land restoration · Bonn challenge · Biodiversity conservation

M. Thummarukudy
G20 Global Land Initiative, United Nations Convention to Combat Desertification, Berlin, Germany
e-mail: mthummarukudy@unccd.int

D. Mittal (✉)
Indira Paryavaran Bhawan, Ministry of Environment, Forest and Climate Change, New Delhi, Delhi, India
e-mail: dmittal.ifs@gov.in

D. Mittal, D. K. Gupta (eds.), *Ecological Restoration in India: Perspective and Practices*, https://doi.org/10.1007/978-981-96-8943-9_3

3.1 Introduction

Land is the foundation of human existence on Earth, and we depend on land for our habitation, establishing our infrastructure, and most of our food. Globally, land can be categorized into many groups, and one way to group them is to differentiate between forest land, agricultural land, rangeland, wetlands, areas used for human settlement and infrastructure, and areas used for mining. Human activity has already touched upon up to 70% of all global land, and as both the numbers and standard of living of human beings increase, there is continuing pressure on land.

The Food and Agricultural Organization (FAO) defines land degradation as "the reduction in the capability of the land to produce benefits from a particular land use under a specified form of land management." The UN Convention to Combat Desertification (UNCCD) has a more elaborate definition where "Land degradation means reduction or loss, in arid, semi-arid and dry subhumid areas, of the biological or economic productivity and complexity of rainfed cropland, irrigated cropland or range, pasture, forest and woodlands resulting from land uses or from a process or combination of processes, including processes arising from human activities and habitation patterns such as soil erosion caused by wind and/or water; deterioration of the physical, chemical and biological or economic properties of soil; and long-term loss of natural vegetation."

One significant category of land is agricultural land, which includes arable fields, pastures, and plantations. These lands are crucial for food production and economic stability. However, agricultural land is highly susceptible to degradation due to unsustainable farming practices, overgrazing, and deforestation. Soil erosion, one of the most common forms of degradation, occurs when the protective cover of vegetation is removed, exposing the soil to wind and water forces. Nutrient depletion is another major concern, caused by continuous cropping without adequate soil replenishment. Inappropriate use of fertilizers and pesticides can further harm the soil's biological health, leading to a loss of fertility and productivity.

Forest lands, another critical category, are often degraded through deforestation and illegal logging. Forests provide a range of ecosystem services, including carbon sequestration, water regulation, and biodiversity conservation. Yet, they are frequently cleared for agriculture, urbanization, and infrastructure development. Deforestation not only diminishes biodiversity but also increases soil erosion and disrupts water cycles. Forest degradation is exacerbated by climate change, which heightens the risk of wildfires, pest infestations, and diseases. The loss of forest lands has far-reaching implications, including reduced resilience to climate shocks and loss of habitat for countless species.

Urban and industrial lands constitute another category that is, by its very nature, degraded land. The construction of buildings, roads, and factories often leads to soil sealing, where the natural soil surface is covered with impermeable materials like concrete and asphalt. This process disrupts natural water infiltration, increases surface run-off, and contributes to urban flooding. Additionally, industrial activities can result in soil contamination from hazardous waste, chemicals, and heavy metals, rendering the land unfit for agricultural or ecological purposes. Urban sprawl also encroaches upon agricultural and forest lands, further compounding the problem of land degradation.

Wetlands, which include marshes, swamps, and bogs, represent a unique and vital category of land. These ecosystems are essential for water purification, flood control, and biodiversity conservation. However, wetlands are increasingly threatened by drainage for agriculture, pollution, and infrastructure development. The conversion of wetlands into agricultural or urban areas leads to a loss of their ecological functions, such as carbon storage and habitat provision for aquatic species. Pollution from industrial effluents, agricultural runoff, and untreated sewage further degrades wetland quality, often resulting in the loss of aquatic flora and fauna.

Rangelands, commonly used for livestock grazing, are another category vulnerable to degradation. Overgrazing is a primary driver, leading to the loss of vegetation cover and soil compaction. This, in turn, reduces the land's water-holding capacity and increases susceptibility to erosion. Climate change exacerbates the degradation of rangelands through prolonged droughts and desertification. As rangelands become less productive, the livelihoods of pastoralist communities that depend on them are severely impacted.

Land degradation is often the result of complex interactions between human activities and natural processes. For instance, climate change acts as a multiplier, intensifying the effects of other degradation drivers. Prolonged droughts, rising temperatures, and extreme weather events accelerate soil erosion, reduce vegetation cover, and disrupt ecosystems. Human-induced factors such as unsustainable land use, deforestation, and pollution further exacerbate these natural processes, creating a vicious cycle of degradation.

3.2 Land Restoration: Global Initiatives

The challenges that land degradation causes to the economy and ecology are being increasingly understood around the world, and there have been many global initiatives for land restoration. In addition, increasingly there is a realization that biodiversity protection, climate change adaptation and mitigation, and disaster risk reduction could all have land restoration as a nature-based solution. Commitments are also being made under global processes through the UN Convention on Biological Diversity and the UN Framework Convention on Climate Change for land restoration.

The Bonn Challenge, established in 2011, set ambitious targets to restore 150 million hectares of degraded land by 2020 and 350 million hectares by 2030. To date, more than 210 million hectares have been pledged by over 70 entities worldwide, supported by regional programs such as AFR100 in Africa and Initiative 20 × 20 in Latin America. The UN Decade on Ecosystem Restoration (2021–2030), jointly led by the United Nations Environment Programme (UNEP) and the Food and Agriculture Organization (FAO), underscores the need for large-scale restoration efforts to achieve goals related to climate mitigation, water and food security, and biodiversity conservation. This initiative aligns closely with the Sustainable Development Goals (SDGs) and promotes nature-based solutions as a cornerstone of restoration strategies.

The New York Declaration on Forests, launched in 2014, outlines commitments to halve deforestation by 2020 and eliminate it by 2030, alongside efforts to restore degraded forests. This declaration emphasizes the importance of public and private sector collaboration to achieve its ambitious targets. Similarly, the Trillion Trees Initiative, spearheaded by BirdLife International, WWF, and the Wildlife Conservation Society, focuses on protecting and restoring one trillion trees globally. This initiative prioritizes biodiversity conservation and carbon sequestration, particularly through the restoration of ecological corridors. Another significant global effort is the Global Mangrove Alliance, established in 2017, which aims to increase global mangrove cover by 20% by 2030. This alliance highlights the vital role of coastal ecosystems in climate resilience, biodiversity, and disaster mitigation.

Regional commitments further strengthen these global efforts by addressing specific challenges and opportunities in different parts of the world. The African Forest Landscape Restoration Initiative (AFR100) seeks to restore 100 million hectares of degraded land across Africa by 2030, with countries such as Ethiopia pledging 15 million hectares under this program. Similarly, Initiative 20 × 20 in Latin America and the Caribbean initially targeted the restoration of 20 million hectares by 2020 but expanded its goals to 52 million hectares by 2030, demonstrating the growing momentum in the region. In Europe, the Caucasus, and Central Asia, the ECCA30 initiative focuses on restoring 30 million hectares of degraded forest landscapes by 2030, employing Forest Landscape Restoration (FLR) principles to guide its actions.

The Great Green Wall Initiative is another ambitious regional effort, envisioning a green belt across the Sahel region to restore 100 million hectares of land. Since 2021, the Great Green Wall Accelerator has mobilized over $19 billion in funding to fast-track this initiative, addressing desertification and supporting sustainable livelihoods in one of the world's most vulnerable regions. In the Middle East, the Green Initiative led by Saudi Arabia aims to restore 200 million hectares of land, plant 50 billion trees, and significantly reduce carbon emissions, showcasing a strong commitment to environmental sustainability in arid and semi-arid landscapes.

3.3 G20 Global Land Restoration Initiative

The leaders of the G20 met in Riyadh, Saudi Arabia, in November 2020 and came out with a Leaders' Declaration, where they expressed "…[their] conviction that coordinated global action, solidarity, and multilateral cooperation are more necessary today than ever to overcome the current challenges and realize opportunities of the 21st century for all by empowering people, safeguarding the planet, and shaping new frontiers."

Within the framework of ensuring sustainable development, the Leaders' Declaration also stated that "Building on existing initiatives, we share the ambition to achieve a 50 per cent reduction of degraded land by 2040, on a voluntary basis." In this context, they launched the "Global Initiative on Reducing Land Degradation and Enhancing Conservation of Terrestrial Habitats to prevent, halt, and reverse land degradation."

The commitment of the G20 leaders was built upon the communiqué from the G20 Environmental Ministers Meeting, also held in Riyadh in September 2020, where they recognized "the urgent need for bold, coordinated and collective initiatives on land protection and restoration and on sustainable land management and use to strengthen existing efforts, such as the Bonn Challenge, the United Nations Decade on Ecosystem Restoration, and other multilateral initiatives, as well as to promote new ones." They also supported the "launch of a Global Initiative on Reducing Land Degradation and Enhancing Conservation of Terrestrial Habitats which aims to support existing efforts to prevent, halt, and reverse land degradation and habitat loss through sharing of knowledge and best practices on protecting, conserving, sustainably managing, restoring, and rehabilitating degraded land, and by showcasing and disseminating publicly available data and information on degraded lands and conservation/restoration efforts. The initiative will also contribute to capacity building and encourage greater private sector support and general public engagement in land restoration efforts. The initiative focuses on complementing and supporting existing efforts while striving to avoid any duplication of efforts. The Initiative will seek synergies with existing relevant initiatives including the UN Decade on Ecosystem Restoration and the implementation of the post-2020 Global Biodiversity Framework. We further acknowledge the benefits of the existing efforts and commitments such as those developed for Land Degradation Neutrality (LDN), Forest Landscape Restoration, Ecosystem-based Adaptation, and Disaster Risk Reduction as well as targets under the Convention of Biological Diversity and the United Nations Convention to Combat Desertification."

The key objectives of the Global Land Initiative are as follows:

1. *Conserving land and halting habitat loss, habitat fragmentation, and land degradation*, notably through sharing knowledge and best practices on conservation incentives, including strengthening physical and functional connectivity and the quality of protected areas, enhancing land conservation, promoting sustainable land management practices, promoting active fire management, and implementing other policies and best practices to enhance land conservation and reduce land degradation;
2. *Promoting integrated, sustainable, and resilient land and landscape management* through nature-based solutions and ecosystem-based approaches; financing mechanisms; urban and rural land use planning; stronger implementation of international agreements and local environmental governance and laws; the empowerment of Indigenous peoples and local communities (including women, youth, and smallholders) in land management; secure land tenure, property, and land use rights in accordance with national legislation; as well as support for sustainable land and water management policies and sustainable agricultural practices, including traditional practices, in order to maintain and enhance ecosystem functionality; and
3. *Restoring degraded land* through sustainable and locally or regionally appropriate reforestation, afforestation, natural regeneration/revegetation, restoration of ecosystem services, sustainable agricultural practices, and deployment of nature-

based solutions and ecosystem-based approaches for biodiversity conservation, among other things, in order to restore ecosystem functionality in a landscape context. This objective will seek to prevent the transfer of degradation (i.e., avoid restoring one area to the detriment of another).

3.4 National Commitments for Land Restoration

While there are many international processes, the operational unit for land restoration is local, and it is important that there are robust processes for land restoration at the national level. Countries have been committing to land restoration under the three Rio Conventions.

A compilation of the Global Land Restoration Commitments by the International Union for Conservation of Nature (IUCN) obtained data from 140 countries that have contributed to at least one of the four global processes (UNCCD's Land Degradation Neutrality, UNCBD's National Biodiversity Strategy and Action Plan, UNFCCC's Nationally Determined Contributions, and the Bonn Challenge). Land Degradation Neutrality Targets topped the list with 614 million hectares, followed by 291 million hectares under NDCs and 205 million hectares under the Bonn Challenge. Under the Kunming-Montreal Global Biodiversity Framework, countries have committed 30% of areas to be under effective restoration by 2030, but the new National Biodiversity Strategy and Action Plans, which will translate these global commitments into specific national actions, are still being prepared.

Ecological Restoration of Mined-Out Areas for Achieving Land Degradation Neutrality: Opportunities and Challenges

4

Vivek Kumar Choudhary, Sudeshna Mazumdar-Leighton, and Cherukuri Raghavendra Babu

Abstract

Millions of hectares of productive land are being degraded and deforested every year globally. Mined-out degradation is the key area of concern that threatens global ecosystems and climate resilience. Ecological restoration of mined-out areas not only boosts our commitment to the UN Decade on Ecological Restoration 2020–2030 but also opens avenues for many opportunities and challenges. Restoring degraded land can lead to the recovery of ecosystems and strengthen ecological services, including carbon sequestration, food security, livelihood options, water filtration, and habitat provision for diverse biota. Ecological degradation can be instrumental in lowering natural calamities and disasters such as landslides, floods, extreme weather, and droughts. At the same time, restoring land on a large scale is a herculean task as it requires significant resources and coordination among different stakeholders, such as policymakers, experts, funding agencies, and local communities. This chapter describes the status of the mined-out area and its restoration in India, particularly in the mineral-rich state of Odisha.

V. K. Choudhary (✉)
Plant-Biotic Interactions Group, Department of Botany, University of Delhi, New Delhi, India

Centre for Environmental Management of Degraded Ecosystems (CEMDE), University of Delhi, New Delhi, India

S. Mazumdar-Leighton
Plant-Biotic Interactions Group, Department of Botany, University of Delhi, New Delhi, India

C. R. Babu
Centre for Environmental Management of Degraded Ecosystems (CEMDE), University of Delhi, New Delhi, India

D. Mittal, D. K. Gupta (eds.), *Ecological Restoration in India: Perspective and Practices*, https://doi.org/10.1007/978-981-96-8943-9_4

Keywords

Ecological restoration · Natural succession · Mined-out area · Wildlife · Biodiversity

4.1 Introduction

Mining in India has an ancient origin. It has coevolved and developed along with human civilization. The importance of mining is indicated in the following quote from Kautilya's Arthasastra (350–275 BC):

> *The wealth of the state is its source in the mining and metallurgical industry; the power of the state comes out of these mineral resources.*

After economic liberalization and globalization in the 1990s, India has emerged as a leading country in the international mineral market. India is a significant player in mining and metals. India stands fourth among the top 20 global mineral producers (Hazra et al. 2013). Its mineral production in terms of USD was 26,042 million in the year 2010, which was 5.6% of the total global production (Hodge 2012). This mining boom has shifted related technologies from conventional or manual types to modern, highly efficient technologies for the extraction of mineral resources. These improvements in mining technology are essential for national growth and development as well as global business. Unfortunately, all milestones of production and profits related to mining were achieved without due concern for biodiversity conservation and ecology or environmental sustainability for local people living in and around the mines. Substantial mining leads to drastic alterations of natural landscapes, above- and belowground biodiversity, soil productivity, water regimes, and drainage patterns, as well as air quality (Kuter 2013). Along with climate change and threats to biodiversity, the degradation of the environment is a global concern in the twenty-first century (Hansen et al. 2010). Mining-related destruction is a considerable challenge for the government, mining authorities, and the scientific community, with a need to heal and restore degraded lands into natural productive ecosystems (without compromising systematic and scientific mining essential for the economic growth and development of the nation).

4.2 Overview of Mining in India with Special Reference to Odisha

4.2.1 Impacts of Mining on Society and Natural Ecosystems

The adverse effects of mining activities on society include environmental pollution, ecological degradation, and changes in socioeconomic status. These aspects have been addressed by many authors in India and abroad (Felleson 1999; Cooke and Johnson 2002; Lee 2003; Wong 2003; Navarro et al. 2013; Gomes and Favas 2006;

Sun et al. 2006; Chopin and Alloway 2007; Moreno et al. 2007; Panda and Barik 2011; Saha et al. 2011; Sahoo 2013). Mining-related environmental problems may be one of the prime reasons for India's poor environmental performance index on the global stage (Hsu et al. 2014). It is an unfortunate truth that the mineral-rich areas of Odisha are superimposed with biodiversity-rich dense forest areas, fragile water catchment areas of major rivers, habitats of indigenous tribal communities, and endemic flora and fauna of the state (Patra et al. 2008). The local peoples include some of the most primitive tribes, who are totally dependent on forests, forest services, and agriculture for their livelihoods and survival (Panda and Barik 2011). Impacts of mining can be grouped into two broad categories: (A) impact of mining on ecology and environmental status and (B) impact of mining on socioeconomic status.

4.2.1.1 Mining and Economy

The exploitation of mineral resources and the development of mining-related activities are integral parts of the economic development of a country (Bogdetsky et al. 2005; Dash 2007; Ofosu-Mensah 2011). Mining empowers society by generating revenue for the state exchequer, strengthening foreign trade, and generating employment and livelihood opportunities for local peoples (Akabzaa and Darimani 2001; Bogdetsky et al. 2005; Ofosu-Mensah 2011). The mining and mine-related industry in India is the second-largest sector after agriculture (Dash 2007). It is one of the largest providers of employment and contributes about 2.6% of India's GDP. The Government of India controlled Public Sector Undertakings (PSUs) such as Steel Authority of India Limited (SAIL), the National Mineral Development Corporation (NMDC), and other small PSUs produce about 33% of total iron ore in the country. The remaining 67% of total production is shared by the private sector, which includes major companies like Tata Steel (TISCO), Jindal Steel (JSP), and other minor companies (Annual Report 2014). The leading iron ore-producing states like Odisha, Jharkhand, Goa, Chhattisgarh, and Karnataka, produced approximately 97% of the total production during the year 2011–2012 (Annual Report 2014).

The state of Odisha is also endowed with rich natural resources, including dense forests with rich biodiversity, as well as marine and river water resources (Panda and Barik 2011). Odisha leads among the states/union territories in having 16.92% of the total mineral reserves of the country and producing 10.21% of total nation minerals (Dash 2007). According to the Ministry of Mines, Government of India (Annual Report 2014), Odisha produces 10–14% of all valuable minerals, such as 93% chromite, 92% nickel ore, 90% platinum group of metals, 69% cobalt ore, 52% bauxite, 51% titaniferous magnetite, 44% manganese ore, 22% pyrophyllite, 33% iron ore (hematite), 20% each of mica and sillimanite, 24% each of fireclay, coal and garnet, 20% vanadium ore, 6% garnet and 5% zircon resources. The other important minerals that occur in the state are china clay, fireclay, garnet, graphite, limestone, quartz/silica sand, talc/steatite, titanium minerals, and zircon. Mining contributes significantly to Odisha's GDP. Iron ore is among the most important minerals in terms of exports (Sahoo 2013).

4.2.1.2 Impacts of Mining on Local Ecology and Environment

Negative impacts of mining are felt at every single stage of the mining operation, starting from pre-mining exploration to post-mining closure (Patra et al. 2008; Panda and Barik 2011). Mining involves activities such as the excavation of artificial pits and the destruction of rocks and hillocks using heavy machines and explosives, leading to regional land degradation and changes in land use patterns (Meher 2009). The change in land use patterns disturbs local climate by affecting ecosystem services (Reid et al. 2005). The degradation of land, water, and air due to mining is the driving force behind the degradation of ecology and environment at the local as well as global levels. Villages closest to mines appear to bear the highest pain and pay a great environmental cost (Pattanayak et al. 2010). The impact of mining on ecology and environmental status can be described as:

4.2.1.3 Impacts of Mining on Degradation of Lands and Landscapes

Mineral production and mining activities produce enormous quantities of waste and pollutants in the form of overburden dumps (OBDs) and tailings, acidic drainage, and fine dust, which cause serious degradation of land and forests (Sahu and Dash 2011). According to UNEP (1992), land degradation can be identified simply by "temporary or permanent lowering of the productive capacity of land." Degradation of land due to disturbances in the physical, chemical, and biological features of topsoil is the inevitable outcome of all mining activities (Cooke and Johnson 2002). The level of degradation is highest where opencast iron ore mines are operational (Cooke and Johnson 2002; Yan et al. 2013). Mining-generated land degradation happens due to the cumulative effect of several biological and physicochemical disasters, including loss of biodiversity in aboveground vegetation and belowground microbial communities (Kundu and Ghose 1997; Del Val et al. 1999), low mycorrhizal densities (Reddell and Milnes 1992), and loss of soil profile, mineral cycling, and fertility (Harris and Birch 1989; Ghose 2004b). Nitrogen levels and water retention capacity of soils reduce (Antonopoulos et al. 1998), high soil compaction occurs (Schwenke et al. 2000), and soil erosion/runoff increases (McIntosh and Barnhisel 1993; Leavitt et al. 2000) as a result of mining. Degraded mined-out areas associated with heavy metal contaminants are usually poor in nutritional profile, lack moisture-retaining capacities, and do not possess sufficient surface soil to support vegetation establishment (Vangronsveld and Cunningham 1998; Eisler 2004).

Odisha's food production decreased by 40% in the last 50 years, and around 9,00,000 ha of land suffers from crop loss every year due to different calamities fueled by mining-associated climate change (Envis Newsletter 2006, 2011). It has been recorded that the majority of land types diverted for mining and related activities are either native forests, agricultural lands, or free multipurpose lands (especially grasslands) that are used by local communities. Mining produces high amounts of dust and acidic drain, which ultimately destroy water bodies and agriculture-related irrigation water (Lin et al. 2005). Soil erosion is a common phenomenon in mining areas, which makes land unstable and infertile. Apart from core mining, other mining-related activities like ore transporting, construction of approach roads and other infrastructure, and setting of crushers result in large patches of forest land degradation (Datar et al. 2011). Mining activities not only cause loss of habitat and environmental deterioration but also disturb a country's natural beauty (Ehrlich 1995).

4.2.1.4 Impacts of Mining on Water Pollution and Water Catchment Areas

Mining changes the landscape of the local area. Mining can convert high hillocks into deep voids within weeks or months. Mining operations make land surfaces barren (without vegetation) and exposed to the air and water. This alters water retention capacity and water leaching capacity, leads to soil erosion, and pollutes surrounding water bodies around the area (Nordstrom et al. 2000). Similarly, perennial streams and tributaries passing through the mining areas can become severely polluted, biologically dead, and often change into seasonal streams due to the destruction of water catchment areas (Panda and Barik 2011). Like industrial effluents, mining activities can adversely affect the quality of water in nearby water bodies by lowering pH and increasing heavy metal concentrations and pollutants (Nordstrom et al. 2000; Baker and Banfield 2003; Lin et al. 2005). Polluted waters make it tough for aquatic flora and fauna to survive, as well as impart negative effects on socioeconomic conditions of native people and their health (Patra et al. 2008; Das 2015). For example, breeding and species composition of fish are badly affected by mining-generated polluted waters (Parulekar et al. 1986; Mol and Ouboter 2004). This affects viable livelihood options for fishermen, causing them to become daily wage earners or migrant laborers (Oraon 2012; Panda 2014).

4.2.1.5 Impacts of Mining on Climate Change

Climate change has made a substantial impact on seasonal and rainfall patterns in Odisha in the last few decades (Wysham 2003a, 2003b; Jogesh and Dubash 2014). Northwestern parts of Odisha, where major mining districts are located, have become more vulnerable to drought (HDR, Government of Orissa 2004). According to a study conducted by the Institute for Policy Studies (Wysham 2003b), Odisha's contribution to global greenhouse gas emissions may rise up to one percent as a result of mining-related activities. The study further pronounced Odisha as a "Global skyrocket greenhouse emitter," releasing toxic global warming agents like tetrafluoromethane and hexafluoroethane, which are by-products of aluminum smelting practices in Odisha. Lost CO_2 uptake by forest vegetation and water bodies that are cleared during mining contributes to global warming. According to Justice M.B. Shah Commission Report 2014 on Odisha's illegal mining of iron and manganese ore (Shah Commission Report 2014), 77% of mining happens in pristine forest areas of Odisha, and 172 out of 192 project areas operate in dense forest areas. These native forests and aquatic life are very critical for carbon sequestration (i.e., absorbing and fixing atmospheric carbon dioxide into biomass and maintaining a healthy balance between CO_2 emissions and CO_2 uptake). Heavy machines and vehicles used to extract and transport ores are fueled by hydrocarbons (mainly diesel) and emit substantial amounts of greenhouse gas into the environment.

4.2.1.6 Impacts of Iron Ore Mining on Wildlife and Biodiversity

Odisha has 34.23% of the total green cover in India, which includes 31.38% forest and 2.85% area outside of forest (http://www.odisha.gov.in/forest_environment). According to Odisha Forest and Environment Department records, remote sensing data suggests that 7073 sq. kms out of a total 48,855 sq. kms consists of a very dense forest. The open forest cover is only 21,394 sq. kms. Iron ore reserves lie beneath the continuous green patch of the Toda and Saranda Reserve Forest areas, with "Sal" (*Shorea robusta*) dominated dense semi-deciduous forest (Champion and Seth 1968). At present the forest areas are still reserved, but plant biodiversity and density have reduced drastically due to mining and mining-related activities. The natural forest in the area has shifted from lush green cover to patches of red steps due to mine benches (Wickham et al. 2007; Hogan et al. 2012). The Toda range of Saranda reserve forest is traditionally home to numerous wildlife, insects, and rare plant species (Singh 2010). However, mining and related activities have altered the habitat, removing dense vegetation and topsoil, destroying water resources, and generating noise and other pollutants (Kumari and Sahu 2010). Habitat fragmentation caused by mining is a common phenomenon (Holden and Jacobson 2012). Habitat fragmentation created due to mines is known to cause difficulties in the dispersal and movement of native species from one patch to another, reduced home range, shrinking of interior-to-edge ratio, and cutting off migratory routes (Valladares et al. 2006). Mining causes damage to wildlife along with habitat loss due to deforestation and degradation of land by excavation and redistribution of soil surfaces. The degraded soil microbial communities are severely affected by heavy metal concentrations (Del Val et al. 1999). Aquatic systems such as rivers, streams, lakes, ponds, or marshes are polluted or drained, causing aquatic flora and fauna to vanish. The food chains and food webs of a local ecosystem are destroyed. Exotic and invasive species colonize in fragmented patches and establish themselves successfully (Lindenmayer and Fischer 2006; Bennett and Saunders 2010).

4.2.1.7 Impacts of Mining on Social Community

Local communities are the ultimate stakeholders in mining areas. Mining empowers local communities and improves the local economy for the moment by providing livelihood and compensation for destroying productive land. However, mining affects local people by destroying their sustainability, ecosystem, health, and environment (Pattanayak et al. 2010; Das 2015). Mining in Odisha has increased the incidence of chronic diseases, depletion and pollution of water sources, and displacement of local populations (Wysham 2003a, b). Polluted and contaminated water remains one of the major causes of the low average life span and high mortality in mining areas (Suresh et al. 2007; Adamu et al. 2015; Das 2015). Some studies also suggest that the prevalence of diseases such as malaria, respiratory infections, and skin diseases is high in mining zones (Yeboah 2008). Until very recently, communities living in mining areas (especially in Sundargarh district) were totally dependent upon native forests and forest services for their livelihood. People met their daily requirements through agriculture, fishing, hunting, and livestock management. But due to the degradation of forests, their socioeconomic condition and lifestyle have completely changed (CEMDE reports 2012; Oraon 2012; Dalei and Gupt 2014). In every tribal village, there used to be a "Baida" (traditional Ayurvedic practitioner) who cured diseases and provided traditional treatments to indigenous peoples based on native herbs and trees. But due to the destruction of native forests, herbal plants disappeared and "Baidas" slowly shifted to other occupations (Khatua and Stanley 2006).

The mining industry can be considered as one of the worst sectors with respect to gender equality and socioeconomic status of women (Bhanumathi 2002). For example, in terms of employability and opportunities, only one-fourth of the total workforce in mining sectors were females (Kitula 2006). This reflects that mining jobs are gender-oriented and provide more opportunities to men than women.

4.3 Extent of Land Degradation in India with Special Reference to Odisha

More than half of the arable land in our country has been degraded (Sehgal and Abrol 1994). Estimated land degradation due to mining is about 36 lakh ha of land, which is 0.11% of the total land area of the country (Singh 2004). In Odisha, 50% of deforestation (which accounts for 17% of the deforestation of the country) estimated between the years 1980 and 2005 occurred due to mining-related activities (Hota 2012). Sahu and Dash (2011) stated that during the year 2004–2005, about 13,546 ha of land was affected due to mining operations, in which iron ore mining affected 1544 ha. The current area affected by mining is not very clear. However, its negative impact has reached beyond the mining areas. Total mining lease areas of 0.8 mha may be considered degraded land (Annual reports 2011–12, MoEF-CC, India). In Odisha, the total area covered by 605 leases was 99,931.55 ha. Of this, iron ore mines contribute around 21.10% (Khatua and Stanley 2006). Sundargarh has the largest number of mining leases and the second-largest area under mining (~20.03%).

The practice of ecological restoration in India is picking up momentum due to heightened awareness among local communities, environmentalists, mining authorities, and the government. However, the extent of restoration work undertaken, its aims, and its methodology pose several questions that need to be addressed for proper ecological restoration in the country. India does not have a detailed inventory of rehabilitation and ecological restoration efforts for abandoned mined-out areas. According to Bhushan (2008), around 510 abandoned mines of major minerals and coal exist in the country. However, this figure is likely to be much higher if all unrecorded and illegal mines are added. According to the Annual Reports (2006) of the Ministry of Mines, Government of India, degraded mined-out areas of only 53 mines covering 660 hectares had been reclaimed until 2005. Various reviews of literature related to the mining sector of India are available (Ghose 2004a, b; Ghose and Kumar 2004; Pandey et al. 2009; Prasad 2004; Singh and Shukla 2008; Sheoran et al. 2010; Datar et al. 2011). A comprehensive study by Datar et al. (2011) involving 57 case studies from 1995 to 2011 reported that four plant species (*Pongamia pinnata*, *Dalbergia sissoo*, *Albizia* lebbeck, and *Azadirachta indica*) are the most popular choices for the rehabilitation process in all kinds of mines. Various post-mining revegetation program sites located at Sundargarh and neighboring districts in Odisha also reveal that most sites were planted with exotic species like *Pongamia pinnata, Cassia siamea,* and *Acacia auriculiformis* (Patra et al. 2008). The introduction of exotic/invasive species can threaten native species and affect the biodiversity and functioning of the ecosystem being restored (Berger 1993; Simberloff 2000; Powell et al. 2011).

4.3.1 Soil Remediation at Degraded Mined-Out Areas: Ex Situ *and* In Situ

Degraded and highly polluted mined-out areas can be mitigated and restored by physical, chemical, and biological techniques (McEldowney et al. 1993). Depending upon the extent and types of soil degradation by various contaminants, different methodologies for soil remediation have been recommended (Mulligan et al. 2001). These methods are based upon (i) the containment of contaminants, (ii) their in situ stabilization, and ultimately (iii) their removal from the contaminated site (Gremion et al. 2004). All soil remediation techniques are broadly divided into ex situ and in situ methods.

4.3.1.1 Ex Situ Methods

These methods include soil excavation, storage, and transportation, as well as washing and chemical treatment of the degraded soils (Gohre and Paszkowski 2006). These ex situ methods remove pollutants and help in the mitigation of degraded soils but are labor-intensive, environmentally invasive, and unfit for plant growth and establishment. Besides, they are also toxic for almost all soil biota (Gaur and Adholeya 2004; Khan 2005). In addition, these approaches are costly (Chaudhry et al. 2005; Mulligan et al. 2001; Padmavathiamma and Li 2007). Glass (1999) has summarized the high-cost values of ex situ remediation methods. The costs of these methods are estimated to range between USD 75 and 425 per ton for nitrification and USD 20–200 per ton for landfilling and chemical treatments (Mulligan et al. 2001). In contrast, in situ methods cost only USD 10–40 per ton of soil (Mulligan et al. 2001). Ex situ methods generate toxic substances as by-products containing trace metals and other hazardous pollutants, which need further treatment for proper remediation (Khan et al. 2004). Furthermore, ex situ remediation technologies are not suitable for large areas of degraded and contaminated mining sites (Mulligan et al. 2001; Khan et al. 2004). In addition, ex situ remediation techniques irrevocably alter the physical, chemical, and biological properties of the soil being remediated (Khan 2005).

4.3.1.2 In Situ Methods

The most common in situ method involves "phytoremediation." It is a form of bioremediation. Bioremediation is a broader term that involves the use of living organisms (for example, animals, plants, insects, or microorganisms) and their products to treat an environmental problem. Phytoremediation is a more specific form of bioremediation where plants are used to solve environmental problems (Jan et al. 2014). Phytoremediation is one of the most versatile and popular methods of soil remediation in which plants and their associated microbial communities are used for environmental cleanup (Baker et al. 1994; Cunningham et al. 1997; Brooks 1998; Salt et al. 1995, 1998; Jan et al. 2014). Phytoremediation is said to be an efficient remediation technology for a range of organic and inorganic (including heavy metal) pollutants using the naturally occurring processes by which plants and their rhizospheric microorganisms reduce soil contaminants (Pilon-Smits 2005). Phytoremediation is broadly categorized under five types of remediation strategies:

(i) Phytoextraction: Removal and sequestration of heavy metals or organics from soil into aboveground harvestable parts of a plant (Lasat 2002; Khan 2005)
(ii) Phytodegradation: Degradation of contaminants into harmless components by plants and their associated microbes (Newman and Reynolds, 2004)
(iii) Rhizofilteration: Involves absorption/filtration of soluble heavy metals by plant roots (Dushenkov et al. 1995)
(iv) Phytostabilization: Using plant vegetation for immobilizing and reducing the bioavailability of contaminants, thereby preventing pollutant runoffs to nearby water bodies (Bolan et al. 2011)
(v) Phytovolatilization: Emission of contaminants as volatiles into the atmosphere due to enzymatic reduction of solid pollutants by processes driven by plants and their rhizospheric microbes (Chaney et al. 1997)

In phytoremediation, plants either remove contaminants or render them as harmless components of soil systems (Salt et al. 1995; Chaney et al. 1997; Flathman and Lanza, 1998; Khan 2005). Phytoremediation can remediate a whole range of pollutants. Examples include the remediation of inorganic pollutants like nitrates and phosphates (Horne 2000); trace elements such as Cu, Fe, and Zn (Lytle and Cronan 1998); nonessential elements such as Cr, Cd, Hg, Se, and Pb (Salt et al. 1995; Blaylock and Huang 2000; Horne 2000); and radioactive isotopes such as ^{238}U and ^{137}Cs (Dushenkov and Kapulink 2000; Dushenkov 2003). In situ approaches for soil remediation are preferred over chemical and physical approaches, being "environmentally friendly" or "green" remediation (Comis 1996; Pilon-Smits 2005). However, phytoremediation is considered a relatively slow process (Prasad and de Oliveira Freitas 1999; Prasad and Freitas 2003).

Ecological restoration of degraded mined-out areas and/or deforested landscapes to their natural state of ecosystems is different from afforestation, reforestation, ecological rehabilitation, and other *in situ methods*. Afforestation and reforestation create new ecosystems mostly of monoculture plantations or nonnative species plantations, which may not render a wide range of ecological services; similarly, ecological rehabilitations restore the degraded/deforested landscapes not to the level of natural state but halfway and generate some ecological services.

4.4 Ecologically Restored Forests Promote Resilience and Sustenance to Local Ecosystems

Planning, strategy, and execution for ecological restoration of mined-out areas are site-specific and vary from one site to another according to objective, geographical, and climate conditions. However, ecological restoration of degraded mined-out areas can be carried out through the following steps: (i) selection of native plant species and assembling them into ecological communities and "planting these assemblages of species in the site based on ecological succession principle"; (ii) inoculation of the species with plant growth-promoting rhizobacteria (PGPR) and arbuscular mycorrhizal fungi (AMF) that promote the establishment and growth of plants; (iii) monitoring of habitat response for ecosystem redevelopment; and (iv) evaluation of ecosystem services of the restored ecosystem. The biological inputs used in the ecological restoration of degraded terrestrial ecosystems are typically (i) grasses, (ii) legumes, (iii) woody plants (iv) soil microbes (PGPR and AMF), (v) soil invertebrates, and (vi) pollinators and seed dispersal agents.

4.4.1 Brief Overview of Ecological Restoration of Degraded Mined-Out Areas

There are several definitions of ecological restoration (Sweeney 2000). The most globally accepted and relevant definition was perhaps given by the Society for Ecological Restoration (SER 2004; Gann et al. 2019). According to SER (2004), ecological restoration is defined as "an intentional activity that initiates or accelerates the recovery of an ecosystem with respect to its health, integrity and sustainability and attempts to accelerate the natural recovery process artificially to achieve the otherwise long term goals in a short time." Ecological restoration is basically a human-assisted activity that involves multiple processes to bring back a disturbed ecosystem to an undisturbed state (Palmer and Filoso 2009; Gann et al. 2019). Colloquially, it also refers to the recovery and healing of degraded ecological systems. An ecologically restored system should be multidimensional and capable of meeting the requirements of the local human populations (Falk et al. 2006, page 244). Anthropogenic ecological requirements include economic services such as pure water and pollutant amelioration, nutrient cycling, biomass productivity, carbon sequestration, decreasing heat islands, noise reduction, reduction of GHG, habitat recreation, and water recharge (Clewell and Aronson 2007; del Moral et al. 2007; Naeem 2006). Although enhancing species biodiversity is a major goal in many restoration projects, sometimes the goal of restoring ecological functioning takes priority (del Moral et al. 2007; Palmer et al. 2007; Palmer and Filoso 2009). Dealing with the degradation of soil health is a major uphill task in most ecological restoration projects. But for each problem, there are immediate and long-term solutions available (Dobson et al. 1997).

There is a sharp difference between restoration and other related methodologies. According to Bradshaw (1984), "rehabilitation" occurs when only some of the original ecosystem functions and only some of the original species are recovered, whereas in "reclamation," replacement of the original native ecosystem takes place with another productive ecosystem at the degraded sites. Natural revegetation of degraded sites that occurs through natural colonization without any human assistance may take more than 100 years (Dobson et al. 1997; Mendez and Maier 2008). Natural forest ecosystems serve as model systems for planning any eco-restoration approach, and the same site can also serve as a reference for the evaluation of the project (Moore et al. 1999). A reference site could be an actual site or a document like a flora, or both (SER 2004; Gann et al. 2019). The establishment of diverse aboveground plant communities is often used to evaluate the restoration of natural habitats and the establishment of functional ecosystems with related processes (Halle et al. 2004; Ruiz-Jaen and Mitchell Aide 2005a, b). Clewell and Aronson (2007) have also suggested that the greater the variety and diversity of species, the greater the chance that a system will be restored and will have adequate ecological function.

The success of an ecological restoration program depends on critical analysis of the environment (soil and climate) and careful selection of plant species to be introduced into the degraded environments (Hansen et al. 2010). According to SER (2004), some features considered for choosing a plant species for introduction to a site undergoing ecological restoration are (i) species should be local and indigenous to the area, (ii) of economic importance and provide ecological services, (iii) better adapted to poor stressed soils, (iv) the biomass of debris or leaf litter should be biodegradable, and (v) increase soil organic matter (SOM) content. Dobson et al. 1997, suggested long- and short-term strategies for any successful ecological restoration process.

4.4.1.1 Survey of Degraded Area and Generating Base Line Data for Selection of Plant Species

Proper survey, selection, and identification of suitable plant species and their propagules for establishing aboveground biodiversity are very important (Sharma et al. 2000; Halofsky and McCormick 2005a, b). The introduction of local plant communities, especially keystone species, ensures the rebuilding of specific associations between plants and other biotic components such as pollinators, invertebrates, mycorrhizae, and prokaryotes in a restored ecosystem (Aronson et al. 1993; Dorner 2002). Since the loss of keystone species has a higher impact on an ecosystem's functioning than the loss of other species (Garibaldi and Turner 2004), the identification of native keystone species is one of the most important tasks for the successful initiation of ecological restoration programs (Aronson et al. 1993). The major targets or final outcomes of ecological restoration programs usually involve the reestablishment of local native flora and/or high-economic-value plant species.

Ecological restoration programs usually involve the removal of exotic, invading alien species. According to the eighth Article of the United Nations Convention on Biological Diversity of 1992 (CBD 1992), "Prevent the introduction of, control or eradicate those alien species which threaten ecosystems, habitats or species." Introduction of nonindigenous species can overwhelm indigenous plant species because of high aggressiveness and tolerance of weeds to hostile soil and abiotic environments (Berger 1993; Skeel and Gibson 1996; Siniscalco et al. 1997; Smith et al. 2005; Singh and Singh 2006; Wu et al. 2008). Nonindigenous, usually invasive species are thus recognized as a major threat to biological diversity essential for a functioning ecosystem (Babu et al. 2009; Pejchar and Mooney 2009). Indigenous flora can be identified by exploring the flora of native forests of surrounding areas (Sharma and Sunderraj 2005; Brunet 2007). Native flora can be restored by introducing plant saplings from the nearby forest (Hagen 2003) and redistributing indigenous topsoils for natural germination of seeds, spores, and plant parts (propagules) (Zhang et al. 2001).

4.4.1.2 Preparation of Sites for Ecological Restoration

Preparation of sites for ecological restoration involves proper landscaping, weed eradication, and protection from disturbances like grazing. Landscaping involves several activities, including proper contouring of the site undergoing ecological restoration. Contouring can be done to reduce steep slope angles, with which the restorers can reduce water and wind erosion and help in water retention and sapling stabilization. Landscaping may also involve recontouring the site so that it matches the surrounding terrain for predominantly aesthetic reasons (Gunn and Bailey 1993). Degraded sites at mined-out areas are considered disturbed, fragmented, and unattended habitats, more prone to invasion by weeds and invasive plants as compared to other well-managed ecosystems and habitats. If the landscape is more heterogeneous, it may be inhabited by more diverse communities that provide more competition and resistance to weedy invasions (Crawley 1987).

For successful ecological restoration processes, the site needs to be protected from unwanted anthropogenic activities and animal grazing. Grazing can be a significant factor for revegetation of both planted and naturally revegetated sites (Wood et al. 1995). Grazing is often considered a detrimental factor to planted vegetation (Muzzi and Fabbri 2007; Koch et al. 2004). The targeted restoration can be achieved by minimizing or excluding grazing activities at sites undergoing restoration (Davis et al. 1985; Davis et al. 1993).

4.4.1.3 Management of Exotic and Weed Species

Degraded sites are often infested with invasive species that hinder the development of native flora and outcompete it. These exotic plants compete with native plant species for nutrition, light, and space and affect indigenous species diversity, soil ecology, and dynamics (Booth et al. 2003; Hulme 2003; Babu et al. 2009; Love et al. 2009; Dogra et al. 2010). There are about 173 documented invasive exotic species, which can be grouped into 117 genera and further into 44 families spread across the country (Reddy 2008). Some of the most successful invasive plant species that have invaded several thousand hectares of land and forest are *Lantana camara, Parthenium hysterophorus,* and *Chromolaena odorata* (Love et al. 2009; Kannan et al. 2013). Many methods, like slashing/chopping, burning, manual/mechanical grubbing, chemical control, and biological controls, are practiced for control of invasive species in degraded lands and forest ecosystems of India with various advantages and disadvantages (Love et al. 2009). A very good method using indigenous manual/mechanical practices has been developed by a team of researchers at CEMDE, University of Delhi, known as the "cut-rootstock" method (Love et al. 2009) for the removal and management of invasive species and weeds.

4.4.1.4 Soil Amendment at Degraded Sites

Topsoil surfaces in mined-out areas are often severely degraded. There is low cohesion of soil aggregates, poor nutrition from aboveground vegetation, low soil organic matter (SOM), low water-retaining capacity, high concentrations of toxic metals, and extremely low pH. Most of these deteriorations occur due to acidic drainage from oxidation and weathering of exposed sulfides like iron pyrite and FeS_2 (Cooke and Johnson 2002; Johnson and Hallberg 2005; Alcolea et al. 2012; Asensio et al. 2013). Degraded soils are nutritionally and physically poor, which prevents the sustenance of plant growth if left unamended (Byron and Bradshaw 1991). Treatment of degraded soils with fertilizers generally shows positive effects on vegetation growth (Singh et al. 2000; Singh and Singh 2001). However, this can also be problematic as it can promote fast growth of exotic and weedy species over introduced desirable native plant species (O'Dell and Claassen 2006). Chemical fertilizers are less preferred as soil amendments for ecological restoration due to high costs. Soil amended with chemical fertilizers can also reduce the number and diversity of bacterial biomass compared to soils amended with organic matter, such as sewage sludge composts (Zaman et al. 2002). Current trends show that amendment practices have shifted from chemical fertilizers to organic fertilizers for promoting plant growth and efficient nutrient cycling in agroecosystems and revegetation of mined-out areas (Zaman et al. 2002; Prasad and Freitas 2003; Sheoran et al. 2010). Two important sources of soil amendments are discussed below:

4.4.1.5 Soil Amendments with Native Forest Soil and Organic Matter

Native forest soil inoculums are composed of healthy, nutritionally rich topsoils. They are rich in seeds and propagules of indigenous plant species and soil microbes like bacteria and symbiotic fungi (Holmes 2001; Rokich et al. 2000). The application of topsoil treatment can increase microbial metabolic activities. For example, Rivera et al. (2014) found that topsoil application enhanced plant species richness of restored areas in relation to controls. Due to the spreading topsoil at embankment revegetation sites, 45% activity of soil β-glucosidases, 57% activity of soil phosphatases, and 60% activity of soil respiration increased (Rivera et al. 2014). Soil amended with native forest soil can support introduced saplings to establish at an early stage. Topsoil applications can improve the recovery of microbial populations (Moynahan et al. 2002), improve soil properties and vegetation coverage (Tormo et al. 2007), and reestablish the functionality of a degraded ecosystem (Rivera et al. 2014). Martinez-Ruiz and Fernandez-Santos (2005) studied natural succession at mined-out areas after topsoil amendment application and observed that revegetation was achieved rapidly. However, the species composition of vegetation at restored sites differed from reference areas.

Cow dung inoculums (or "Gobar") have always been used as manure and as a very effective, universal organic amendment in Indian agroecosystems. Cow dung inoculums worked wonders in the restabilization of greenery and healing of the degraded soils at mined-out areas in the Tri-State Mining Region (states: Missouri, Kansas, and Oklahoma) of the USA (Baker et al. 2011). Degraded and nutritionally poor soils amended with organic matter showed significant increases in levels of phosphorus, nitrogen, and carbon and improved water retention capacity in soils (Rotkittikhun et al. 2006; Sen and Kumar 2012; Davis and Whiting 2013). Organic matter, including cow dung manure, increases colonization and stabilization of microflora and microfauna, thereby improving soil texture, soil enzyme activities, and nitrification (Mujumdar 1960; Karaca 2004; Jordao et al. 2006; Iovieno et al. 2009). Organic matter (such as biosolids, sewage effluents, pig or cow manure, composts, animal wastes, and food wastes) are rich sources of carbon, microbes, and essential plant nutrients such as nitrogen and phosphorus that promote microbial activities in soil (Simek et al. 1999; Enwall et al. 2005; Palmroth 2006). Reid and Neath (2005a, b) demonstrated that the addition of organic matter to soil reduced problems related to poor soil depth, poor fertility, degraded structure, and low water-holding capacity. Organic amendments also improved the extent of land cover and biomass of native seeded grasses and legumes (Noyd et al. 1995; Noyd et al. 1996; Davis and Whiting 2013). In addition, organic amendments such as manures helped to immobilize zinc (Zn), cadmium (Cd), and lead (Pb) ions in a metal-contaminated alluvial soil (Pierzynski and Schwab 1993).

4.4.1.6 Restoration via Natural Succession

Transforming highly degraded, barren mined-out areas into productive lands covered with grasses and legumes, prior to introductions of woody and/or nonwoody plant species, has been a key strategy for successful ecological restoration in India (Babu et al. 1990; Babu et al. 1993; Jha et al. 1995; CEMDE reports 2012; Babu et al. 2015). Such planning for the progression of pioneer communities and other groups is being increasingly recognized (Baasch et al. 2012). Grasses as pioneer species can enrich soil qualities by retaining water and accumulating organic biomass. Its rhizospheres help in the colonization of belowground biota. Grasses are quick-growing, provide biomass, have the ability to survive on toxic waste material, and are tolerant to adverse pH, extremely low moisture, and low-nutrient conditions (Babu et al. 1990; Maiti and Maiti 2014; Babu et al. 2015). Nyamadzawo et al. (2008) demonstrated efficient enrichment of soil seeded with grasses and subsequently planted with woody tree saplings. Planting bamboo clumps has also been shown to have a positive effect on soil health with respect to nutrition and the status of microbial biomass (Singh and Singh 1999). Since degraded soils lack bioavailable nitrogen, seeding leguminous species along with grasses yields remarkable recovery and stabilization of ecosystems (Babu et al. 1990; Babu et al. 1993; Jha et al. 1995; Singh et al. 2006; Brunet 2007; Maiti and Maiti 2014). Some rapid-growing plant species like *Argyrolobium flaccidum, Astragalus graveolens, Crotalaria* sp., *Desmodium* sp., *Indigofera gangetica, Lespedeza stenocarpa, Stylosanthes humilis, Stylosanthes hamata,* and *Sesbania sesban* are known to be excellent candidate legumes with proven abilities to fix nitrogen and produce high biomass in mined-out degraded areas (Jha et al. 1995; Caravaca et al. 2002; Madejon et al. 2006; Maiti and Maiti 2014). Plants with high tolerance to drought and metal contamination are introduced first, and then less tolerant species are used. Soil structure and plant composition of such aboveground vegetation improve with time (Baig 1992).

4.4.1.7 Monitoring Ecological Restoration Project

Determination of the success of ecological restoration efforts requires monitoring and evaluation of key indicators (Ruiz-Jaen and Mitchell Aide 2005a, b). In plant community recovery processes, species richness has been widely used as an indicator of restoration efficiency (Willems 2001; Lindborg and Eriksson 2004; Cristofoli et al. 2010). Temporal measurements of vegetation cover at restoration areas have also been used to provide inferences on ecosystem function (Clewell and Aronson 2007). Various environmental factors like soil pH and soil organic matter are good indicators of soil restoration processes (Ruiz-Jaen and Aide 2006). More information on soil health factors that can be monitored during ecological restoration programs is provided in the subsequent paragraphs. The rate of recovery in ecological restoration or ecological rehabilitation pathways is inversely related to the structural and functional complexity of the ecosystem selected as a reference ecosystem (Aronson et al. 1993). If evaluation trends lead toward the reference condition, then it confirms that the restoration is following its intended trajectory. Three approaches were suggested by SER (2004) on ecological restoration for evaluation and monitoring: (i) direct comparison with reference sites, (ii) attribute analysis, and (iii) trajectory analysis. In direct comparison, selected parameters (up to 30), including biota and environmental factors, are monitored or measured in the reference sites and restoration sites. However, in some instances, direct comparisons could create ambiguity in the interpretation of results, especially if some of the comparisons are close and others are not (SER 2004). In attribute analysis, attributes can be assessed by quantitative and semiquantitative data from scheduled monitoring and other inventories (SER 2004). Trajectory analysis is still being developed and may be considered the most promising method based on the interpretation of large sets of comparative data generated at restoration sites with respect to established trends (SER 2004).

4.5 Lessons Learned from Successful Ecological Restoration Projects

For the first time in India and perhaps in the world, 250 acres of limestone mined-out area in the tropical region of Odisha (Purnapani) has been restored to a three-story rainforest ecosystem with a top canopy reaching 80 feet within 12 years (Fig. 4.1a, b).

This site is exemplary on many counts—within a span of 12 years, a barren mined-out area was restored to a three-story tropical forest ecosystem with a top

Fig. 4.1 Mined-out sites before restoration (**a**) and the same site after restoration (**b**) at PL & DQ Mines of SAIL in Purnapani, Sundergarh, Odisha, India

canopy reaching 80 ft. in height, capable of providing sustainable livelihoods to local communities besides other ecosystem services. This success can serve as a perfect model for replication elsewhere and act as a demonstration plot for restoration practitioners and other stakeholders on how ecological transformation of degraded landscapes into naturally productive forest ecosystems can be achieved, leading to sustainable development goals. The forest developed at mined-out area harbors about 200 native tree species, some of which are host plants of tasar

Fig. 4.2 (**a**) Field training of tasar silkworm cultivation by experts of the State Sericulture Department, Odisha, while figure (**b**) showing tasar silkworm cultivation by local SHG members at restored sites at Purnapani, Odisha

silkworms and lac insects, and others produce nectar-bearing flowers, and still other species yield minor forest produce such as tendu leaves, sal leaves and seeds, mahua flowers and seeds, Amla, Chironji, ber, diospyros, soap nut, and others. The local communities harvest edible mushrooms, tasar silk cocoons, and honey from restored ecosystems. The biological productivity of the 40 m deep mine void water reservoir has been enhanced, and the reservoir harbors fish with each fish weighing about 8–15 kg. About 10 lakh fingerlings of four fish species, such as Rohu and Katala, have been introduced into the restored mining void in the current year. Locals have been harvesting fishes and fishing has become an alternative livelihood for locals and provides protein-rich food. The local communities (self-help groups) have been trained in rearing tasar silkworms (Fig. 4.2a, b), lac insect culture, fish cultivation, and apiculture, and the village community is now sustainably utilizing the restored ecosystem for enhancing income as well as socioeconomic status. The ecological services rendered by restored ecosystems at Purnapani, Odisha, include (i) an increase in groundwater table through recharging of groundwater; (ii) CO_2 sequestration through storage of carbon in biomass and soils; (iii) imparting climate resilience; (iv) promoting local cloud formation and precipitation; and (v) serving as a habitat for wildlife. (vi) promoting recreational and educational activities. The restored site serves as a model for replication in the ecological restoration of mined-out areas to the original natural forest ecosystem that used to exist before mining. The restored ecosystem, after the extraction of georesources, has not only brought back the ecological services lost during mining but also provided sustainable livelihoods and imparted climate resilience. As such, the Purnapani model is an excellent sustainable development model that can be adopted across the country and elsewhere.

References

Adamu C, Nganje T, Edet A (2015) Heavy metal contamination and health risk assessment associated with abandoned barite mines in Cross River State, southeastern Nigeria. Environ Nanotechnol Monitor Manag 3:10–21

Akabzaa T, Darimani A (2001) Impact of mining sector investment in Ghana: a study of the Tarkwa mining region. Draft report prepared for the structural adjustment participatory review international network. SAPRIN, Washington, DC

Alcolea A, Vazquez M, Caparros A, Ibarra I, Garcia C, Linares R, Rodriguez R (2012) Heavy metal removal of intermittent acid mine drainage with an open limestone channel. Miner Eng 26:86–98

Annual Report, MoM (2014) Annual Report 2013–14, Ministry of Mines. Retrieved from Ministry of Mines, Government of India Website: http://mines.nic.in/writereaddata/UploadFile/Annual%20Report%202013-2014.pdf

Antonopoulos Z, Vasssilis C, Wyseure G (1998) Modeling of water and nitrogen dynamics on an undisturbed soil and a restored soil after open-cast mining. Agric Water Manag 37:21–40

Aronson J, Floret C, Floch E, Ovalle C, Pontanier R (1993) Restoration and rehabilitation of degraded ecosystems in arid and semi-arid lands. I. A view from the south. Restor Ecol 1:8–17

Asensio V, Vega FA, Singh BR, Covelo EF (2013) Effects of tree vegetation and waste amendments on the fractionation of Cr, Cu, Ni, Pb and Zn in polluted mine soils. Sci Total Environ 443:446–453

Baasch A, Kirmer A, Tischew S (2012) Nine years of vegetation development in a postmining site: effects of spontaneous and assisted site recovery. J Appl Ecol 49(1):251–260

Babu CR, Jha PK, Nair S (1990) Biological reclamation of derelict ecosystems: problems and prospects. In: Dhar BB (ed) Environmental management of mining operations. Ashish Publications, New Delhi, pp 327–336

Babu CR, Subramaniam B, Nair S, Jha PK, Bhattacharya A, Kishore L, Natarajan K (1993) Biological nitrogen fixation technologies for ecological rehabilitation of degraded soils and natural ecosystems. Proc Indian Natl Sci Acad Part B Biol Sci 59:359–366

Babu S, Love A, Babu CR (2009) Ecological restoration of lantana-invaded landscapes in Corbett Tiger Reserve. Indian Ecol Restorat 27:467–477

Babu CR, Choudhary VK, Kumar V (2015) Tropical grassland ecosysytem and climate change. In: Vijay D, Srivastava MK, Malaviya DR, Roy MM, Mahanta SK, Singh JB, Maity, Ghosh PK (eds) Sustainable use of grassland resources for forage production, biodiversity and environment protection. Proceedings-23rd international grassland congress. Keynote lectures. Range Management Society of India, Jhansi, pp 371–373

Baig M (1992) Natural revegetation of coal mine spoils in the Rocky Mountains of Alberta and its significance for species selection in land restoration. Mt Res Dev 12(3):285–300

Baker BJ, Banfield JF (2003) Microbial communities in acid mine drainage. FEMS Microbiol Ecol 44:139–152

Baker A, McGrath S, Sidoli C, Reeves R (1994) The possibility of in situ heavy metal decontamination of polluted soils using crops of metal-accumulating plants. Resour Conserv Recycl 11:41–49

Baker LR, White PM, Pierzynski GM (2011) Changes in microbial properties after manure, lime, and bentonite application to a heavy metal-contaminated mine waste. Appl Soil Ecol 48:1–10

Bennett AF, Saunders DA (2010) Habitat fragmentation and landscape change. Conservation biology for all, vol 93, pp 1544–1550

Berger JJ (1993) Ecological restoration and nonindigenous plant species: a review. Restor Ecol 1:74–82

Bhanumathi K (2002) The status of women affected by mining in India. In: McDonald I, Rowland C (eds) Tunnel vision: women, mining and communities. Oxfam Community Aid Abroad, Fitzroy, pp 20–25

Bhushan C (2008) Rich, lands, poor people: the socio-environmental challenges of mining in India. India Econ Rev 5:44–53

Blaylock MJ, Huang JW (2000) Phytoextraction of metals. In: Raskin I, Ensley BD (eds) Phytoremediation of toxic metals: using plants to clean-up the environment. Wiley, New York, pp 53–70

Bogdetsky V, Ibraev K, Abdyrakhmanova J (2005) Mining industry as a source of economic growth in Kyrgyzstan. Bishek, World Bank

Bolan NS, Park JE, Robinson B, Naidu R, Huh KY (2011) Phytostabilization: a green approach to contaminant containment. Adv Agron 112:145–204

Booth MS, Caldwell MM, Stark JM (2003) Overlapping resource use in three Great Basin species: implications for community invasibility and vegetation dynamics. J Ecol 91:36–48

Bradshaw A (1984) Ecological principles and land reclamation practice. Landsc Plan 11:35–48

Brooks RR (1998) Plants that hyperaccumulate heavy metals: their role in phytoremediation, microbiology, archaeology, mineral exploration and phytomining. CAB international, Wallingford

Brunet J (2007) Plant colonization in heterogeneous landscapes: an 80-year perspective on restoration of broadleaved forest vegetation. J Appl Ecol 44:563–572

Byron K, Bradshaw A (1991) The potential value of sewage sludge in land reclamation. In: Hall JE (ed) Alternative uses for sewage sludge. Elsevier, Pergamon, Oxford, pp 1–20

Caravaca F, Barea J, Figueroa D, Roldan A (2002) Assessing the effectiveness of mycorrhizal inoculation and soil compost addition for enhancing reafforestation with *Olea europaea subsp. sylvestris* through changes in soil biological and physical parameters. Appl Soil Ecol 20:107–118

CBD (1992) Convention on biological diversity 1992, United Nations. Available on https://www.cbd.int/doc/legal/cbd-en.pdf

CEMDE Report. (2012). Final technical report on DBT-programme support in environmental biotechnology-restoration ecology 2010-2011 project sanction no. BT/PR5630/ BCE/08/372/2004

Champion SH, and Seth SK. (1968). A revised survey of the forest types of India. New Delhi: Oxford and IBH Publishers

Chaney RL, Malik M, Li YM, Brown SL, Brewer EP, Angle JS, Baker AJ (1997) Phytoremediation of soil metals. Curr Opin Biotechnol 8:279–284

Chaudhry Q, Blom-Zandstra M, Gupta SK, Joner E (2005) Utilising the synergy between plants and rhizosphere microorganisms to enhance breakdown of organic pollutants in the environment. Environ Sci Pollut Res 12:34–48
Chopin E, Alloway B (2007) Trace element partitioning and soil particle characterisation around mining and smelting areas at Tharsis, Riotinto and Huelva, SW Spain. Sci Total Environ 373:488–500
Clewell AF, Aronson J (2007) Reference models and developmental trajectories. In: Clewell AF, Aronson J (eds) Ecological restoration: principles, values, and structure of an emerging profession. Island Press, Washington, DC, pp 75–87
Comis D (1996) Green remediation: using plants to clean the soil. J Soil Water Conserv 51:184–187
Cooke JA, Johnson MS (2002) Ecological restoration of land with particular reference to the mining of metals and industrial minerals: a review of theory and practice. Environ Rev 10:41–71
Crawley MJ (1987) What makes a community invasible? In: Gray AJ, Crawley MJ, Edwards PJ (eds) Colonization, succession and stability. Blackwell Scientifc, Oxford, UK, pp 429–453
Cristofoli S, Monty A, Mahy G (2010) Historical landscape structure affects plant species richness in wet heathlands with complex landscape dynamics. Landsc Urban Plan 98:92–98
Cunningham S, Shann J, Crowley DE, Anderson TA (1997) Phytoremediation of contaminated water and soil. J Environ Qual 28:760–766
Dalei NN, Gupt Y (2014) Livelihood sustainability of Forest dependent communities in a mine-spoiled area. Int J Ecol Econom Stat 35:30–47
Das N (2015) Socio-economic impact of mining on rural communities: a study of the Ib Valley Coalfield in Odisha (Doctoral Thesis). Retrieved from http://ethesis.nitrkl.ac.in/6913/1/Nabanita_509HS304_PhD_2015.pdf
Dash L (2007) Economics of mining in Orissa. Orissa Review, 69–76. Retrieved from http://www.orissa.gov.in/e-magazine/orissareview/nov-2007/engpdf/pages69-76.pdf
Datar A, Audet P, Mulligan D (2011) Post-mined land rehabilitation in India: cataloguing plant species used in land revegetation. In: Vikuckis N, Nichols O (eds) Proceedings of the 2011 workshop on Australian mine rehabilitation Australian mine rehabilitation workshop 2011. JKTech, Adelaide, pp 197–208
Davis JG, Whiting D (2013) Gardening services, Colorado State University, US Department of Agriculture and Colorado Countries Cooperating. Fact sheet number 7, p 235
Davis B, Lakhani K, Brown M, Park D (1985) Early seral communities in a limestone quarry: an experimental study of treatment effects on cover and richness of vegetation. J Appl Ecol 22:473–490
Davis B, Lakhani K, Brown M (1993) Experiments on the effects of fertilizer and rabbit grazing treatments upon the vegetation of a limestone quarry floor. J Appl Ecol 30(4):615–628
del Moral R, Walker LR, Bakker JP (2007) Insights gained from succession for the restoration of landscape structure and function. In: Walker LR, Walker J, Hobbs RJ (eds) Linking restoration and ecological succession. Springer, New York, pp 19–44
Del Val C, Barea J, Azcon-Aguilar C (1999) Diversity of arbuscular mycorrhizal fungus populations in heavy-metal-contaminated soils. Appl Environ Microbiol 65:718–723
Dobson AP, Bradshaw AD, Baker AJM (1997) Hopes for the future: restoration ecology and conservation biology. Science 277:515–522
Dogra KS, Sood SK, Dobhal PK, Sharma S (2010) Alien plant invasion and their impact on indigenous species diversity at global scale: a review. J Ecol Nat Environ 2:175–186
Dorner J (2002) An introduction to using native plants in restoration projects. Center for Urban Horticulture. University of Washington, Seattle. Retrieved from http://www.nps.gov/plants/restore/pubs/introatplant
Dushenkov S (2003) Trends in phytoremediation of radionuclides. Plant Soil 249:167–175
Dushenkov S, Kapulink Y (2000) Phytofilitration of metals. In: Raskin, Ensley BD (eds) Phytoremediation of toxic metals-using plants to clean-up the environment. John Wiley & Sons, New York, pp 89–106
Dushenkov V, Kumar PN, Motto H, Raskin I (1995) Rhizofiltration: the use of plants to remove heavy metals from aqueous streams. Environ Sci Technol 29:1239–1245

Ehrlich PR (1995) The scale of the human enterprise and biodiversity loss. Oxford University Press, New York
Eisler R, (2004). Gold concentrations in abiotic materials, plants, and animals: a synoptic review. Environmental Monitoring and Assessment 90:73–88
Envis Newsletter (2006) Center for Environment studies. Forest and Environment Department, Government of Odisha, Bhubneswer 4(1), Retrieved from http://www.cesorissa.org/PDF/newsletter4.pdf
Envis Newsletter (2011) Center for Environment studies. Forest and Environment Department, Government of Odisha, Bhubneswer 25, 1–8. Retrieved from http://www.cesorissa.org/pdf/newsletter25.pdf
Enwall K, Philippot L, Hallin S (2005) Activity and composition of the denitrifying bacterial community respond differently to long-term fertilization. Appl Environ Microbiol 71:8335–8343
Falk DA, Palmer MA, and Zedler JB (2006). Foundations of Restoration Ecology. Washington, DC: Island Press
Felleson DA (1999) Iron ore and taconite mine reclamation and revegetation practices on the Mesabi Range in northeastern Minnesota. University of Minnesota, Department of Horticultural Science. Retrieved from the University of Minnesota Digital Conservancy, http://hdl.handle.net/11299/59 502
Flathman PE, and Lanza GR (1998). Phytoremediation: current views on an emerging green technology. Journal of Soil Contamination 7:415–432
Gann GD, McDonald T, Walder B, Aronson J, Nelson CR, Jonson J et al (2019) International principles and standards for the practice of ecological restoration. Restor Ecol 27(S1):S1–S46
Garibaldi A, Turner N (2004) Cultural keystone species: implications for ecological conservation and restoration. Ecol Soc 9(3):1
Gaur A, Adholeya A (2004) Prospects of arbuscular mycorrhizal fungi in phytoremediation of heavy metal contaminated soils. Curr Sci 86(4):528–534
Ghose MK (2004a) Effect of opencast mining on soil fertility. J Sci Ind Res 63(12):1006–1009
Ghose MK (2004b) Restoration and revegetation strategies for degraded mine land for sustainable mine closure. Land Contamin Reclamat 12(4):363–378. https://doi.org/10.2462/09670513.632
Ghose MK, Kumar A (2004) Mineral industries and their environmental aspects in Indian context. Indian J Eng Mater Sci 11(5):433–437
Glass DJ (1999) US and international markets for phytoremediation, 1999-2000. Retrieved from D. Glass Associates website: http://www.dglassassociates.com/INFO/phy99exc.htm
Gohre V, Paszkowski U (2006) Contribution of the arbuscular mycorrhizal symbiosis to heavy metal phytoremediation. Planta 223(6):1115–1122. https://doi.org/10.1007/s00425-006-0225-0
Gomes MEP, Favas PJ (2006) Mineralogical controls on mine drainage of the abandoned Ervedosa tin mine in North-Eastern Portugal. Appl Geochem 21(8):1322–1334. https://doi.org/10.1016/j.apgeochem.2006.06.007
Gremion F, Chatzinotas A, Kaufmann K, Von Sigler W, Harms H (2004) Impacts of heavy metal contamination and phytoremediation on a microbial community during a twelve-month microcosm experiment. FEMS Microbiol Ecol 48(2):273–283. https://doi.org/10.1016/j.femsec.2004.02.004
Gunn J, Bailey D (1993) Limestone quarrying and quarry reclamation in Britain. Environ Geol 21(3):167–172. https://doi.org/10.1007/BF00775301
Hagen D (2003) Assisted recovery of disturbed arctic and alpine vegetation: an integrated approach. Retrieved from Fakultet for naturvitenskap og teknologi. [Faculty of Natural Sciences and Technology](123253)
Halle S, Fattorini M, Temperton V, Hobbs R, Nuttle T, Halle S (2004) Advances in restoration ecology: insights from aquatic and terrestrial ecosystems. In: Temperton VM, Hobbs RJ, Nuttle T, Hale S (eds) Assembly rules and restoration ecology: bridging the gap between theory and practice. Island Press, Washington, D.C., pp 10–33
Halofsky JE, McCormick LH (2005a) Establishment and growth of experimental grass species mixtures on coal mine sites reclaimed with municipal biosolids. Environ Manag 35(5):569–578

Halofsky JE, McCormick LH (2005b) Effects of unseeded areas on species richness of coal mines reclaimed with municipal biosolids. Restor Ecol 13(4):630–638. https://doi.org/10.1111/j.1526-100X.2005.00081.x

Hansen L, Hoffman J, Drews C, Mielbrecht E (2010) Designing climate-smart conservation: guidance and case studies. Conserv Biol 24(1):63–69. https://doi.org/10.1111/j.1523-1739.2009.01404.x

Harris J, Birch P (1989) Soil microbial activity in opencast coal mine restorations. Soil Use Manag 5(4):155–160. https://doi.org/10.1111/j.1475-2743.1989.tb00777.x

Hazra AK, Chauhan SBS, Gupta A, Sharma E (2013) Development of Indian Mining Industry-The way forward, Non-fuel minerals 2013. FICCI Mines and Metals Division, Federation of Indian Chambers of Commerce and Industry. Retrieved from http://www.ficci.com/spdocument/20317/MiningIndustry.pdf

Human Development Report (2004). Orissa Human Development Report, Planning and Coordination Department, Government of Orissa Retrieved from http://planningcommission.nic.in/plans/stateplan/sdr_pdf/shdr_ori04.pdf

Hodge A (2012) Mining's contribution to sustainable development. International Council on Mining and Metal, London. Avalable on http://hub.icmm.com/document/3716

Hogan DM, Labiosa W, Pearlstine L, Hallac D, Strong D, Hearn P, Bernknopf R (2012) Estimating the cumulative ecological effect of local scale landscape changes in South Florida. Environ Manag 49(2):502–515. https://doi.org/10.1007/s00267-011-9771-8

Holden WN, Jacobson RD (2012) Mining and natural Hazard vulnerability in the Philippines: digging to development or digging to disaster? Anthem Press

Holmes PM (2001) Shrubland restoration following woody alien invasion and mining: effects of topsoil depth, seed source, and fertilizer addition. Restor Ecol 9(1):71–84. https://doi.org/10.1046/j.1526-100x.2001.009001071.x

Horne AJ (2000) Phytoremediation by constructed wetlands. In: Terry N, Banuelos G (eds) Phytoremediation of contaminated soil and water. CRC Press, Boca Raton, pp 13–40

Hota P (2012) Extractions of mineral resources and local climate change: the case of Orissa in India. Retrieved from http://www.planetunderpressure2012.net/

Hulme PE (2003) Biological invasions: winning the science battles but losing the conservation war? Oryx 37(20):178–193. https://doi.org/10.1017/S003060530300036X

Hsu A, et al. (2014). Environmental Performance Index. New Haven, CT: Yale University. Retrived from http://www.epi.yale.edu

Iovieno P, Morra L, Leone A, Pagano L, Alfani A (2009) Effect of organic and mineral fertilizers on soil respiration and enzyme activities of two Mediterranean horticultural soils. Biol Fertil Soils 45(5):555–561. https://doi.org/10.1007/s00374-009-0365-z

Jan AT, Ali A, Haq QMR (2014) Phytoremediation: A promising strategy on the crossroads of remediation. In: Hakeem K, Sabir M, Ozturk M, Mumet A (eds) Soil remediation and plants: prospects and challenges. Elsevier, Waltham, pp 1–36

Jha PK, Nair S, Gopinathan MC, Babu CR (1995) Suitability of rhizobia-inoculated wild legumes *Argyrolobium flaccidum, Astragalus graveolens, Indigofera gangetica* and *Lespedeza stenocarpa* in providing a vegetational cover in an unreclaimed limestone quarry. Plant Soil 177(2):139–149. https://doi.org/10.1007/BF00010120

Jogesh A, Dubash NK (2014) Mainstreaming climate change in state development planning: an analysis of the odisha climate change action plan. Centre for Policy Research (CPR), New Delhi. Retrieved from http://www.cprindia.org/sites/default/files/working_papers/jogesh__dubash_mainstreaming_climate_in_state_planning_odisha_climate_plan_feb_2014.pdf

Johnson DB, Hallberg KB (2005) Acid mine drainage remediation options: a review. Sci Total Environ 338(1–2):3–14. https://doi.org/10.1016/j.scitotenv.2004.09.002

Jordao C, Nascentes C, Cecon P, Fontes R, Pereira J (2006) Heavy metal availability in soil amended with composted urban solid wastes. Environ Monit Assess 112(1–3):309–326. https://doi.org/10.1007/s10661-006-1072-y

Kannan R, Shackleton CM, Uma Shaanker R (2013) Playing with the forest: invasive alien plants, policy and protected areas in India. Curr Sci 104(9):1159–1165

Karaca A (2004) Effect of organic wastes on the extractability of cadmium, copper, nickel, and zinc in soil. Geoderma 122(2–4):297–303. https://doi.org/10.1016/j.geoderma.2004.01.016

Khan AG (2005) Role of soil microbes in the rhizospheres of plants growing on trace metal contaminated soils in phytoremediation. J Trace Elem Med Biol 18(4):355–364. https://doi.org/10.1016/j.jtemb.2005.02.006

Khan FI, Husain T, Hejazi R (2004) An overview and analysis of site remediation technologies. J Environ Manag 71(2):95–122. https://doi.org/10.1016/j.jenvman.2004.02.003

Khatua S, Stanley W (2006) Ecological debt: a case study from Orissa, India. Retrieved from http://www.deudaecologica.org/publicaciones/Chapter5(125-168).pdf

Kitula A (2006) The environmental and socio-economic impacts of mining on local livelihoods in Tanzania: a case study of Geita District. J Clean Prod 14(3):405–414. https://doi.org/10.1016/j.jclepro.2004.01.012

Kumari J, and Sahu R, (2010). Bio diversity loss due to mining activities at Chiria mines, Dist. W. Singhbhum, Jharkhand. Asian Journal of Environmental Science 5(1):1–4

Kundu N, and Ghose M, (1997). Soil profile characteristic in rajmahal coalfield area. Indian Journal of soil and water conservation 25(1):28–32

Kuter N (2013). Reclamation of Degraded Landscapes due to Opencast Mining. In M. Ozyavuz (Ed.) Advances Landscape Architecture. INTECH. DOI: https://doi.org/10.5772/55796

Koch JM, Richardson J, Lamont BB (2004) Grazing by kangaroos limits the establishment of the grass trees *Xanthorrhoea gracilis* and *X. preissii* in restored bauxite mines in eucalypt forest of southwestern Australia. Restor Ecol 12(2):297–305. https://doi.org/10.1111/j.1061-2971.2004.00335.x

Lasat MM (2002) Phytoextraction of toxic metals: a review of biological mechanisms. J Environ Qual 31(1):109–120

Leavitt KJ, Fernandez GC, Nowak RS (2000) Plant establishment on angle of repose mine waste dumps. J Range Manag 53(4):442–452. https://doi.org/10.2307/4003758

Lee CH (2003) Assessment of contamination load on water, soil and sediment affected by the Kongjujeil mine drainage, Republic of Korea. Environ Geol 44(5):501–515

Lin C, Tong X, Lu W, Yan L, Wu Y, Nie C, Chu C, Long J (2005) Environmental impacts of surface mining on mined lands, affected streams and agricultural lands in the Dabaoshan mine region, southern China. Land Degrad Dev 16(5):463–474. https://doi.org/10.1002/ldr.675

Lindborg R, Eriksson O (2004) Effects of restoration on plant species richness and composition in Scandinavian semi-natural grasslands. Restor Ecol 12(3):318–326. https://doi.org/10.1111/j.1061-2971.2004.00334.x

Lindenmayer DB, Fischer J (2006) Habitat fragmentation and landscape change: an ecological and conservation synthesis. Island Press, Washington, D.C.

Love A, Babu S, Babu C (2009) Management of *Lantana*, an invasive alien

Lytle DE, Cronan CS (1998) Comparative soil CO2 evolution, litter decay, and root dynamics in clearcut and uncut spruce-fir forest. For Ecol Manag 103(2–3):121–128. https://doi.org/10.1016/S0378-1127(97)00182-5

Madejon E, De Mora AP, Felipe E, Burgos P, Cabrera F (2006) Soil amendments reduce trace element solubility in a contaminated soil and allow regrowth of natural vegetation. Environ Pollut 139(1):40–52. https://doi.org/10.1016/j.envpol.2005.04.034

Maiti D, Maiti SK (2014) Ecorestoration of waste dump by the establishment of grass-legume cover. Int J Sci Technol Res 3(3):37–41

Martinez-Ruiz C, Fernandez-Santos B (2005) Natural revegetation on topsoiled mining-spoils according to the exposure. Acta Oecol 28(3):231–238. https://doi.org/10.1016/j.actao.2005.05.001

McEldowney S, Hardman DJ, Waite S (1993) Pollution: ecology and biotreatment. Longman Scientific & Technical, New York

McIntosh J, Barnhisel R (1993) Erodibility and sediment yield by natural rainfall from reconstructed mine soils. Soil Sci 156(2):118–126

Meher R (2009). Globalization, Displacement and the Livelihood Issues of Tribal and Agriculture Dependent Poor People The Case of Mineral-based Industries in India. Journal of Developing Societies 25(4), 457–480. https://doi.org/10.1177/0169796X0902500403

Mendez MO, and Maier RM (2008). Phytostabilization of mine tailings in arid and semiarid environments- An emerging remediation technology. Environmental Health Perspectives 116(3):278. https://doi.org/10.1289/ehp.10608

Mol JH, Ouboter PE (2004) Downstream effects of erosion from small-scale gold mining on the instream habitat and fish community of a small neotropical rainforest stream. Conserv Biol 18(1):201–214

Moore MM, Wallace Covington W, Fule PZ (1999) Reference conditions and ecological restoration: a southwestern ponderosa pine perspective. Ecol Appl 9(4):1266–1277

Moreno T, Oldroyd A, McDonald I, Gibbons W (2007) Preferential fractionation of trace metals–metalloids into PM10 resuspended from contaminated gold mine tailings at Rodalquilar, Spain. Water Air Soil Pollut 179(1–4):93–105

Moynahan OS, Zabinski CA, and Gannon JE (2002). Microbial Community Structure and Carbon-Utilization Diversity in a Mine Tailings Revegetation Study. Restoration Ecology 10(1):77–87

Mujumdar N (1960) Cow-dung as manure. Econ Weekly 12:743–744

Mulligan C, Yong R, Gibbs B (2001) Remediation technologies for metal-contaminated soils and groundwater: an evaluation. Eng Geol 60(1):193–207

Muzzi E, Fabbri T (2007) Revegetation of mineral clay soils: shrub and tree species compared. Land Degrad Dev 18(4):441–451

Naeem S (2006) Biodiversity and ecosystem functioning in restored ecosystems: extracting principles for a synthetic perspective. Foundat Restorat Ecol 88:114

Navarro CA, von Bernath D, Jerez CA (2013) Heavy metal resistance strategies of acidophilic bacteria and their acquisition: importance for biomining and bioremediation. Biol Res 46(4):363–371

Newman LA, and Reynolds CM (2004). Phytodegradation of organic compounds. Current opinion in Biotechnology 15(3):225–230

Nordstrom DK, Alpers CN, Ptacek CJ, Blowes DW (2000) Negative pH and extremely acidic mine waters from Iron Mountain, California. Environ Sci Technol 34(2):254–258

Noyd RK, Pfleger F, Norland MR, Sadowsky MJ (1995) Native prairie grasses and microbial community responses to reclamation of taconite iron ore tailing. Can J Bot 73(10):1645–1654

Noyd RK, Pfleger F, Norland MR (1996) Field responses to added organic matter, arbuscular mycorrhizal fungi, and fertilizer in reclamation of taconite iron ore tailing. Plant Soil 179(1):89–97

Nyamadzawo G, Shukla M, Lal R (2008) Spatial variability of total soil carbon and nitrogen stocks for some reclaimed minesoils of Southeastern Ohio. Land Degrad Dev 19(3):275–288

O'Dell RE, Claassen VP (2006) Relative performance of native and exotic grass species in response to amendment of drastically disturbed serpentine substrates. J Appl Ecol 43(5):898–908

Ofosu-Mensah EA (2011) Gold mining and the socio-economic development of Obuasi in Adanse. Afr J History Cult 3(4):54–64

Oraon V (2012) Change patterns of tribal livelihood: a case study in Sundergarh District, Odisha. In National Institute of Technology Rourkela India. e thesis available at http://ethesis.nitrkl.ac.in/3379/1/VIJAY_FINAL_REPORT_MAY12TH.pdf

Padmavathiamma PK, Li LY (2007) Phytoremediation technology: hyper-accumulation metals in plants. Water Air Soil Pollut 184(1–4):105–126

Palmer MA, Filoso S (2009) Restoration of ecosystem services for environmental markets. Science 325(5940):575

Palmer M, Allan JD, Meyer J, Bernhardt ES (2007) River restoration in the twenty-first century: data and experiential knowledge to inform future efforts. Restor Ecol 15(3):472–481

Palmroth M (2006) Enhancement of in situ remediation of hydrocarbon contaminated soil. Tampereen teknillinen yliopisto. Julkaisu-Tampere University of Technology. Publication, p 595

Panda G (2014) Mining sector and its impact on Odisha economy: proper utilization of limited resources using forecasting techniques. J Business Manag Sci 2:41–45

Panda S, Barik A (2011) Impact of iron ore mines generated pollutants on peripheral environment and its effective management—a case study of Koira region, Odisha. In Proceedings of the XI international seminar on mineral processing technology (MPT-2010) (Vol. 2, No. Section 14). Allied Publishers, New Delhi, pp 1147–1156

Pandey VC, Abhilash P, Singh N (2009) The Indian perspective of utilizing fly ash in phytoremediation, phytomanagement and biomass production. J Environ Manag 90(10):2943–2958

Parulekar A, Ansari Z, Ingole B (1986) Effect of mining activities on the clam fisheries and bottom fauna of Goa estuaries. Proc Anim Sci 95(3):325–339

Patra HS, Mishra BK, Sahu B, Dash P, Mohapatra P (2008) Impact of mining in scheduled area of Orissa, A Case Study from Keonjhar. http://vasundharaodisha.org/Research%20Reports/Impact%20of%20Mining%20in%20Schduled%20Area%20of%20Orissa.pdf

Pattanayak S, Saha S, Sahu P, Sills E, Singha A, Yang J (2010) Mine over matter? Health, wealth and forests in a mining area of Orissa. Indian Growth Develop Rev 3:166–185

Pejchar L, Mooney HA (2009) Invasive species, ecosystem services and human well-being. Trends Ecol Evol 24(9):497–504

Pierzynski GM, Schwab AP (1993) Bioavailability of zinc, cadmium, and lead in a metal-contaminated alluvial soil. J Environ Qual 22(2):247–254

Pilon-Smits E (2005) Phytoremediation. Ann Rev Plant Biol 56:15–39. https://doi.org/10.1146/annurev.arplant.56.032604.144214

Powell KI, Chase JM, Knight TM (2011) A synthesis of plant invasion effects on biodiversity across spatial scales. Am J Bot 98(3):539–548

Prasad M (2004) Phytoremediation of metals and radionuclides in the environment: the case for natural hyperaccumulators, metal transporters, soil-amending chelators and transgenic plants. In: Heavy metal stress in plants. Springer, Berlin, pp 345–391

Prasad MNV, de Oliveira Freitas HM (1999) Feasible biotechnological and bioremediation strategies for serpentine soils and mine spoils. Electron J Biotechnol 2:7–8

Prasad MNV, Freitas HDO (2003) Metal hyperaccumulation in plants—biodiversity prospecting forphytoremediation technology. Electron J Biotechnol 6(3):110–146

Reddell P, Milnes A (1992) Mycorrhizas and other specialized nutrient-acquisition strategies: their occurrence in woodland plants from Kakadu and their role in rehabilitation of waste rock dumps at a local uranium mine. Aust J Bot 40(2):223–242

Reddy CS (2008). Catalogue of invasive alien flora of India. Life Science Journal 5(2):84–89

Reid NB, Naeth MA (2005a) Establishment of a vegetation cover on tundra kimberlite mine tailings: 1. A greenhouse study. Restor Ecol 13:594–601

Reid NB, Naeth MA (2005b) Establishment of a vegetation cover on tundra kimberlite mine tailings: 2. A field study. Restor Ecol 13:602–608

Rivera D, Mejías V, Jáuregui BM, Costa-Tenorio M, López-Archilla AI, Peco B (2014) Spreading topsoil encourages ecological restoration on embankments: soil fertility, microbial activity and vegetation cover. PLoS One 9(7):e101413

Rokich DP, Dixon KW, Sivasithamparam K, Meney KA (2000) Topsoil handling and storage effects on woodland restoration in Western Australia. Restor Ecol 8(2):196–208

Rotkittikhun P, Kruatrachue M, Chaiyarat R, Ngernsansaruay C, Pokethitiyook P, Paijitprapaporn A, Baker A (2006) Uptake and accumulation of lead by plants from the Bo Ngam lead mine area in Thailand. Environ Pollut 144(2):681–688

Ruiz-Jaen MC, Aide TM (2006) An integrated approach for measuring urban forest restoration success. Urban Forest Urban Greening 4(2):55–68

Ruiz-Jaen MC, Mitchell Aide T (2005a) Restoration success: how is it being measured? Restorat Ecol 13(3):569–577

Ruiz-Jaen MC, Mitchell Aide T (2005b) Vegetation structure, species diversity, and ecosystem processes as measures of restoration success. For Ecol Manag 218(1):159–173

Saha S, Pattanayak SK, Sills EO, Singha AK (2011) Under-mining health: environmental justice and mining in India. Health Place 17(1):140–148

Sahoo M (2013) Socio-economic impact of mining on living in mining areas: a case study in Joda block of Keonjhar district of Odisha. Int J Arts Comm Literat 1(2)

Sahu H, Dash S (2011) Land degradation due to mining in India and its mitigation measures. In: Second international conference on environmental science and technology, February 26–28, 2011. Singapore (IEEE)
Salt DE, Prince RC, Pickering IJ, Raskin I (1995) Mechanisms of cadmium mobility and accumulation in Indian mustard. Plant Physiol 109(4):1427–1433
Salt DE, Smith R, Raskin I (1998) Phytoremediation. Annu Rev Plant Biol 49(1):643–668
Sehgal J, Abrol I (1994) Soil degradation in India: status and impact. Oxford & IBH Publishing Co., New Delhi
Schwenke GD, Mulligan DR, Bell LC. (2000). Soil stripping and replacement for the rehabilitation of bauxite-mined land at Weipa. I. Initial changes to soil organic matter and related parameters. Australian Journal of Soil Research, 38(2):345–369
Sen S, Kumar V (2012) The effect of different soil amendments on growth of mung bean (*Vigna radiata* L.) in coal mine overburden dumps. Int J Emerg Technol Adv Eng 2:264–270
SER (2004) The Society for Ecological Restoration Primer on ecological restoration. Society for Ecological Restoration, CSON, Arizona, USA. http://www.ser.org/resources/resources-detail-view/ser-international-primer-on-ecologicalrestoratation
Shah commission reports. (2014). The report of Justice M B Shah Commission of Enquiry on illegal iron and manganese ore mining in Odisha. http://mines.nic.in/writereaddata/UploadFile/First%20Report
Sharma D, Sunderraj SW (2005) Species selection for improving disturbed habitats in Western India. Curr Sci 88(3):462–467
Sharma K, Kumar S, Gough L (2000) Rehabilitation of lands mined for limestone in the Indian desert. Land Degrad Dev 11(6):563–574
Sheoran V, Sheoran A, Poonia P (2010) Soil reclamation of abandoned mine land by revegetation: a review. Int J Soil Sediment Water 3(2):13
Simberloff D (2000) Introduced species: the threat to biodiversity and what can be done. Action Bioscience original article http://www.actionbioscience.org/biodiversity/simberloff.html
Simek M, Hopkins D, Kalcik J, Picek T, Santruckova H, Stana J, Travnik K (1999) Biological and chemical properties of arable soils affected by long-term organic and inorganic fertilizer applications. Biol Fertil Soils 29(3):300–308
Singh S (2004) Biological reclamation of degraded mined land a sustainability indicator. EniviroNews 10:1–3. http://isebindia.com/01_04/04-01-2.html
Singh AP (2010) Butterfly diversity in tropical moist deciduous sal forests of Ankua Reserve Forest, Koina Range, Saranda Division, West Singhbhum District, Jharkhand, India. J Threat Taxa 1130–1139
Singh AN, Singh J (1999) Biomass, net primary production and impact of bamboo plantation on soil redevelopment in a dry tropical region. Forest Ecol Manag 119(1):195–207
Singh G, Shukla AK (2008) Environmental issues with best practice examples ofresponsible mining in India. In 21st World Mining Congress and Expo Krakow, Poland. 7–12 September 2008, pp 255–265, Kraków, Poland
Singh A, Singh J (2001) Comparative growth behaviour and leaf nutrient status of native trees planted on mine spoil with and without nutrient amendment. Ann Bot 87(6):777–787
Singh A, Singh J (2006) Experiments on ecological restoration of coal mine spoil using native trees in a dry tropical environment, India: a synthesis. New For 31(1):25–39
Singh A, Jha A, Singh J (2000) Effect of nutrient enrichment on native tropical trees planted on Singrauli Coalfields, India. Restorat Ecol 8(1):80–86
Singh A, Zeng D, Chen F (2006) Effect of young woody plantations on carbon and nutrient accretion rates in a redeveloping soil on coalmine spoil in a dry tropical environment, India. Land Degradat Develop 17(1):13–21
Siniscalco C, Peyron A, Pata S, Reyneri A, Ciotti A (1997) Use of wild plant herbaceous species for revegetation. In: Symposium on plant biotechnology as a tool for the exploitation of mountain lands 457, Torino, Italy, pp 371–378
Skeel VA, Gibson DJ (1996) Physiological performance of *Andropogon gerardii*, *Panicum virgatum*, and *Sorghastrum nutans* on reclaimed mine spoil. Restor Ecol 4(4):355–367

Smith BM, Diaz A, Winder L, Daniels R (2005) The effect of provenance on the establishment and performance of *Lotus corniculatus* L. in a re-creation environment. Biol Conserv 125(1):37–46

Sun Y, Xie Z, Li J, Xu J, Chen Z, Naidu R (2006) Assessment of toxicity of heavy metal contaminated soils by the toxicity characteristic leaching procedure. Environ Geochem Health 28(1):73–78

Suresh S, Dinakar N, Prasad T, Nagajyothi P, Damodharam T, and Nagaraju A, (2007). Effects of a barite mine on ground water quality in Andhra Pradesh, India. Mine Water and the Environment 26(2):119–123

Sweeney S (2000) Different means, shared ends: environmental restoration and restoration ecology. Acta Universitas Palackianae Olomucensis Facultas Rerum Naturalium. Biologica 38:129–139

Tormo J, Bochet E, García-Fayos P (2007) Roadfill revegetation in semiarid Mediterranean environments. Part II: topsoiling, species selection, and hydroseeding. Restor Ecol 15:97–102

UNEP (1992). Global biodiversity strategy. Guidelines for action to save, study and use earth's biotic wealth sustainably and equitably. World Resources Institute, Washington DC, USA

Valladares G, Salvo A, Cagnolo L (2006) Habitat fragmentation effects on trophic processes of insect-plant food webs. Conserv Biol 20:212–217

Vangronsveld J, Cunningham SD (1998) Introduction to the concepts. In: Vangronsveld J, Cunningham SD (eds) Metal-contaminated soils: in situ inactivation and Phytorestoration. Springer, pp 1–4

Wickham J, Riitters KH, Wade T, Coan M, Homer C (2007) The effect of Appalachian mountaintop mining on interior forest. Landsc Ecol 22:179–187

Willems J (2001) Problems, approaches, and results in restoration of Dutch calcareous grassland during the last 30 years. Restor Ecol 9:147–154

Wong M (2003) Ecological restoration of mine degraded soils, with emphasis on metal contaminated soils. Chemosphere 50:775–780

Wood MK, Buchanan BA, Skeet W (1995) Shrub preference and utilization by big game on New Mexico reclaimed mine land. J Range Manag 48:431–437

Wu C, Lai S, Xu Y, Xu T, Lei J (2008) Application of 17 wild plant species in afforesting the abandoned quarry. Acta Hortic 501

Wysham D (2003a) Sustainable development south and north "climate change policy coherence in global trade and financial flows". Sustainable energy and economy network paper. ENV/EPOC/RD(2003) 8, available at: http://www.oecd.org/dataoecd/31/18/2510875.pdf

Wysham D (2003b) Destroying Orissa, fueling climate change: a joint project of the World Bank, Transnational Corporations, and G-7 governments. http://islandnet.com/plethora/mai/orissa.html

Yan D, Zhao F, Sun OJ (2013) Assessment of vegetation establishment on tailings dam at an iron ore mining site of suburban Beijing, China, 7 years after reclamation with contrasting site treatment methods. Environ Manag 52:748–757

Yeboah JY (2008) Environmental and health impact of mining on surrounding communities: a case study of Anglogold Ashanti in Obuasi. Thesis (no-299) (Department of Geography and Rural Development, Kwame Nkrumah University of Science and Technology)

Zaman M, Di H, Sakamoto K, Goto S, Hayashi H, Inubushi K (2002) Effects of sewage sludge compost and chemical fertilizer application on microbial biomass and N mineralization rates. Soil Sci Plant Nutr 48:195–201

Zhang Z, Shu W, Lan C, Wong M (2001) Soil seed bank as an input of seed source in revegetation of lead/zinc mine tailings. Restor Ecol 9:378–385

Ecological Restoration in India for Socio-Economic Well-Being

5

Pankaj Satija

Abstract

Ecological restoration is critical for improving socio-economic well-being by addressing environmental deterioration and promoting long-term development. In India, restoration activities improve livelihoods, public health, biodiversity conservation, and climate resilience. This chapter looks at the socio-economic advantages of ecological restoration, such as job creation, improved health, and community empowerment. It focuses on a variety of case studies, including the rejuvenation of the Yamuna River and mangrove restoration in the Sundarbans, to demonstrate how ecological interventions reduce climate risks, increase biodiversity, and benefit local economies. Restoration initiatives in metropolitan areas, such as the Sabarmati Riverfront development, exhibit better quality of life through green infrastructure. It emphasizes the combination of traditional ecological knowledge and contemporary approaches, as demonstrated by watershed management in Jharkhand and biodiversity protection in Odisha. Policy gaps, financing restrictions, and competing land uses are examined, as well as future options that emphasize multidisciplinary collaboration, strong governance, and public–private partnerships.

Keywords

Ecological restoration · Socio-economic well-being · Biodiversity conservation · Climate resilience · Community empowerment · Sustainable development · Ecosystem services

P. Satija (✉)
Ferro Alloys & Minerals Division, Tata Steel Limited, Bhubaneswar, India

D. Mittal, D. K. Gupta (eds.), *Ecological Restoration in India: Perspective and Practices*, https://doi.org/10.1007/978-981-96-8943-9_5

5.1 Introduction

5.1.1 Socio-Economic Benefits of Ecological Restoration

Socio-Economic Aspects of Ecological Restoration in India

The socio-economic aspects of ecological restoration in India encompass a range of dimensions, including livelihoods, health, economic stability, and social equity. One of the primary socio-economic benefits of ecological restoration in India is the creation of employment opportunities. Restoration projects, such as afforestation, wetland management, and river rejuvenation, require significant labour input, generating jobs, particularly for rural communities and marginalized groups. For example, the Mahatma Gandhi National Rural Employment Guarantee Act (MGNREGA) has incorporated ecological restoration activities, providing financial security to millions while simultaneously restoring ecosystems (Ministry of Rural Development).

Moreover, healthy ecosystems provide essential services that support livelihoods. Forests, wetlands, and rivers offer resources such as timber, non-timber forest products, fish, and clean water. Restoration enhances these ecosystem services, ensuring sustainable resource availability for local communities. For instance, the restoration of mangroves along India's coastlines not only protects against storm surges but also supports fisheries and provides raw materials for local industries (Dasgupta et al. 2021). Socio-economic well-being through ecological restoration can be achieved through different methods, which are further elaborated below.

5.2 Methods of Socioeconomic Well Being through Ecorestoration

5.2.1 Health and Well-Being

Restoration of ecosystems has both direct and indirect benefits for public health. Healthy ecosystems reduce the prevalence of vector-borne diseases such as malaria and dengue by regulating the habitats of mosquitoes and other disease vectors (World Health Organization [WHO] 2022). Additionally, clean air and water—outcomes of successful restoration projects are fundamental to public health. For instance, the rejuvenation of the Yamuna River in Delhi aims to reduce water pollution, thereby decreasing the incidence of waterborne diseases (Central Pollution Control Board [CPCB] 2023). Green spaces in urban areas, such as parks and restored wetlands, contribute to mental well-being and provide recreational opportunities. These spaces are crucial for the physical and psychological health of urban populations, who often experience the stresses of city life (Gupta and Sharma 2020).

5.2.1.1 Yamuna River Rejuvenation - Case Study

The *Yamuna Action Plan (YAP)* (Nakamura 2004), a bilateral project between the governments of Japan and India, was introduced by the Central Government in 1993 to aid State Governments in their efforts to combat the Yamuna River pollution issue. The Yamuna Action Plan is among India's biggest initiatives for river restoration. By installing public restrooms and sewers and conducting environmental and hygiene education in 15 cities across the three states of Delhi, Haryana, and Uttar Pradesh (UP) that border the Yamuna River Basin, the goal was to lessen the pollutants that untreated sewage and human waste add to the river. This would also help to improve water conservation in the river and the cleanliness of the cities surrounding the river basin.

The projects undertaken under YAP include the collection and redirection of raw sewage, the establishment of sewage treatment plants, the construction of affordable sanitation facilities, the establishment of electric or upgraded wood crematoria, and the development of riverfront properties. The plan included 21 towns from the three states of Uttar Pradesh, Haryana, and Delhi. A sewage treatment capacity of 753.25 million litres per day was created under the plan, at a cost of Rs. 872.15 crore (Japan International Cooperation Agency [JICA] 2004).

To ensure that only treated water enters the Yamuna and clears the city's main drains, the Delhi government has created a five-point action plan to be implemented by 2025.

1. To collect and remove plastic waste in one location, floating booms will be installed to clean drains in various zones. In addition, an in-situ treatment technique will be applied, where water is cleaned on-site rather than being transported to a different location, as shown in Fig. 5.1.
2. Near these drains, dams and weirs are to be contructed. The goal of constructing a dam is to raise the water's depth and cause its fine particles to settle on the ground and from the top of the dam, clean water overflows and flows forward.
3. Aeration devices will be installed to dissolve oxygen in the water and make it cleaner.
4. In various areas, floating wetlands will be erected to collect and hold dirt from the water. These will be constructed from long-lasting, reasonably priced bamboo. To help clean water by absorbing pollutants, wetland plants will be planted.
5. The drain zones will undergo chemical dosing to lower the wastewater's phosphate content. One of the main contaminants that cause the Yamuna to foam is phosphate.

Fig. 5.1 (**a**) Polluted state of Yamuna River; (**b**) rejuvenation of Yamuna River

5.2.2 Social Equity and Community Empowerment

Ecological restoration projects can promote social equity by involving local communities in the planning and implementation processes. When communities participate actively, they not only gain economically but also build a sense of ownership and stewardship over local natural resources. This empowerment can strengthen social cohesion and resilience against environmental shocks (Kumar and Singh 2011). Programmes that integrate traditional ecological knowledge with modern restoration techniques respect and preserve indigenous cultures and practices. In regions like the Western Ghats, involving indigenous tribes in restoration efforts has proven effective, as these communities possess deep ecological knowledge and a vested interest in the health of their environment (Das 2010).

5.2.2.1 Integrated Watershed Management in the Sarunala Catchment (Case Study)

The livelihoods of local tribal communities in Jharkhand, a state renowned for its mineral wealth, are in jeopardy due to an ongoing water crisis. Droughts, poverty, and migration are the results of ecosystem damage caused by unplanned mining, deforestation, and a lack of conservation practices. To restore ecosystems and improve water security, the non-governmental organizations (NGOs) Integrated River Basin Management Society (IRBMS), Jharkhand Vikash Parishad (JVP), and Association for India's Development (AID) collaborated to introduce Integrated Watershed Management in the Sarunala catchment (Integrated River Basin Management Society, Jharkhand Vikash Parishad, and Association for India's Development 2004). This initiative offers hope for ecological restoration and sustainable development in the area by involving local youth (World Bank 2007).

Part of the Bansloi River catchment, the Sarunala watershed is primarily in the hilly regions of the Rajmahal Hills, which are noted for their basalt trap formations. The project began in Nipania village, where IRBMS, AID, and JVP teams regularly interacted with locals to share technical concepts and illustrate in an understandable way the benefits of watershed management (Integrated River Basin Management Society, Jharkhand Vikash Parishad, and Association for India's Development 2004). Utilizing data from remote sensing and sophisticated geo-informatics tools, an objective assessment of the watershed's current state was conducted (Kumar and Singh 2011). The participation of stakeholders in the participatory watershed planning and implementation process was facilitated by regular community meetings (Das 2010). Early on, there were difficulties because the IRBMS, JVP, and the tribal community were unfamiliar with the concepts and the traditional way of life of the tribe (World Bank 2007).

The programme started to grow quickly in several villages with the help of local youth and JVP's volunteers. Notably, the watershed development programme (Fig. 5.2) spread to 75 villages in the Sarunala catchment thanks to the enthusiastic involvement of local youth, especially young girls and women. The Pahariya community, which lives on the level hillsides and ridgelines of the watersheds, was one of the hardest-hit communities. They had to walk kilometres downhill to collect

Fig. 5.2 Community engagement in planning process for watershed management

water from the few wells that remained during dry seasons when natural springs had dried up owing to deforestation and soil erosion, which had a severe impact on their way of life. Pahariya villages were given priority treatment in order to address this, with initiatives centred on the ridge areas showing promising results.

With the initiative's advancement, several villages implemented customized soil and water conservation strategies. While young volunteers in the village started creating and distributing seed balls to promote quick reforestation of degraded areas, the Pahariya community constructed stone guard walls to prevent soil erosion on their lands. The local community understood and accepted the building of gully plugging structures, check dams on small streams and nalas, and farm pond excavation in agricultural lands. The local hydrological cycle and ecosystem saw substantial changes because of these efforts over time, having a profound effect on people's lives and means of subsistence.

The youth population in the area has demonstrated a crucial role in propelling these initiatives ahead. Every year, they plan neighbourhood activities like farmers' markets and football and archery tournaments. In the Sarunala catchment villages, one noteworthy occasion that has grown in popularity is the Jal Jatra, a watershed awareness rally. It is encouraging to note that local panchayats and government block development officials have started taking part in these initiatives as well, having seen the positive changes they brought about. Like-minded local NGOs in other districts and river catchment areas have begun to replicate similar endeavours as word of the initiative's success and experience has spread. The joint endeavour of IRBMS, JVP, and AID provides a positive example of how young people and community members can be empowered to take an active role in watershed management and ecology restoration.

5.2.3 Economic Resilience and Climate Change Adaptation

Ecological restoration enhances the resilience of local economies to climate change. Restored ecosystems, such as wetlands and mangroves, act as natural buffers against extreme weather events like floods and cyclones. These natural defences can significantly reduce the economic damage to infrastructure, agriculture, and communities, thereby stabilizing local economies. For instance, the restoration of mangrove forests in Sundarbans and Odisha has been instrumental in protecting coastal villages from the devastating impacts of cyclones.

5.2.3.1 Mangrove Restoration in Sundarbans (case study)

The *Sundari Mangrove Restoration Project*: The Sundari Project (2023) in India is probably one of the most well-known biodiversity conservation projects through eco-restoration. Mangroves are essential natural ecosystems because they safeguard coastal ecosystems from natural disasters like typhoons, store a wealth of biodiversity, aid in the sequestration of carbon dioxide, and provide employment opportunities for nearby communities. However, because of disruptions brought on by humans and climate change, they are constantly degrading.

The project's goal is to plant a variety of native species and provide support for alternative local livelihoods in order to restore 4000 hectares of urban degraded mangroves in the Indian Sundarbans (Fig. 5.3), and it also has alignment with the Sustainable Development Goals (SDGs), where the project integrates environmental and community concerns with the management of natural resources. The project, which aims to improve local livelihoods for the more than 1.5 million residents of the 24th South Parganas District in West Bengal spread across 400 communities, was designed with a community-based approach in mind. Since the project's design phase, community landowners and representatives have been involved and engaged in the decision-making process. To develop afforestation, reforestation, and restoration activities, including monitoring and surveillance, community members have been hired and trained.

As per the project's alignment with Sustainable Development Goals, it aligns with SDG13—mangroves play a positive role in climate change mitigation; SDG15—the project combats desertification, stops and reverses land degradation, stops the loss of biodiversity, and manages forests sustainably; SDG8—the project employs local people to monitor and implement activities such as coordinating

Fig. 5.3 (**a**) Destroyed mangrove forest in West Bengal, India; (**b**, **c**) plantation of mangrove trees

plantations and maintaining nurseries and sites; SDG14—mangroves play an essential role in preventing flooding of coastal areas and serve as commercial fishing areas; and SDG5—more than 50% of the workers participating in the project are women.

The project hopes to restore 4000 hectares of mangrove and plant 19 mangrove species, while currently 600,000 trees have been planted, thereby protecting 1434 fauna species indirectly and employing more than 900 people.

5.2.4 Agriculture and Food Security

Healthy ecosystems are vital for sustainable agriculture and food security. Soil restoration projects improve soil health, enhance water retention, and increase agricultural productivity. Initiatives such as agroforestry and watershed management support farmers by providing diverse crops and mitigating the effects of droughts and floods.

5.2.4.1 Coastal Salinity Prevention Project (Case Study)

The success of the *Coastal Salinity Prevention Project* (Tata Trusts 2025) in Kodinar, Gujarat, has improved water availability and quality, leading to increased agricultural output and food security for local communities. In the 1990s, the Ambuja Cement Foundation (ACF) observed that the salinity had seeped inland in the Gujarat coastline up to 15 kilometres, and the total dissolved solids (TDS) was found to be as high as 4000 mg/litre for certain seacoast villages. The situation worsened owing to the intensive agricultural patterns in the area and the over-exploitation of groundwater. As a result, the local community, as shown in Fig. 5.4, in Kodinar had no potable water available, and agricultural yield was adversely hit, impacting the major source of community livelihood.

Fig. 5.4 Securing agriculture and food security for farmers in Kodinar, Gujarat

The solution was found by addressing salinity in a cost-effective way:

1. Surface water harvesting structure: Construction of check dams, revival of traditional water bodies, farm ponds, pond deepening, and interlinking canals.
2. Groundwater recharge: Construction of nala bunds, farm ponds, percolation well, tube well recharge.
3. In-situ moisture conservation: Drainage line treatment, nala plugs, etc.
4. Creating water reservoirs from mined-out pits: ACF initiated the conversion of mined-out pits into water reservoirs as a major sustainability initiative to address salinity. These pits were a low-cost solution for the storage of rainwater.

As an impact, in an independent study conducted, the programme was reported to deliver 13 times the social return on investment. In the initial days of the programme, ACF partnered with the Tata Trusts and pooled their resources for funding, technical expertise, and innovation to impact more villages in a robust, sustainable manner. In six years, they were able to demonstrate a robust model and impact. The government was convinced and agreed to fund close to 60–80% of all future initiatives of this nature. Following the success of the project, ACF now works closely with the government, NGOs, corporations, and development agencies with an aim to facilitate the pooling of resources and technical knowledge. One such collaborative initiative is the 'Coastal Salinity Prevention Cell', with ACF as its founding member along with Aga Khan Rural Support Program India and Tata Trusts. Established in 2008, the cell acts as the advisory body to address salinity in the coastal regions of Gujarat.

5.2.5 Biodiversity Conservation

Ecological restoration promotes biodiversity conservation, which is crucial for maintaining ecosystem services that support human well-being. Biodiverse ecosystems are more resilient to environmental changes and provide a wide array of resources, such as medicinal plants and pollinators for crops. In regions like the Western Ghats and the Sundarbans, restoration projects aim to protect endangered species and restore habitat connectivity, thereby supporting both conservation and local livelihoods that depend on biodiversity.

Fig. 5.5 (**a**) Asola Bhatti mines before eco-restoration; (**b**, **c**) Asola Bhatti in its present state

5.2.5.1 Biodiversity Conservation is Asola Bhatti (Case Study)

One of the prominent examples of eco-restoration for biodiversity conservation is *Asola Bhatti* (Fig. 5.5). The Asola Bhatti (Goel and Rath 2022) nature sanctuary, which is home to over 200 species of birds, animals, and plants, is currently rebuilding its enormous flora and fauna after once being a hub of the mining industry. The Asola Bhatti Wildlife Sanctuary stretches over 6874 acres, extending across the border between Delhi and Haryana. There are massive deposits of quartzite on the ridge that were used as building material. In 1986, the region was designated as a wildlife sanctuary, and in 1991, additional regions were included. Up until the late 1990s, the illegal mining business prospered from the minerals it extracted.

In addition to mining, the sanctuary was protected by stopping the grazing of cattle and the cutting down of trees. One strategy used to repair the sanctuary from the ravages of mining was reforestation. Steps have been taken to prevent unlawful entrance, including intensive patrolling. A few endeavours have been made to reestablish Asola Bhatti Nature Sanctuary, such as planting saplings, rainwater harvesting, restoration of degraded land and habitat, agroforestry, herbal plantations, use of solar energy, and organic farming. More than 193 different species of birds, 80 different kinds of butterflies, and other animals like leopards, blue bulls, nilgai bulls, and black bucks may all be found in Asola Bhatti. In the near future, the Forest Department is also working to create strategies for ecotourism while keeping biodiversity in mind.

5.2.5.2 Regeneration of Forests at the Noamundi Mines of Tata Steel (Case Study)

Another case study, as shown in Fig. 5.6, is the *regeneration of forests at the Noamundi mines of Tata Steel* (Husain et al. 2020). The Noamundi iron ore mine is naturally endowed with a lower stripping ratio, and the overburden produced has been utilized in the reclamation/restoration of the mine. The restoration was scientifically engineered over many decades, while the implementation of the

Fig. 5.6 (**a**, **b**) Noamundi Hill 1 and 2 before; (**c**, **d**) after reclamation

Fig. 5.7 Niche nesting in Tata Steel mining locations

afforestation plan was a joint exercise with local stakeholders. The company has covered a 126-hectare area across which it adopted progressive reclamation.

The afforestation has made the restored mined-out area a habitat for local birds and animal species today. Local stakeholders use these regenerated forests as a source of livelihood, extracting traditional medicinal plants and minimizing the impact on other natural forests. The plantation of native species in the dump helps reclaim the mine to its original position, which is in harmony with the surrounding virgin forest. Developing local forests as part of a progressive reclamation plan helps in achieving the goals of the final mine closure plan. Environmental amelioration and human well-being strategies for reducing the rate of degradation, fragmentation, and loss of all natural habitats have been done in 2020.

Also, *niche nesting* (Husain et al. 2020) was taken as one of the projects identified under the Biodiversity Management Plan (BMP) (Fig. 5.7) to improve bird biodiversity at all the mining locations of Tata Steel. The location of the nest and its monitoring were done in the geographic information system (GIS) platform. A niche is a space selected by any animal or plant species for feeding, resting sleeping, nesting, breeding, and growing plants. A niche creates a secure space for a species with less competition from other species and the freedom to grow.

With guidance from the International Union for Conservation of Nature (IUCN), Tata Steel has developed the nest boxes and placed them at various locations in reclaimed as well as domestic areas of the mining site to provide an essential nesting area for members of many species of birds.

5.2.5.3 Regenerating Forests at Kakudia Dump of the Sukinda Mine in Jajpur, Odisha (Case Study)

Similarly, Tata Steel has worked hard in *regenerating forests through the restoration of the 100 ha Kakudia dump* of the Sukinda mine in Jajpur, Odisha. The dump was reclaimed on a set of standard operating procedures for reclamation and rehabilitation through a scientifically engineered process of afforestation, and the entire plantation was carried out with the involvement of local stakeholders. The very purpose was to set an example of recreating the original land use, i.e. natural ecosystem, by bringing back potential natural species of the area by enhancing the quality of ecological services.

Fig. 5.8 (**a**, **b**) Dump before eco-restoration; (**c**–**g**) after eco-restoration, flora growth in reclaimed dump area; (**h**–**o**) revival of fauna presence in restored Kakudia dump at Sukinda

A total of approximately 5,27,760 numbers of saplings were planted. All these conscious steps and regular monitoring have resulted in creating the dense forest on the Kakudia dump with enriched local biodiversity and an improved ecosystem. Because of the conversion of the mining land into a thick biomass cover and a good habitat for the local species, the following social, environmental, and governance benefits and outcomes have been observed:

1. Food security for the local flora and fauna.
2. Food security for the local villagers.
3. Employment for the local villagers because of them being used for horticulture and plantation purposes.
4. Increased sustainability in terms of employment, climate change, and social responsibility.

During the survey after reclamation in Fig. 5.8, among flora, 30 species of trees, 6 species of shrubs, 14 species of herbs, 3 species of climbers, 3 species of grass, and 1 species of bamboo were recorded. Around six species of mammals were recorded during the survey through direct and indirect signs. Among avifauna, around 18 species of birds were observed. Among butterflies, 20 species of butterflies belonging to 5 families were recorded. The common emigrant was recorded in maximum number from the study site, as this species feeds on plants belonging to the Cassia genus, and the study site has plenty of Cassia siamea in plantation areas.

5.2.6 Tourism and Recreation Opportunities

Ecological restoration can boost ecotourism and recreational opportunities, contributing to local economies. Restored landscapes attract tourists, providing income and employment opportunities for local communities. National parks, wildlife sanctuaries, and nature reserves that undergo restoration can draw visitors for activities such as wildlife viewing, hiking, and nature photography.

5.2.6.1 Restoration of the Chilika Lake in Odisha (Case Study)

The *restoration of the Chilika Lake in Odisha* (Finlayson et al. 2020), for instance, has revitalized the local tourism industry by attracting bird watchers and nature enthusiasts from around the world. Chilika (Fig. 5.9) is a hotspot of biodiversity that is home to several endangered species that are listed on the IUCN Red List of Threatened Species. It is a unique combination of marine, brackish, and freshwater ecosystems with estuarine characteristics. More than a million migratory birds winter there, making it a sight to behold. More than 0.15 million fishermen live in and around the lagoon, where their livelihood is supported by the highly productive ecosystem and abundant fisheries. Numerous issues and threats have been plaguing

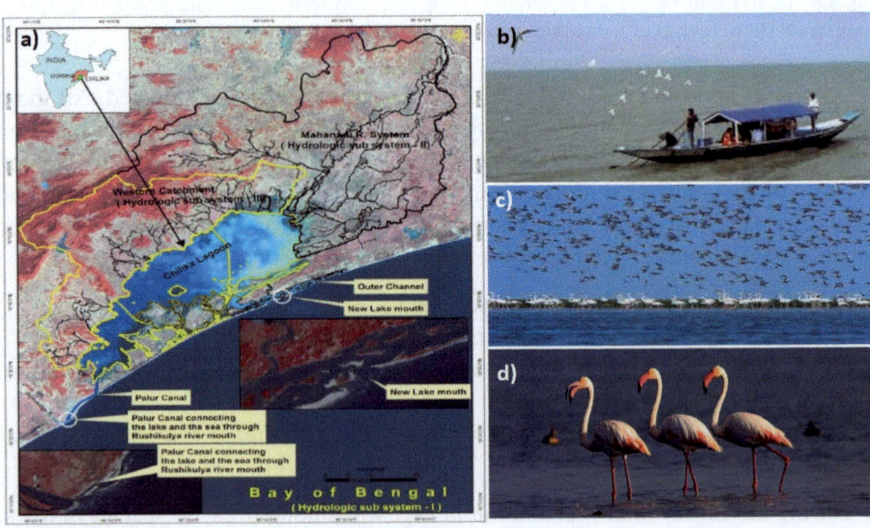

Fig. 5.9 (**a**) Chilika Lake with location of New Lake mouth; (**b**–**d**) birdwatching by tourists in Chilika Lake

the Chilika ecosystem, including siltation, shrinking of the water spread area, clogging of the inlet channel, shifting of the mouth that connects to the sea, a drop in salinity and fisheries resources, the spread of invasive species that live in freshwater environments, and a general loss of biodiversity. Chilika Development Authority (CDA) was established in 1992 to oversee the restoration and general development of the lagoon due to concerns regarding the deterioration of the lagoon ecosystem.

In order to tackle this issue, CDA hired top national institutions, including the National Institute of Oceanography (NIO) in Goa, to conduct studies on the lead channel's bathymetry, long-shore sediment transport along the shore, and the inlet's wave climate. The hydrological and two-dimensional mathematical model studies were conducted at the Central Water and Power Research Station (CWPRS), Pune. The studies found that the shoal formation along the lead channel and the constant mouth shifting, which resulted in a significant hydraulic head loss, significantly reduced the tidal influx into the lagoon. On September 23, 2000, an artificial mouth was opened in accordance with the CWPRS's recommendations, resulting in an 18-kilometre reduction in the outflow channel's length.

In addition to revitalizing the lagoon ecosystem, the opening of the artificial mouth and the desiltation of the lead channel greatly helped the fishermen, whose average annual income increased by more than Rs. 50,000 per family. The lagoon's ecosystem has improved overall as a result of the hydrological interventions made to restore it, which have also had a positive impact on biodiversity, water quality, and fishery resources. Further, Chilika is one of the largest communal centres for exotic and migratory birds in India. Ecotourism resorts have been successfully running in Chilika, where there is a footfall of many tourists interested in birdwatching, boat riding, and other leisure activities.

5.2.6.2 Ecorestoration by Coal India for Tourism

Another example of eco-restoration, as shown in Fig. 5.10, for tourism purposes is the activity of *pisciculture, creating restaurant and boating in old quarries of Southeastern Coalfields Limited (SECL) of Coal India* (Southeastern Coalfields Limited 2024).

The amenities have been built on a 26-acre water-filled site that was once Quarry No. 6 of Bishrampur Open Cast (OC) of SECL in the Surajpur District of

Fig. 5.10 (**a**) Pisciculture and (**b**) floating restaurant at Bishrampur OC, SECL

Chhattisgarh, created by the state fisheries department of Chhattisgarh, with funding from the Bishrampur Area's Mine Closure Fund of Rs. 197.00 lakh. There are 32 pisciculture cages, 1 floating restaurant, 2 motorboats, and floating rafts available as facilities. The district administration has given the self-help group of the local villagers complete control over the setup for upkeep and operation, sustaining the lives of roughly 30 to 40 people in neighbouring villages. A respectable yield of fish (Pangash) is regularly released for commercial sale of about 800 quintals each year. Boating, floating restaurants, and other water sports also generate revenue. The location has grown in popularity among the locals as a boating and tourism destination, drawing an average of 100 visitors every day.

Coal India Limited (CIL) is currently transforming its closed mines into eco-parks (Southeastern Coalfields Limited 2024), which are becoming well-liked travel destinations for ecotourists. The local population is finding that these tourist attractions and eco-parks provide them with a means of subsistence. Plans are underway to create additional eco-parks and eco-restoration sites in the mining areas owned by CIL, adding to the 30 existing eco-parks that are currently drawing consistent foot traffic.

Gunjanpark, Eastern Coalfields Limited (ECL); Gokul eco-cultural park, Bharat Coking Coal Limited (BCCL); Kenapara ecotourism site and Ananya Vatika, SECL; Krishnashila eco-restoration site and Mudwani eco-parks, Northern Coalfields Limited (NCL); Ananta medicinal garden, Mahanadi Coalfields Limited (MCL); Bal Gangadhar Tilak eco-park, Western Coalfields Limited (WCL); and Chandra Sekhar Azad eco-park, Central Coalfields Limited (CCL) are some eco-restored mining locations that encourage coal mine tourism.

In its several mines, CIL is also implementing innovative methods, including Miyawaki planting, seed casting with drones, and seed ball planting. The reclamation of mined-out areas, overburden dumps, and other locations happens simultaneously with their delinking from active mining zones. Selected species for biological reclamation are chosen after consulting with state and federally funded expert agencies. Currently, about 33% of the land is covered in greenery, and remote sensing is used to monitor land reclamation and restoration.

5.2.7 Urban Development and Quality of Life

In urban areas, ecological restoration contributes to improved quality of life by enhancing green spaces, reducing urban heat island effects, and improving air and water quality. Projects that restore urban rivers, parks, and wetlands provide recreational spaces, reduce pollution, and mitigate the impacts of climate change.

5.2.7.1 Rejuvenation of the Sabarmati River (Study Study)

The *rejuvenation of the Sabarmati River* (Fig. 5.11) *in Ahmedabad (The Project—Sabarmati Riverfront)* is a notable example, where the restoration of the riverfront has transformed it into a vibrant public space, improving the quality of life for city residents. In Ahmedabad, India, the Sabarmati Riverfront development has been developed along the banks of the Sabarmati River. Constructed in 2005, the idea was first conceived in the 1960s. Phase 1 of the project began in 2012 and gradually

Fig. 5.11 (a) View of the slums on the bank of the Sabarmati River before construction; (b–d) rejuvenation of the Sabarmati River in Ahmedabad city

opened the waterfront to the public while other facilities were being built. The three main goals of the initiative were sustainable development, social infrastructure, and environmental enhancement. In 2020, Phase 2 was approved.

In 2005, work on the project started. The sewage system, land reclamation, and heavy engineering were finished at a cost of ₹900 crore (US$110 million). The lower promenades spanning 11.5 km on both sides have been finished, with a portion of the route opening to the public on August 15, 2012. Since then, a number of facilities have been actively constructed. By 2014, ₹1152 crore had been invested in the project, and ₹1400 crore had been spent by November 2019.

Both governmental and private development are carried out on the recovered area. Nearly 14% of the recovered land is intended to be used for commercial and residential purposes, with the remaining over 85% going towards public infrastructure, parks, sports facilities, and gardens. The barriers built on both banks enhance the environment by lessening bank erosion and flooding in the city's low-lying districts.

Several amenities and recreational areas, such as boating, parks and gardens, markets, shopping complexes, and sports complexes, have been built with the planning of building residential and commercial development.

5.2.8 Education and Capacity Building

Restoration projects (Fig. 5.12) offer opportunities for education and capacity building within communities. Training programmes in sustainable land management, conservation techniques, and biodiversity monitoring empower local people with skills and knowledge that can be applied beyond the immediate restoration projects. This capacity building not only enhances the effectiveness of restoration efforts but also provides long-term benefits by creating a skilled workforce capable of addressing future environmental challenges.

5.2.8.1 Tata Steel's Endeavour to Conserve Sukinda Ecorace (Case Study)

A case study for biodiversity conservation through eco-restoration by education and capacity building of farmers and communities is *Tata Steel's endeavour to conserve Sukinda Ecorace* (Tata Steel 2024), an indigenous silkworm breed of Sukinda, located in the Jajpur district of Odisha.

Fig. 5.12 Revival process of Sukinda Ecorace silkworm

Out of eight breeds of silkworms available in Odisha, hundreds of farmers in the region depend on tasar farming of Sukinda Ecorace for their livelihood. However, (since 2007 the Tasar Rearers' Cooperative Society of Sukinda had reared Sukinda Ecorace), it is now rarely seen, as another breed—Daba TV—has almost replaced it due to higher yields. The natural habitat of the superior Sukinda Ecorace breed is the Sukinda region, where the forest type is tropical moist deciduous, and the soil type is red loamy.

The reasons for the extinction of this breed can be attributed to over-exploitation/over-collection, degradation of natural habitat, deforestation, pollution, and increasing inbreeding depression. While the experts at the Pilot Project Centre (PPC), Sukinda, under the Directorate of Textile, Government of Odisha, and the Regional Extension Centre of the Central Silk Board (CSB), Baripada, under the Government of India, have started working on the conservation of the species, Tata Steel has come forward to join hands with them to contribute its bit to save this superior indigenous breed and promote biodiversity. Areas that need urgent intervention are being worked out jointly, particularly identifying forest patches in the region where this breed is naturally available. Facilitating more cocoon grainages and raising nurseries for its primary food plants, like Arjun (*Terminalia arjuna*) and Asana (*Terminalia tomentosa*), are also being explored.

The process involved not only awareness among farmers for adaptation and commercial rearing but also involvement of local communities in the conservation process. From 550 cocoons, farmers have successfully multiplied to 3212 numbers of cocoons (2200 numbers of good-quality cocoons and 1012 numbers of flimsy cocoons). The entire lot at present is preserved by the farmers themselves. This will not only help in the revival of the Sukinda Ecorace breed, thereby protecting the livelihood of farmers, but also enhance the biodiversity of the area.

5.3 Challenges and Future Directions

Despite the evident benefits, ecological restoration in India faces significant challenges. Funding constraints, policy inconsistencies, and the lack of a comprehensive restoration framework hinder progress. Additionally, land tenure issues and competing land-use interests can complicate restoration efforts. To overcome these challenges, a multi-faceted approach is necessary. Integrating ecological restoration into national and regional development plans, increasing investment in restoration projects, and fostering public–private partnerships can enhance the scope and scale of restoration efforts. Additionally, continuous monitoring and adaptive management are essential to ensure the long-term success of restoration projects.

Education and awareness campaigns can also play a crucial role. By educating the public about the importance of ecological restoration, support for restoration initiatives can be increased, leading to more community-driven projects and broader societal benefits.

Policy and Governance Effective ecological restoration requires robust policy frameworks and governance structures. Policies that support restoration efforts through incentives, regulations, and funding mechanisms are crucial for success. The Indian government's initiatives, such as the National Afforestation Programme and the Green India Mission, highlight the importance of policy support in scaling up restoration activities. Strengthening governance at local levels ensures better implementation, monitoring, and maintenance of restoration projects.

Interdisciplinary Collaboration Ecological restoration benefits from interdisciplinary collaboration, integrating ecological science, social sciences, economics, and traditional knowledge. Collaborative efforts involving scientists, local communities, government agencies, and non-governmental organizations ensure that restoration projects are scientifically sound, socially inclusive, and economically viable. Successful projects often result from such multi-stakeholder approaches, which combine diverse expertise and perspectives.

The socio-economic aspects of ecological restoration in India underscore its potential to create a more sustainable and equitable society. Through job creation, improved public health, enhanced social equity, community empowerment, economic resilience, enhanced food security, conserved biodiversity, built community capacity, boosted tourism, improved urban living conditions, and fostered effective policy and governance, ecological restoration can drive significant socio-economic progress. However, addressing the challenges requires coordinated efforts across governmental, non-governmental, and community organizations. As India continues to develop, integrating ecological restoration into its growth paradigm is not only an environmental necessity but also a socio-economic imperative.

Conflict-of-Interest Statement The author declares that they have no conflict of interest.

References/Credits

Asola Bhati Wildlife Santuary. https://abwls.eforest.delhi.gov.in/

Asola Bhatti: Once a mining industry, now a healing biodiversity, 2022, Judith Mariya Antony, The Patriot. https://thepatriot.in/delhi-ncr/asola-bhatti-once-a-mining-industry-now-a-healing-biodiversity-26699

Central Pollution Control Board (2023) Yamuna river rejuvenation report. Retrieved from https://cpcb.nic.in

Chilika Lake: A Beacon for Wetland Conservation. Kushagra Meshram. 2022. https://www.nature-infocus.in/environment/chilika-lake-a-beacon-for-wetland-conservation

Coastal Salinity Prevention and Mitigation. Tata Trusts. https://www.tatatrusts.org/our-work/livelihood/agriculture-practices/coastal-salinity-prevention-and-mitigation#:~:text=In%20an%20effort%20to%20resolve,Cell%20(CSPC)%20in%20Ahmedabad.

Coastal Salinity Prevention Cell. Peer Water Exchange. https://peerwater.org/en/organizations/124-Coastal-Salinity-Prevention-Cell

Das P (2010) The role of community participation in ecological restoration projects: a study in the Western Ghats. Ecological Sustainability Press

Dasgupta P, Bose R, Sinha S (2021) Mangrove restoration and coastal resilience in India. Environ Res J 45(2):134–148

Delhi government prepares 5-point action plan to clean Yamuna by 2025. India Today. Pankaj Jain. 7th April 2023. https://www.indiatoday.in/india/delhi/story/delhi-government-5-point-action-plan-to-clean-yamuna-by-2025-2357222-2023-04-07

Eco-Restoration, Chilika Development Authority, Department of Forest and Environment, Government of Odisha. https://www.chilika.com/eco-restoration.php

Empowering Nature, Nurturing Growth: Coal Sector Transforming Landscapes Through Sustainable Greening Initiatives for the Benefit of Coal Communities. Coal India, Ministry of Coal. Government of India. https://pib.gov.in/PressReleasePage.aspx?PRID=2007973#:~:text=Implementation%20of%20the%20Miyawaki%20technique,such%20as%20fertilization%20and%20watering.

Finlayson CM, Rastogi G, Mishra DR, Pattnaik AK (2020) Ecology, conservation, and restoration of Chilika Lagoon, India. In: Wetlands: ecology, conservation and management. https://doi.org/10.1007/978-3-030-33424-6

Forest Regeneration & Niche Nesting for Bird Conservation at Tata Steel. Hishmi Jamil Husain. https://www.tatasteel.com/media/11655/forest-regeneration-niche-nesting-for-bird-conservation-at-tata-steel.pdf

Goel V, Rath R (2022) Vantage. J Thematic Anal 3(2):104–116. https://journalvantage-maitreyi.com/images/Volume-3/9.%20Asola%20Bhatti%20Wildlife%20Sanctuary-A%20Case%20Study%20of%20Ecological%20Restoration_Its%20Success%20and%20Present%20Status%201.pdf

Government Of India, Ministry of Environment And Forests, Rajya Sabha, Question No 13.07.2009, Answered On Yamuna Action Plan. https://sansad.in/getFile/annex/217/Au892.pdf?source=pqars#:~:text=Yamuna%20Action%20Plan%20(YAP)%20was,of%20pollution%20of%20river%20Yamuna.

Gupta R, Sharma M (2020) Urban green spaces and mental health: a case study in India. J Environ Psychol 38(1):22–35

Husain HJ, Tata Steel, International Union for Conservation of Nature (2020) Forest regeneration & niche nesting for bird conservation at Tata Steel. Harnessing Nature Magazine 2:29–31

INDIA (Sundarbans): 16 million mangrove trees to protect local communities. https://livelihoods.eu/portfolio/news-india/

India: Chilika Lake Restoration, Orissa. Society for Ecological Restoration. https://old.ser-rrc.org/project/india-chilika-lake-restoration-orissa/#:~:text=Opening%20of%20a%20new%20mouth,before%20opening%20the%20new%20mouth.

Integrated River Basin Management Society. https://www.irbmsindia.com
Integrated River Basin Management Society, Jharkhand Vikash Parishad, Association for India's Development (2004) Integrated watershed management in the Sarunala catchment. Retrieved from https://www.irbms.org/sarunala-catchment
Internal Company Archives, Tata Steel Limited.
Japan International Cooperation Agency (JICA) (2004) Yamuna Action Plan Project: External evaluation report. Retrieved from https://www.jica.go.jp/Resource/english/our_work/evaluation/oda_loan/post/2005/pdf/2-33_full.pdf
Kumar R, Singh P (2011) Watershed management and its role in sustainable development in Jharkhand. Environ Manag J 28(2):97–110
Management Plan of Asola Bhatti Wildlife Santuary, (Govt. of NCT Delhi) 2015-16 to 2024-25, Resource Survey and Management Division, Forest Research Institute Dehradun. https://forest.delhi.gov.in/sites/default/files/generic_multiple_files/management_plan_of_ashola_bhati.pdf
Mine Water Utilization - Every Drop matters Every Drop Matters. Ministry of Coal. Government of India. https://coal.gov.in/en/sustainable-development-cell/mine-water-utilization
Ministry of Rural Development (n.d.) Vision of Mahatma Gandhi NREGA. Retrieved from https://nrega.nic.in/MGNREGA_new/Nrega_home.aspx
Nakamura C (2004) Yamuna Action Plan Project: External evaluation report. Japan International Cooperation Agency (JICA). Retrieved from https://www.jica.go.jp/Resource/english/our_work/evaluation/oda_loan/post/2005/pdf/2-33_full.pdf
Reviving the Sabarmati: Drishti Foundation Trust's Visionary River Cleaning Initiative. Dristi Foundation Trust. https://drishtifoundation.org/sabarmati-river/#:~:text=This%20helped%20in%20controlling%20the,green%20belts%20along%20the%20riverbanks.
Roadmap to Conserve Indigenous Silkworm Breed in Odisha's Sukinda. 2022. Times of India. https://timesofindia.indiatimes.com/roadmap-to-conserve-indigenous-silkworm-breed-in-odishas-sukinda/articleshow/93407253.cms
South Eastern Coalfields Ltd to use "Miyawaki" Plantation Method to Enhance Green Cover (2024) With Rs.4 crore Investment Pilot Project to be Implemented in SECL Gevra Area. Coal India, Ministry of Coal. Government of India. https://pib.gov.in/PressReleaseIframePage.aspx?PRID=1983156
Sundari.life. https://sundari.life/
Sundari – Mangrove restoration project. Ecoact. https://eco-act.com/project/sundari-mangrove-restoration-india/#:~:text=The%20Indian%20Sundarbans%20are%20home,species%20and%20supporting%20local%20livelihoods.
Tata Trusts P (2025) Coastal Salinity Prevention and Mitigation – Tata Trusts. Tata Trusts. https://www.tatatrusts.org/our-work/livelihood/agriculture-practices/coastal-salinity-prevention-and-mitigation
The story of Water Management in Kodinar. Ambuja Cement Foundation. https://www.ambujafoundation.org/uploads/our-publication/ReversingtheTide.pdf
The Sabarmati story: A river is more than a usable resource. Darshan Desai. 2022. https://questionofcities.org/the-sabarmati-story-a-river-is-more-than-a-usable-resource/
Tata Steel. (2024). Tata Steel's Sukinda Chromite mine to help conserve indigenous Tasar silkworm breed Sukinda EcoRace. Tata Steel. https://www.tatasteel.com/newsroom/press-releases/india/2016/tata-steel-s-sukinda-chromite-mine-to-help-conserve-indigenous-tasar-silkworm-breed-sukinda-ecorace/
Tata Steel's Sukinda Chromite Mine to Help Conserve Indigenous Tasar Silkworm Breed Sukinda Ecorace. https://www.tatasteel.com/media/newsroom/press-releases/india/2016/tata-steel-s-sukinda-chromite-mine-to-help-conserve-indigenous-tasar-silkworm-breed-sukinda-ecorace/
The project – Sabarmati Riverfront (n.d.). https://sabarmatiriverfront.com/the-project/
Versha Goel, Rita Rath. Asola Bhatti Wildlife Sanctuary-A Case Study of Ecological Restoration: Its Success and Present Status. Journal of Thematic Analysis, 2022; 3(2): Page 104–116, https://journalvantage-maitreyi.com/images/Volume-3/9.%20Asola%20Bhatti%20Wildlife%20

Sanctuary-A%20Case%20Study%20of%20Ecological%20Restoration_Its%20Success%20and%20Present%20Status%201.pdf
Vision of Mahatma Gandhi NREGA. https://nrega.nic.in/MGNREGA_new/Nrega_home.as
World Bank (2007) Ecological restoration and sustainable development in India: a case study of the Western Ghats. World Bank Publications
World Health Organization (2022) Ecosystem restoration and public health: a global perspective. WHO Environmental Reports. Retrieved from. https://www.who.int
Yamuna Action Plan. https://en.wikipedia.org/wiki/Yamuna_Action_Plan
Yamuna Action Plan Project, External Evaluator: Chiaki Nakamura, Field Survey: September 2004. https://www.jica.go.jp/Resource/english/our_work/evaluation/oda_loan/post/2005/pdf/2-33_full.pdf
Youth Empowerment in Watershed Management in Jharkhand, India. Sujit Choudhury. 7th Aug 2023. IUCN. https://iucn.org/story/202308/youth-empowerment-watershed-management-jharkhand-india

Picture References/Credits

(a) Explained: How Difficult it is to Clean Yamuna? Ayanabha Banerjee. 2024. Themooknayak. https://en.themooknayak.com/environment/explained-how-difficult-it-is-to-clean-yamuna; (b) Yet another Yamuna action plan. Damini Nath. 2016. The Hindu. https://www.thehindu.com/news/cities/Delhi/yet-another-yamuna-action-plan/article8571844.ece
Integrated River Basin Management Society. www.irbmsindia.com
(a) Pinpointing priority areas for mangrove conservation in the Sundarbans. Mongabay. 2023. EastMojo. https://www.eastmojo.com/national/2023/02/15/pinpointing-priority-areas-for-mangrove-conservation-in-the-sundarbans/; (b) Sundari—Mangrove restoration project. Ecoact. https://eco-act.com/project/sundari-mangrove-restoration-india/#:~:text=The%20Indian%20Sundarbans%20are%20home,species%20and%20supporting%20local%20livelihoods; (c) India (Sundarbans): 16 million mangrove trees to protect local communities. https://livelihoods.eu/portfolio/news-india/
CSR: Water as an impact multiplier. The CSR Journal. https://www.google.com/url?sa=i&url=https%3A%2F%2Fthecsrjournal.in%2Fcsr-water-impact-multiplier%2F&psig=AOvVaw3YXKLoV0S87JHpLMAxFp1p&ust=1726116674077000&source=images&cd=vfe&opi=89978449&ved=0CBUQjhxqFwoTCMDelvKLuogDFQAAAAAdAAAAABAE
(a) Bhatti Mines (2004). Ravi Agarwal. https://www.raviagarwal.com/2004/09/bhatti-mines-2004/; (b) Delhi to turn mining pits of Asola sanctuary into reservoirs. Urban Update. https://urbanupdate.in/tag/bhatti-mines/; (c) Asola Bhatti: Once a mining industry, now a healing biodiversity. Judith Mariya Antony. 2022. The Patriot. https://thepatriot.in/delhi-ncr/asola-bhatti-once-a-mining-industry-now-a-healing-biodiversity-26699
Internal Company Archives, Tata Steel Limited (Fig. 5.6).
Internal Company Archives, Tata Steel Limited (Fig. 5.7).
Internal Company Archives, Tata Steel Limited (Fig. 5.8).
Internal Company Archives, Tata Steel Limited (Fig. 5.9).
(a) Surya Mohanty, Subhrendu Sekhar Mishra, Muntaz Khan, Rajeeb K. Mohanty, Anil Mohapatra, A. Pattanaik. 2015. Ichthyofaunal diversity of Chilika Lake, Odisha, India: An inventory, assessment of biodiversity status and comprehensive systematic checklist (1916–2014). Check List. Volume 11, Number 6, Article 1817. https://www.researchgate.net/figure/Location-map-of-Chilika-Lake-and-its-connection-with-river-systems-and-the-Bay-of-Bengal_fig1_287277817; (b) A Deep Dive into Chilika Lake's Unique Ecosystem and Conservation Efforts. https://queenoftreasures.com/2023/06/22/a-deep-dive-into-chilika-lakes-unique-ecosystem-and-conservation-efforts/; (c) Chilika bird population surges. PTI. 2019. The

Hindu. https://www.thehindubusinessline.com/news/variety/chilika-bird-population-surges/article25926013.ece; (d) Chilika At a Glance. Chilika Development Authority, Department of Forest and Environment, Government of Odisha. https://www.chilika.com/chilika-at-a-glance.php

(a) & (b) Mine Water Utilization – Every Drop matters Every Drop Matters. Ministry of Coal. Government of India. https://coal.gov.in/en/sustainable-development-cell/mine-water-utilization

https://en.wikipedia.org/wiki/Sabarmati_Riverfront

Mangroves of India: Diversity, Conservation and Restoration Prospects

6

Karthigeyan Kaliyamurthy and Ashiho Asosii Mao

Abstract

Mangroves are plant communities that thrive in the harsh, salty conditions of tropical and subtropical intertidal zones around the world. Mangroves are ecosystems of paramount ecological importance and socio-economic value, providing a wealth of benefits, including shoreline protection, nursery habitats for fisheries, and carbon sequestration. They are one of the world's most significant reservoirs of biological and genetic diversity. Despite covering a mere 0.7% of the world's tropical forested area, mangrove deforestation is a significant contributor to global CO_2 emissions, accounting for around 10% of emissions from deforestation worldwide. Mangrove forests are facing severe threats from climate change, sea-level rise, and human activities in coastal regions. This chapter highlights the diversity of mangroves in India, their present status, distribution across the coastal states, threatened species of mangroves in India and the significant threats faced by the Indian mangroves. This chapter also provides a comprehensive overview of the current state of legal protection, restoration efforts, and conservation strategies for mangrove ecosystems in India. It highlights the existing legislative frameworks and policies in place to safeguard these vital ecosystems, as well as ongoing rehabilitation and restoration activities. Furthermore, this chapter offers recommendations and suggestions for effective conservation and management practices to ensure the long-term protection and sustainability of mangroves.

Keywords

Flora · Threats · Ecosystem · Intertidal · MISHTI · IUCN · Sundarbans

K. Kaliyamurthy · A. A. Mao (✉)
Headquarters, Botanical Survey of India, Salt Lake City, Kolkata, West Bengal, India
e-mail: aamao@bsi.gov.in

D. Mittal, D. K. Gupta (eds.), *Ecological Restoration in India: Perspective and Practices*, https://doi.org/10.1007/978-981-96-8943-9_6

6.1 Introduction

Mangroves are highly productive, salt-tolerant plant communities found in the tropical and subtropical intertidal regions of the world that are important for the well-being, food security and protection of coastal communities worldwide. They have evolved unique structural and physiological features to thrive in the harsh, saline, and dynamic environment of the coastal zone. Their specialised roots, such as stilt roots and pneumatophores, help them stabilise and breathe in the waterlogged soil; salt glands and various physiological mechanisms enable them to manage salt levels. These remarkable adaptations allow mangroves to flourish in areas where other plants would struggle to survive.

Mangroves are among the world's most dynamic ecosystems that play an important role in the coastal landscape where the communities depend on their rich aquatic resources. As one of the world's most significant reservoirs of biological and genetic diversity, mangroves support a rich diversity of associated species, which provide considerable services to human communities. Mangroves fulfil important socio-economic and environmental functions, including the provision of timber, firewood, and non-timber forest products (NTFPs); coastal protection against the effects of winds, waves, and water currents; conservation of biological diversity; protection of seagrass beds and coral reefs; and provision of habitat, spawning grounds, and nutrients for a variety of fish and shellfish, including many commercial species. Mangrove forests play an important role and act as a bioshield that safeguards coastal communities, infrastructure and ecosystems from devastation. Mangroves possess a complex root system which helps to slow down water flows, enhance sediment deposits and arrest coastal erosion. They also protect the water quality by filtering phosphates, nitrates and other pollutants from the water and improve the water quality before they reach the estuarine and ocean environment, thereby protecting sensitive habitats like seagrasses and coral reefs. Mangrove occurs in areas where the temperature ranges between 20 and 35 °C, experiences high relative humidity and receives annual rainfall between 1000 and 3000 mm (Naskar and Guha Bakshi 1987; Waisel 1972).

Mangrove forests comprise unique floristic elements supporting a wide range of terrestrial, estuarine and near-shore marine ecosystems. Mangroves sequester up to 25.5 million tonnes of carbon per year, about ten times greater than upland tropical forests (Ong 1993; Donato et al. 2011; Hutchinson et al. 2014). Mangroves provide a wealth of ecological benefits, including significant carbon sequestration and storage in tropical coastal ecosystems and many other ecosystem services that provide multiple direct and indirect benefits to coastal communities (Barbier et al. 2011; Gutting et al. 2021). Conservation of mangrove forests is widely recognised as a crucial climate adaptation strategy, essential for mitigating carbon emissions and supporting the resilience of coastal communities. Beyond their legal status as protected reserves or their economic value, mangroves are cherished as a public good by coastal populations, who depend on these ecosystems for their livelihoods (Owuor et al. 2024).

Mangrove ecosystems are found in 123 countries and territories and cover a total of 152,361 km² geographical area of the world, of which Southeast Asia alone has 51,049 km² area under mangrove, constituting 33.5% of the total mangrove cover of the world. Despite their global distribution across the coasts of many countries, mangrove forests comprise a mere 0.7% of the world's total tropical forested areas (Spalding et al. 2010). The Indo-Malesian region, with the highest mangrove species diversity in the world with 48 species, is considered the cradle of the evolution of mangrove forests. The mangroves in the Indian subcontinent are viewed as an extension from the Persian Gulf around the coast of India, Pakistan, and Sri Lanka to Myanmar, including the Andaman and Nicobar Islands (Sanjappa et al. 2011).

6.2 Extent and Diversity of Mangroves in India

India ranks 10th among the 12 mangrove-rich countries, with about 2% of global mangrove cover (Spalding et al. 2010). The major mangrove areas of India are found along the East Coast in the Gangetic Sundarbans of West Bengal, the Mahanadi delta of Odisha, the Godavari-Krishna estuaries of Andhra Pradesh and the Cauvery delta and Gulf of Mannar of Tamil Nadu. No major deltas are found along the west coast, but the smaller estuaries in the Gulf of Khambhat and the Gulf of Kutch of Gujarat, Mumbai, Ratnagiri, Malvan, Devgad and Vijayadurg of Maharashtra, Mandavai-Zuari estuary of Goa, Thiruvananthapuram, Kollam, Kottayam, Ernakulam, Kozhikode and Kannur in Kerala have healthy patches of mangroves.

The Indian mangroves exhibit high species diversity, with a total of 42 true mangrove species under 23 genera belonging to 16 families. In addition, four natural hybrids of mangroves, namely, *Rhizophora* × *annamalayana* (Kathiresan), *Rhizophora* × *lamarckii* Montrouz (Rhizophoraceae), *Sonneratia* × *urama* N.C. Duke and *Sonneratia* × *gulngai* N.C. Duke (Sonneratiaceae), are also reported from India (Raghavan et al. 2016). Among the Indian states and UTs, the highest species diversity is found in the Andaman and Nicobar Islands, with 37 true mangrove species, followed by Odisha with 35 species and West Bengal with 32 species (Karthigeyan and Mao 2022) (Fig. 6.1). The majority of the Indian mangrove forests (56.54%) are found on the east coast compared to only 31.06% along the west coast due to the availability of geomorphologically suitable terrain (Fig. 6.2).

6.3 Major Threats to Mangroves

Over the past century, mangrove forest cover has declined significantly. The alarming loss of mangrove ecosystems can be attributed to various human activities, including the widespread clearing of mangrove forests for aquaculture, tourism, urban development and urban and industrial pollution, leading to degradation of mangrove vegetation (Fig. 6.3). Alarmingly, around 26% of the world's mangrove forests are degraded, primarily due to over-exploitation for fuelwood and timber

A panoramic view of the mangroves along the coast of Neil Island

Mangrove forests bordering the inland vegetation along the estuaries in Andaman Islands

Fig. 6.1 Mangroves in Andaman Islands

Dense formations of *Rhizophora apiculata* and *R* × *annamalayana* in Pichavaram mangroves

Luxuriant mangrove vegetation in Sundarbans

Fig. 6.2 Mangrove vegetation

Fig. 6.3 Degradation of mangrove vegetation due to human activities

production (Valiela et al. 2001). The expansion of aquaculture, particularly shrimp farming, is one of the important reasons for mangrove deforestation, responsible for nearly 38% of global mangrove loss. Additionally, other forms of aquaculture contribute to a further 14% of mangrove clearance (Ellison 2008).

In India alone, over 40% of mangrove area on the western coast has been converted to agriculture and urban development (Upadhyay et al. 2002). Unfortunately, these human-induced losses are expected to persist, and the situation may worsen due to the added pressures of climate change. Rising temperatures, sea-level rise, and increased frequency of extreme weather events will likely exacerbate the stresses on mangrove ecosystems, further threatening their survival. Studies suggest that in a changing climate, sea-level rise and increased storm intensity will affect storm surge areas, with the greatest impact expected from the loss of mangroves. It is estimated that under the current climate and mangrove coverage, 3.5 million people and GDP worth roughly US $400 million are at risk, partially protected by mangroves (Blankespoor et al. 2017).

Coastal and estuarine regions are highly vulnerable to the impacts of climate change, facing several challenges, including rising surface water temperatures, increased frequency of extreme weather events, and sea-level rise. Furthermore, the degradation of vital 'blue carbon' ecosystems like mangroves and salt marshes has exacerbated CO_2 concentrations. As these ecosystems deteriorate, they emit stored CO_2 back into the atmosphere, further fuelling climate change (Agarwal et al. 2023). Mangroves are considered one of the most carbon-dense ecosystems in the tropics, storing on average 1023 mg of carbon per hectare (Donato et al. 2011). It is estimated that mangrove deforestation is a significant contributor to global CO_2 emissions, accounting for around 10% of emissions from deforestation worldwide, despite accounting for just 0.7% of tropical forest area (Donato et al. 2011). This estimate highlights the urgent need to preserve mangrove ecosystems, not only for their unique biodiversity but also for their significant role in regulating the climate.

India's mangrove ecosystems are facing unprecedented biotic pressure, primarily driven by land use changes and the increasing demand for various products such as fodder, fuel wood, fibre, timber, paper, charcoal, and medicine. The western coast of India has witnessed a dramatic transformation, with nearly 40% of its mangrove cover being cleared for agricultural expansion and urban development (Upadhyay et al. 2002). The Indian Sundarbans, a critically important dynamic and fragile ecosystem, is under constant pressure from both natural and human-induced factors. The mangrove forests experience annual fluctuations in area due to natural processes of accretion and erosion (Hazra et al. 2002). However, the increasing local population and intensifying impacts from climate change, hydrological alterations, increasing agricultural activities and shrimp farming and illegal fishing pose serious threats to the mangrove ecosystem. The scarcity of freshwater in the Sundarbans poses a significant threat to the growth and survival of freshwater-loving mangrove species, resulting in limited growth and survival of freshwater-loving mangroves and loss of natural reproduction due to high salinity, causing potential local extinction of these species. According to the IUCN Red List of Ecosystems category and criteria, the Indian Sundarban mangrove ecosystem is classified as Endangered (EN) under subcriterion A3 (Sievers et al. 2020).

The Indian Sundarbans is undergoing a worrying transformation, where the southern region, nearer to the ocean, experiences increased salinity levels, leading to a decline in plant species diversity. The increased salt levels are stifling the growth and survival of several mangrove species, leading to a shift in the ecosystem's composition. There is a decline in the population of freshwater-loving mangroves like *Bruguiera* spp. and *Xylocarpus* spp. and an increasing trend of saline water-loving mangroves such as *Ceriops* spp. and *Avicennia* spp., indicating the ongoing transformation of the Indian Sundarbans (Bhadra et al. 2023). Although *Heritiera fomes* is widely distributed across South Asia, its population in the Indian Sundarbans has suffered substantial declines in recent times, highlighting a pressing conservation concern. This species, once abundant, has faced significant threats leading to its dwindling numbers, underscoring the need for targeted conservation efforts to protect and restore its populations. A significant knowledge gap exists regarding the impacts of mangrove area loss on individual species and populations. It is essential to collate and synthesise comprehensive species-specific information, including the presence of threatened species, to develop and implement effective conservation priorities, such as the designation of critical habitats or marine protected areas or informing policies on resource extraction and coastal development (Polidoro et al. 2010).

6.4 Threatened Mangrove Species in India

Among the Indian mangroves, seven species are included in the list of species at an elevated threat of extinction globally, which includes *Sonneratia griffithii* (critically endangered), *Heritiera fomes* (endangered), *Brownlowia tersa* (near threatened), *Phoenix paludosa* (near threatened), *Sonneratia ovata* (near threatened), *Aegialitis rotundifolia* (near threatened) and *Ceriops decandra* (near threatened) (Polidoro et al. 2010) (Fig. 6.4). The critically endangered *Sonneratia griffithii*, found in fragmented populations across India and Southeast Asia, faces a severe threat due to widespread mangrove loss. The region has witnessed an alarming 80% loss of mangrove cover over the past 60 years, with a significant loss in Malaysia (Ong 2003). The primary reasons for this destruction are the clearance of mangroves for rice cultivation, shrimp aquaculture, and coastal development (Ong 1995). The *Sonneratia griffithii* is already reported as locally extinct in multiple areas across its range of distribution. Alarmingly, fewer than 500 mature individuals of this species are known to remain in India, a critically low population that puts the species at risk of extinction. The species *Sonneratia ovata* and *Brownlowia tersa*, though still considered Near Threatened, are facing a concerning trend of severe population decline at the edges of their ranges (Polidoro et al. 2010). It is estimated that three out of ten mangrove species found primarily in the downstream estuarine and low intertidal regions are in the Threatened or Near Threatened categories. The riverine or freshwater-preferring species, such as the endangered *Heritiera fomes*, which was widespread in the Southeast Asian region, has experienced a significant decline. The natural populations of the *Heritiera fomes* have been severely reduced in Southeast Asia due to coastal development, aquaculture ponds, reduction of freshwater inflow, and expansion of palm and timber plantations (Polidoro et al. 2010). Mangrove species occurring along the landward margins are particularly susceptible to the impacts

Fig. 6.4 Threatened/near threatened and endangered mangroves species

of sea-level rise, especially when coastal development blocks their natural landward migration. This group of vulnerable species includes *Brownlowia tersa*, *Bruguiera sexangula*, *Nypa fruticans*, *Phoenix paludosa*, *Lumnitzera racemosa*, *Lumnitzera littorea*, *Sonneratia caseolaris*, *Sonneratia lanceolata*, and *Xylocarpus granatum*, which are all found at the landward margin or upstream in tidal estuaries (Polidoro et al. 2010) (Fig. 6.5). As sea levels continue to rise, these species are vulnerable to population decline and local extinction due to the risk of blocked migration. A total of 23 species of mangroves are listed under various threat categories (Kathiresan 2008) in India, which are found to exhibit severe population decline and need immediate habitat-specific study and monitoring for planning conservation strategies. A list of threatened mangrove species occurring in India is provided in Table 6.1.

The critically low population of the true mangrove tree species *Aglaia cucullata* (Roxb.) Pellegr., with only a few mature individuals remaining in Odisha and West Bengal, raises serious concerns about the long-term viability of this species in the Indian mangroves (Karthigeyan and Mao 2022). This uncertain population status underscores the urgent need for targeted conservation efforts for immediate and effective conservation, including habitat protection to protect and replenish this species, ensuring its survival and potential recovery in the region.

6.5 Current Status of Mangroves in India

Despite the threats and increasing pressure on the mangrove vegetation, there has been a considerable increase in the mangrove cover of India in recent years. India's mangrove conservation efforts have yielded impressive results, with a substantial 21.63% increase in mangrove cover over the past three decades. The country's mangrove area expanded from 4046 to 4992 sq km, representing a steady annual growth rate of 0.72% (Kathiresan 2019; FSI 2021). This notable achievement can be attributed to targeted government initiatives, which focused on intensive conservation and management efforts in selected areas. Additionally, systematic plantation of mangrove species in degraded areas, coupled with rigorous monitoring and protection measures by state forest departments, has contributed to the expansion of mangrove cover, underscoring the effectiveness of collaborative conservation efforts (Fig. 6.6). According to the recent mangrove cover estimate by the Forest Survey of India (FSI 2021), the mangrove cover in the country is 4992 sq km, which is 0.15% of the country's total geographical area. Mangroves are found in nine states and four Union territories along the coast of India. Very dense mangroves comprise 1475 km^2, moderately dense mangroves constitute a total of 1481 km^2, and open mangroves occupy a significant area of 2036 km^2. The state and Union territory-wise mangrove cover from 2011 to 2021 is provided in Table 6.2.

Aglaia cucullata - an extremely rare mangrove tree

Shirakiopsis indica - a locally endangered mangrove tree

Lumnitzera littorea - a rare species found only in the Andaman Islands in India

Xylocarpus granatum - a species vulnerable to habitat loss

Sonneratia caseolaris - a vulnerable mangrove tree

Bruguiera sexangula - a mangrove tree of rare occurrence

Fig. 6.5 Rare/vulnerable mangrove species

Table 6.1 Threatened mangrove species in India

S. No.	Name of species	Family	Threat status in India (Kathiresan 2008)	IUCN status (Polidoro et al. 2010)
1	*Acanthus ebracteatus* Vahl	Acanthaceae	Near threatened	–
2	*Nypa fruticans* (Thunb.) Wurmb	Arecaceae	Endangered	Near threatened
3	*Dolichondrone spathacea* (L.f.) Baill. ex Schumann	Bignoniaceae	Vulnerable	–
4	*Lumnitzera littorea* (Jack) Voight	Combretaceae	Endangered	–
5	*Shirakiopsis indica* (Willd.) Esser	Euphorbiaceae	Endangered	–
6	*Cynometra iripa* Kostel.	Fabaceae	Vulnerable	–
7	*Pemphis acidula* J.R.Forst	Lythraceae	Vulnerable	–
8	*Sonneratia alba* Sm.	Lythraceae	Vulnerable	–
9	*Sonneratia apetala* Buch.-Ham	Lythraceae	Vulnerable	–
10	*Sonneratia caseolaris* (L.) Engl.	Lythraceae	Vulnerable	–
11	*Sonneratia griffithii* Kurz	Lythraceae	Endangered	Critically endangered
12	*Sonneratia ovata* Backer	Lythraceae	–	Near threatened
13	*Heritiera fomes* Buch.-Ham	Malvaceae	Vulnerable	Endangered
14	*Brownlowia tersa* (L.) Kosterm.	Malvaceae	–	Near threatened
15	*Heritiera littoralis* Dryand.	Malvaceae	Vulnerable	–
16	*Xylocarpus granatum* J. Koenig	Meliaceae	Vulnerable	–
17	*Xylocarpus moluccensis* (Lam.) M. Roem.	Meliaceae	Vulnerable	–
18	*Aegialitis rotundifolia* Roxb.	Plumbaginaceae	Vulnerable	Near threatened
19	*Acrostichum speciosum* Willd.	Pteridaceae	Endangered	–
20	*Bruguiera parviflora* Wight & Arn. ex Griff.	Rhizophoraceae	Vulnerable	–
21	*Bruguiera sexangula* (Lour.) Poir.	Rhizophoraceae	Vulnerable	–
22	*Ceriops decandra* (Griff.) Ding Hou	Rhizophoraceae	Vulnerable	Near threatened
23	*Ceriops tagal* (Perr.) C.B. Rob.	Rhizophoraceae	Vulnerable	–
24	*Rhizophora stylosa* Griff.	Rhizophoraceae	Endangered	–
25	*Scyphiphora hydrophyllacea* C.F. Gaertn.	Rubiaceae	Endangered	–

Mangrove nursery raised and maintained by the West Bengal Forest Department

Well grown mangrove saplings raised in Forest Department nursery at Bonnie camp in Sundarbans

Monitoring of the mangrove afforestation programs in Lakshmipur, Sundarbans

Signage on the mangrove plantation in Sundarbans maintained by State Forest Department

Signage highlighting the importance of mangroves kept in Pichavaram mangrove forest - a good method of creating awareness among the tourists

Mangrove nature - walk in Dhaninallah, Andaman Islands- an ecotourism initiative

Fig. 6.6 Mangrove restoration efforts and conservation efforts by State Forest Departments

Table 6.2 Mangrove cover in states and union territories of India (FSI 2021)

S. No.	State/UT	Total (km^2) 2011	Total (km^2) 2013	Total (km^2) 2015	Total (km^2) 2017	Total (km^2) 2019	Total (km^2) 2021
1	Andhra Pradesh	352.00	352.00	367.00	404.00	404.00	405.00
2	Goa	22.00	22.00	26.00	26.00	26.00	27.00
3	Gujarat	1058.00	1103.00	1107.00	1140.00	1177.00	1175.00
4	Karnataka	03.00	03.00	03.00	10.00	10.00	13.00
5	Kerala	06.00	06.00	09.00	09.00	09.00	09.00
6	Maharashtra	186.00	186.00	222.00	304.00	320.00	324.00
7	Odisha	222.00	213.00	231.00	243.00	251.00	259.00
8	Tamil Nadu	39.00	39.00	47.00	49.00	45.00	45.00
9	West Bengal	2155.00	2097.00	2106.00	2114.00	2112.00	2114.00
10	Andaman and Nicobar Islands	617.00	604.00	617.00	617.00	616.00	616.00
11	Daman and Diu	02.00	01.00	03.00	03.00	03.00	03.00
12	Puducherry	01.00	01.63	02.00	02.00	02.00	02.00
	Total	**4663.00**	**4628.00**	**4740.00**	**4921.00**	**4975.00**	**4992.00**

6.6 Restoration and Rehabilitation of Mangroves

Over the past three decades, significant global efforts have been made to restore and rehabilitate degraded mangrove areas. Various approaches have been employed, including the integration of mangroves into coastal defence structures, monoculture plantations, and 'ecological mangrove restoration' (EMR) methods (Ellison et al. 2020). The prevalent approach to mangrove rehabilitation, characterised by short-term, mono-specific plantations, has been ecologically detrimental. This method often disregards crucial species-specific information and habitat characteristics, leading to the misguided introduction of incompatible species into unsuitable environments. As a result, numerous attempts at mangrove rehabilitation resulted in failure as the wrong species were being planted in the wrong environment (Zimmer 2018; Lewis et al. 2019; Raghavan et al. 2020). A more informed, ecologically grounded approach is imperative to ensure the long-term success of mangrove rehabilitation efforts.

Historically, conservation efforts for natural ecosystems like mangroves have focused on establishing protected areas (PAs) in relatively undisturbed locations, an approach that remains prevalent even today (Mace 2014). Encouragingly, there are now over 2500 PAs worldwide that encompass mangrove forests, covering around 39% (approximately 54,000 km^2) of the world's remaining mangroves. Notably, 34 countries have placed more than half of their mangrove areas under protection (Worthington and Spalding 2019; Raghavan et al. 2020). The extensive coverage of mangroves by protected areas represents a strong positive trend in coastal conservation. However, Southeast Asia—a hot spot of mangroves, renowned for its extensive mangrove cover, high species diversity, and rapid deforestation—has only 20% of its mangrove areas under protection (Raghavan et al. 2020).

A promising new estimate reveals that approximately 8120 km^2 of land worldwide is suitable for mangrove restoration, with a significant 6630 km^2 of area considered highly restorable. Furthermore, an additional 1389 km^2 of degraded mangrove areas have been identified as suitable for restoration efforts (Worthington and Spalding 2019). These findings offer a chance to revive and preserve these crucial ecosystems. Community-based rehabilitation, or 'Community based ecological mangrove rehabilitation', has emerged as a highly effective approach to mangrove restoration (Selvam et al. 2003; Datta et al. 2012). This approach leverages local ecological knowledge to bridge gaps in baseline information, such as detailed topography and hydrology data, typically required for restoration efforts (Raghavan et al. 2020). By empowering local communities to take ownership of mangrove rehabilitation, this method fosters a collaborative and inclusive approach to ecosystem restoration. The Pichavaram mangrove restoration project in India stands as a testament to the effectiveness of community-based rehabilitation efforts (Selvam et al. 2003). This successful initiative demonstrates the potential for local communities to take an active role in restoring and preserving vital mangrove ecosystems, highlighting the benefits of collaborative conservation approaches.

India has pioneered a groundbreaking technique in mangrove restoration, showcasing exemplary practices that have yielded remarkable results. Notably, the innovative 'Canal Bank planting' technique, employing a 'fish bone' design, has been successfully implemented in Muthupet and Pichavaram in Tamil Nadu, as well as East Godavari and Krishna districts of Andhra Pradesh (Kathiresan 2018). Community-based co-management has emerged as a highly effective approach to mangrove management in India, with successful initiatives in Andhra Pradesh, Gujarat, Odisha, Tamil Nadu, and West Bengal (Kathiresan 2018). Notably, the joint mangrove management initiatives in Sundarbans, which involve local communities in conservation and restoration efforts, have yielded remarkable success. The M.S. Swaminathan Research Foundation, with financial support from the Indo-Canada Environmental Facility (ICEF), New Delhi, successfully implemented a Joint Mangrove Management (JMM) project, empowering local communities to take an active role in mangrove restoration and conservation. This pioneering initiative engaged 5240 families from 28 villages along India's east coast, planting 6.8 million mangrove saplings and restoring about 1475 ha of mangroves. The Gujarat Ecology Commission implemented a similar project to regenerate and manage mangrove ecosystems across 5000 ha along the Gulf of Kachchh and Gulf of Khambhat through community-based management and regeneration in 5 years from 2001 to 2006 (Kathiresan 2018).

In a significant move to safeguard mangrove ecosystems in Mumbai and adjoining areas, the Government of Maharashtra established the 'Mangrove Cell' in January 2012, followed by the 'Mumbai Mangrove Conservation Unit' in 2013. These initiatives led to the upgradation of mangrove forests on government land from 'protected forests' to 'reserved forests', resulting in the notification of 15,088 ha of mangroves as protected areas across seven districts of Maharashtra. Notably, the Mangrove cell successfully facilitated the plantation of mangroves in over 200 ha of degraded areas within the Greater Mumbai region. As a result of all these combined efforts, the mangrove cover expanded substantially with a remarkable 20% increase between 2013 and 2015 (Vasudevan 2017; Kathiresan 2018).

The Soonabai Pirojsha Godrej Marine Ecology Centre (SPGMEC) in Maharashtra has been at the forefront of mangrove conservation efforts in Vikhroli, Mumbai, adopting a multifaceted approach that encompasses research, monitoring, restoration, education, and community outreach. By engaging local communities and fostering awareness about the importance of mangrove ecosystems, SPGMEC has made a significant role in protecting and preserving these vital forests, serving as a model for effective mangrove conservation in urban settings and spreading awareness about their critical role in maintaining coastal resilience and biodiversity. The AVT Tea Company's Mangrove Interpretation Centre in Kerala serves as a vital hub for environmental education and awareness-raising, providing a unique platform for visitors to learn about the significance of mangrove ecosystems and the importance of conservation efforts (Mahajan 2008; Sahu et al. 2015). The Wildlife Trust of India, in partnership with Apollo Tyres and the Kerala Forest Department, launched the 'Kannur Kandal Project' in Kerala and established a Mangrove Interpretation Centre for mangrove-based research and education and the promotion of mangrove restoration by engaging local communities and government agencies to maintain the long-term health of these critical ecosystems. In the Sundarbans of West Bengal, the Nature Environment and Wildlife Society (NEWS) successfully implemented a comprehensive project, 'Restoration of Mangroves in Sundarbans through Afforestation, Integrated Mangrove-Shrimp Farming, Income Generation, and Community Participation', aimed to not only stabilise and restore the fragile mangrove ecosystems but also enhance the livelihoods of local coastal communities through sustainable afforestation and integrated mangrove-shrimp farming practices (Bhattacharjee et al. 2018).

6.7 Conservation Strategies

Internationally, there are currently 11 treaties for the protection of mangroves, some of which have been in force for over five decades. While these agreements provide a framework for conservation on paper, their effectiveness in safeguarding mangroves in practice remains a concern. These treaties and instruments include the RAMSAR Convention, the Convention on the Prevention of Marine Pollution, CITES, the International Tropical Timber Agreement, the Convention for the Protection and Development of the Marine Environment of the Wider Caribbean Region and the Convention on Biological Diversity (Farnsworth and Ellison 1997; Polidoro et al. 2010). The existing international treaties related to mangroves have significant limitations. They do not provide explicit legal protection for mangrove ecosystems and, more critically, fail to address the specific needs of individual mangrove species. This lack of species-level protection leaves a significant gap in conservation efforts, emphasising the need for new agreements that prioritise the preservation and management of mangrove biodiversity. The disconnect between legal frameworks and effective conservation underscores the need for renewed efforts, increased accountability, and collective action to bridge the gap between intention and implementation (Polidoro et al. 2010).

Mangroves for the Future (MFF) is a pioneering partnership-driven initiative that fosters collaborative investment in coastal ecosystem conservation for sustainable development. Co-led by IUCN and UNDP, MFF serves as a platform to address the complex challenges faced by the coastal ecosystems and communities. The initiative aims to promote a holistic approach to coastal management, enhancing the resilience of coastal communities that depend on these ecosystems. MFF initially focused on supporting the worst-affected countries after the 2004 Indian Ocean tsunami, namely, India, Indonesia, the Maldives, Seychelles, Sri Lanka, and Thailand. Recognising the broader regional need, MFF has since expanded to include Bangladesh, Cambodia, Myanmar, Pakistan, and Vietnam, further strengthening its commitment to coastal ecosystem conservation and community resilience (http://www.mangrovesforthefuture.org).

As of 2023, a significant milestone has been reached, with 97 countries incorporating coastal and marine ecosystems, including mangroves, into their Nationally Determined Contributions (NDCs) under the Paris Agreement. Furthermore, 61 countries have explicitly identified the conservation or restoration of blue carbon ecosystems, such as mangroves, as a key measure for both mitigating and adapting to climate change. This widespread recognition of the critical role that coastal ecosystems play in addressing the climate crisis is a promising step towards safeguarding these vital ecosystems and leveraging their potential for climate action. The Global Mangrove Alliance (GMA) has set the most ambitious global goals to date on mangroves to be achieved by 2030. These goals include halting loss of mangrove ecosystems by reducing net anthropogenic mangrove loss to zero, restoring half of recently lost mangrove area and doubling the area-based protection of mangroves from 40% to 80% (https://www.mangrovealliance.org/about-us/).

Mangroves are highly relevant to the implementation of multiple goals and targets across the Kunming-Montreal Global Biodiversity Framework (GBF) adopted by Parties to the Convention on Biological Diversity (CBD). The GBF's ambitious target of conserving, managing, and restoring 30% of terrestrial, marine, and coastal ecosystems by 2030 presents a significant opportunity for mangrove conservation. It emphasises that conservation should especially focus on areas of importance to biodiversity and ecosystem functions and services (Slobodian and Vidal 2023). By COP16 in 2024, parties are required to revise their National Biodiversity Strategy and Adaptation Plans (NBSAPs) to ensure alignment with the Global Biodiversity Framework (GBF) objectives. This presents a timely opportunity for governments to integrate mangrove conservation and restoration into their updated plans, harnessing the potential of these ecosystems to achieve GBF targets and support biodiversity, ecosystem services, and human well-being (Gil et al. 2023).

The Global Mangrove Watch (GMW) is an online platform that provides remote sensing data and tools for global monitoring of mangroves. Through a collaborative effort with esteemed partners, including Wetlands International, Aberystwyth University, soloEO, The Nature Conservancy, Cambridge University, JAXA, NASA, and many others, this innovative platform provides access to near-real-time data and tools providing unparalleled insights into mangrove health and dynamics worldwide. The GMW offers Parties to the Convention a critical resource to support the integration of mangrove commitments into their NBSAPs revisions and national reports, collectively catalysing ambition and action on mangroves (Gil et al. 2023).

6.8 Conservation Through Traditional Practices: Mangrove Sacred Groves of India

In India, conservation of forest patches in the form of sacred groves has been practised since time immemorial. India's sacred groves represent a remarkable example of traditional conservation, where forest patches are protected by local communities driven by religious and cultural beliefs. These sacred sites have been maintained for centuries, providing a safe shelter for diverse plant and animal species. Mangrove forests have been traditionally protected and conserved in some regions by local communities as sacred groves, based on their spiritual beliefs and religious practices. The mangrove in Shravan Kavadia, Kachchh, is one of the most important sacred groves of Gujarat. Harvesting, logging or collection of firewood has been totally prohibited and considered religiously inauspicious by the local inhabitants. Similarly, the Kagekanu forest patch, situated off the coast of Karwar in Karnataka, is a remarkable example of traditional conservation through sacred groves. Dominated by mangrove species like *Rhizophora mucronata*, *Avicennia officinalis*, and *Kandelia candel*, this mangrove patch is protected by local villagers, enforcing strict regulations on exploitation of resources, allowing only limited, sustainable utilisation for temple purposes (Sahu et al. 2015).

6.9 Legal Protection for Conservation of Mangroves in India

The Government of India has demonstrated a strong commitment to mangrove conservation, establishing a comprehensive legal and regulatory framework to protect these fragile ecosystems. The mangrove forests are recognised as ecologically sensitive areas to ensure the long-term preservation of these vital coastal habitats and the diverse array of species they support. India boasts a pioneering legacy in mangrove forest management, with the Sundarbans being the world's first mangrove ecosystem to be brought under scientific management, with the introduction of a management plan as early as 1892 (Chaudhuri and Choudhury 1994).

In a visionary move, the Government of India established the National Mangrove Committee (NMC) in 1976, recognising the vital importance of mangrove ecosystems. As an advisory body, the NMC has played a pivotal role in providing expert guidance and recommendations, ensuring that mangrove conservation receives the attention it deserves. By acknowledging the significance of mangroves, India has demonstrated a commitment to preserving these critical ecosystems, ensuring the long-term health and resilience of its coastal environments. The Ministry of Environment and Forests has established the Coastal Regulation Zone (CRZ) to govern human activities in coastal areas, ensuring a balance between development and environmental protection. The CRZ categorises coastal zones, extending 500 m landward from the high tide line, into four categories based on their ecological sensitivity and the level of development along the coast. This classification system enables targeted regulations and management strategies to conserve vulnerable ecosystems while also supporting sustainable development in these critical regions. Recognising the vital importance of mangrove ecosystems, in 1987, the Indian government imposed a comprehensive ban on the felling of trees in these sensitive areas (Upadhyay et al. 2002).

Presently, the mangrove habitats of India are protected under the Indian Forest Act, 1927; Wildlife Protection Act, 1972; Water Prevention and Control of Pollution Act, 1974; Forest Conservation Act, 1980; Environmental Protection Act of 1986; National Forest Policy, 1988; Environmental Impact Assessment Notification (EIA) of 1994; Coastal Aquaculture Authority Act, 2005; Wetland (Conservation and Management) Rules, 2017; Coastal Regulation Zone (CRZ) Notification, 2018; National Wetland Management Programme (NWMP); National Lake Conservation Plan (NLCP); and National River Conservation Plan (NRCP) (Raghavan et al. 2016, 2020). The Sundarbans mangrove forests have been protected under the Indian Forest Act since 1927, with the area being designated as a reserved forest (Naskar and Mandal 1999). Nearly a century of protection has helped preserve the Sundarbans' remarkable biodiversity and ecosystem services, serving as a model for the importance of early conservation action.

In the Indian Sundarbans, regular consultations are being held with Forest Protection Committees (FPCs), Ecodevelopment Committees (EDCs), and local Panchayat members to address conservation issues and ensure collaborative decision-making. During site-specific planning for alternate livelihood programmes or infrastructure development, the executive bodies of these committees are engaged in detailed discussions, and village development projects are implemented based on local suggestions. Moreover, Joint Forest Management Committees (FPCs and EDCs), comprising local villagers, are actively participating in mangrove restoration efforts through village afforestation and conservation awareness programmes (Fig. 6.7).

The active participation of women in tree plantation programmes in many parts of the Indian Sundarbans is a significant contributor to the conservation of the Sundarban forest. The Joint Forest Management Committee's (JFMC) active

Fig. 6.7 Mangrove restoration efforts through village afforestation and conservation awareness programmes

presence in the villages has created a conducive environment for women's participation, resulting in a notably higher level of engagement. Furthermore, many villages benefit from the support of various NGOs and self-help groups, which have also contributed to the increased involvement of women in forest conservation efforts (Chandra and Mukopadhyay 2021).

In a significant move to protect India's mangrove ecosystems, the Government of India has launched the ambitious MISHTI (Mangrove Initiative for Shoreline Habitats & Tangible Incomes) scheme. This innovative programme aims to expand mangrove cover along India's coastline and saltpan lands, with a primary focus on the vulnerable Sundarbans delta and Hooghly Estuary in West Bengal. By prioritising mangrove conservation and restoration, the MISHTI scheme seeks to enhance the resilience of these critical ecosystems, promote sustainable livelihoods, and safeguard their rich biodiversity.

6.10 Conclusions and Recommendations

Wetland ecosystems have suffered alarming losses, with a 50% decline since 1900 (Li et al. 2018). The unchecked expansion of coastal development and population growth has made mangroves one of the planet's most endangered ecosystems (Goldberg et al. 2020). The estimated global loss of mangrove cover varies between 0.16% and 0.39% annually (Goldberg et al. 2020). The rampant destruction of mangrove habitats for agricultural and aquaculture expansion poses a significant threat to these vital ecosystems. Ecosystem destruction will reduce the essential products and services that mangroves provide, compromising the well-being and livelihoods of future generations. Unless urgent measures are taken to address the impact of human activities, the alarming rate of mangrove degradation will persist, leading to further loss of these vital ecosystems. The rate of destruction will continue relentlessly, compromising the essential benefits that mangroves provide, including shoreline protection, biodiversity, and fisheries. Preserving the ecological health of existing mangroves requires immediate attention and decisive action. While legislative protection has helped prevent some habitat conversion and supported natural regeneration, its impact is limited by poor enforcement. To safeguard the ecological health of existing mangroves, a comprehensive restructuring of the legislative framework is essential, accompanied by rigorous enforcement and monitoring mechanisms to ensure effective protection (Raghavan et al. 2020).

It is crucial to develop and implement effective conservation and management strategies to safeguard the biodiversity and ecological integrity of this unique ecosystem. The conservation of the Indian Sundarbans requires a holistic approach, recognising the intricate relationships between freshwater flow, salinity levels, and habitat preservation. The influence of salinity on species diversity in the Indian Sundarbans highlights the need to address the sources of increased salinity, such as sea-level rise, freshwater diversion, and human activities. Integrated coastal zone management approaches should be adopted to effectively balance human development and conservation goals, ensuring the sustainability of the mangrove ecosystem. Long-term monitoring programmes are crucial for generating reliable data by regularly assessing the changes in mangrove composition, distribution, and biodiversity (Bhadra et al. 2023). The limited success rate of mangrove restoration and rehabilitation initiatives highlights the need for improved strategies and approaches. It is essential to re-examine current methods and explore innovative solutions to address the complex challenges facing mangrove ecosystems. Since the anthropogenic or climate change impacts are often gradual in mangroves, long-term tracking of hydrological and ecological changes is essential to take effective conservation measures, such as ecosystem design, necessary to prevent catastrophic losses in plant community structure and ecological function (Lewis et al. 2016). Given the site-specific nature of mangrove ecosystems, a tailored approach is necessary for effective conservation.

Effective policies should be formulated with adequate site-specific information rather than global generalisation and should be complemented with economic incentives, such as carbon credits, payments for ecosystem services, deforestation taxes, and certified eco-friendly products. This approach will raise the value of maintaining and protecting the mangroves rather than converting them to other uses (Lee et al. 2019). Updating the IUCN Red List assessments at regular intervals is vital for accurately reflecting the conservation status of the mangrove species (Polidoro et al. 2010). As new data becomes available, the Red List category of the species may change, signalling the success or failure of conservation initiatives. These updates are essential for evaluating the effectiveness of conservation actions, identifying areas for improvement, and adapting management strategies to optimise conservation outcomes.

Restoration efforts must be paired with a comprehensive awareness campaign that educates and engages local communities, stakeholders, and the broader public about the value and vulnerability of mangrove ecosystems. By improving knowledge and awareness of mangrove ecosystems, relevant legislation, policies, and conservation strategies, we can empower local communities to support the long-term protection and sustainability of these vital ecosystems.

References

Agarwal S, Mukherjee P, Sinha M, Lutez JM, Mitra A (2023) Anthropogenic climate change in the mangrove-dominated Indian Sundarbans: spatio-temporal analyses, future trends, and recommendations for mitigation and adaptation. In: Leal Filho W et al (eds) Climate change strategies: handling the challenges of adapting to a changing climate, climate change management. Springer Nature, Cham, pp 249–286

Barbier EB, Hacker SD, Kennedy C, Koch EW, Stier AC, Siliman BR (2011) The value of estuarine and coastal ecosystem services. Ecol Monogr 81(2):169–193. https://doi.org/10.1890/10-1510.1

Bhadra T, Banerjee S, Ghosh S, Saha A, Mukherjee K, Sardar R, Roy S, Das R (2023) Monitoring the mangroves of Indian Sundarbans using geospatial techniques. Int Arch Photogramm Remote Sens Spat Inf Sci XLVIII-1/W2-2023:405–412. Cairo, Egypt

Bhattacharjee B, Karthigeyan K, Dey A, Giri P (2018) Restoration of mangroves at two 'Aila' affected parts of Sundarbans Biosphere Reserve through community participation. In: 28th annual conference of Indian Association for Angiosperm Taxonomy & International Symposium on "Conservation of angiosperm diversity: hidden treasure of today and tomorrow". The Maharaja Sayajirao University of Baroda, Gujarat, pp 3–4

Blankespoor B, Dasguupta S, Lange GM (2017) Mangroves as a protection from storm surges in a changing climate. Ambio 46:478–491

Chandra S, Mukopadhyay M (2021) Involvement of women in forest conservation-a comparative analysis of fringe villages in Sundarban Biosphere Reserve. Ensemble 3(1):69–80

Chaudhuri AB, Choudhury A (1994) Mangroves of the Sundarbans. Volume 1: India. World Conservation Union, Gland, p 247

Datta D, Chattopadhyay RN, Guha P (2012) Community based mangrove management: a review on status and sustainability. J Environ Manag 107:84–95

Donato DC, Kauffman JB, Murdiyarso D, Kurnianto S, Stidham M, Kanninen M (2011) Mangroves among the most carbon-rich forests in the tropics. Nat Geosci 4:293. https://doi.org/10.1038/NGEO1123

Ellison AM (2008) Managing mangroves with benthic biodiversity in mind: moving beyond roving banditry. J Sea Res 59:2–15

Ellison AM, Felson AJ, Friess DA (2020) Mangrove rehabilitation and restoration as experimental adaptive management. Front Mar Sci 7:327. https://doi.org/10.3389/fmars.2020.00327

Farnsworth EJ, Ellison AM (1997) The global conservation status of mangroves. Ambio 26:328–334

FSI (2021) India State of Forest Report 2021. Forest Survey of India (FSI), Dehradun

Gil L, Tol S, Vidal A (2023) Integrating mangrove ecosystems into NBSAPs through the Global Mangrove Watch. https://www.mangrovealliance.org/wp-content/uploads/2023/06/Global-Mangrove-Watch_Integrating-Mangrove-Ecosystems-into-NBSAPs_v13.pdf

Goldberg L, Lagomasino D, Thomas N, Fatoyimbo T (2020) Global declines in human-driven mangrove loss. Glob Chang Biol 26:5844–5855. https://doi.org/10.1111/gcb.15275

Gutting R, Syrbe R, Grunewald K, Mehlig U, Helfer V, Zimmer M (2021) The benefits of combining global and local data-a showcase for valuation and mapping of mangrove climate regulation and food provisioning services within a protected area in Pará, north Brazil. Land 10:432. https://doi.org/10.3390/land10040432

Hazra S, Ghosh T, DasGupta R, Sen G (2002) Sea level and associated changes in the Sundarbans. Sci Cult 68(9–12):309–321

Hutchinson J, Manica A, Swetnam R, Balmford A, Spalding A (2014) Predicting global patterns in mangrove forest biomass. Conserv Lett 7(3):233–240. https://doi.org/10.1111/conl.12060

Karthigeyan K, Mao AA (2022) Mangroves of India- a pictorial field guide. Botanical Survey of India, Kolkata

Kathiresan K (2008) Biodiversity of Mangrove ecosystems. In: Proceedings of mangrove workshop. GEER Foundation, Gujarat, India

Kathiresan K (2018) Mangrove forests of India. Curr Sci 114(5):976–981

Kathiresan K (2019) Why to restore mangroves? An economic assessment. Oceanogr Fish Open Access J 10(5):1–3

Lee SY, Hamilton S, Barbier EB, Primavera J, Lewis RR III (2019) Better restoration policies are needed to conserve mangrove ecosystems. Nat Ecol Evol 3:870–872

Lewis RR, Milbrandt EC, Brown B, Krauss KW, Rovai AS, Beever JW, Flynn LL (2016) Stress in mangrove forests: early detection and pre-emptive rehabilitation are essential for future successful worldwide mangrove forest management. Mar Pollut Bull 109:764–771

Lewis RR, Brown BM, Flynn LL (2019) Methods and criteria for successful mangrove forest rehabilitation. In: Perillo GME, Wolanski E, Cahoon DR, Hopkinson CS (eds) Coastal wetlands: an integrated ecosystem approach, 2nd edn. Elsevier, Amsterdam, pp 863–887. https://doi.org/10.1016/B978-0-444-63893-9.00024-1

Li X, Bellerby R, Craft C, Widney SE (2018) Coastal wetland loss, consequences, and challenges for restoration. Anthr Coasts 1:1–15

Mace GM (2014) Whose conservation? Changes in the perception and goals of nature conservation require a solid scientific basis. Science 345:1558–1559

Mahajan M (2008) A model mangrove conservation initiative in Mumbai Soonabai Pirojsha Godrej Marine Ecology Centre. In: National workshop on "Mangroves in India: biodiversity, protection and environmental services". Institute of Wood Science and Technology, Bangalore, p 52

Naskar K, Guha Bakshi DN (1987) Mangrove swamps of the Sundarbans – an ecological perspective. Naya Prokash, Calcutta, p 263

Naskar KR, Mandal RN (1999) Ecology and biodiversity of Indian mangroves, vol 1. Daya Publishing House, New Delhi

Ong JE (1993) Mangroves – a carbon source and sink. Chemosphere 27:1097–1107

Ong JE (1995) The ecology of mangrove conservation and management. Hydrobiologia 295:343–351

Ong JE (2003) Plants of the Merbok mangrove, Kedah, Malaysia and the urgent need for their conservation. Folia Malays 4:1–18

Owuor M, Santos TMT, Otieno P, Mazzuco ACA, Iheaturu C, Bernardino AF (2024) Flow of mangrove ecosystem services to coastal communities in the Brazilian Amazon. Front Environ Sci 12:1329006. https://doi.org/10.3389/fenvs.2024.1329006

Polidoro BA, Carpenter KE, Collins L, Duke NC, Ellison AM, Ellison JC, Farnsworth EJ, Fernando ES, Kathiresan K, Koedam NE, Livingstone SR, Miyagi T, Moore GE, Nam VN, Ong JE, Primavera JH, Salmo SG, Sanciangco JC, Sukardjo S, Wang Y, Yong JWH (2010) The loss of species: mangrove extinction risk and geographic areas of global concern. PLoS One 5(4):1–10. http://www.mangrovesforthefuture.org/. Accessed on 21.08.2024

Raghavan P, Saxena A, Jayaraj RSC, Mohan PM, Ravichandran K, Saravanan S, Vijayaraghavan A (2016) A review of the mangrove floristics of India. Taiwania 61(3):224–242

Raghavan P, Kathiresan K, Zimmer M, Zhou R, Amir AA, Mohan PM, Rana TS (2020) Three decades of global mangrove conservation – an overview. Malay Nat J 72(4):551–576

Sahu SC, Suresh HS, Murthy IK, Ravindranath NH (2015) Mangrove area assessment in India: implications of loss of mangroves. J Earth Sci Clim Change 6(5):280. https://doi.org/10.4172/2157-7617.1000280

Sanjappa M, Venu P, Arisdason W, Albertson WD (2011) A Review of Mangrove species in India. In: Bhatt JR, Macintosh DJ, Nayar TS, Pandey CN, Nilaratna BP (eds) Towards Conservation and Management of Mangrove Ecosystems in India. IUCN, India, pp 35–50

Selvam V, Ravichandran KK, Gnanappazham L, Navamuniyammal M (2003) Assessment of community-based restoration of Pichavaram mangrove wetland using remote sensing data. Curr Sci 85:794–798

Sievers M, Chowdhury MR, Adame MF, Bhadury P, Bhargava R, Buelow C, Friess DA, Ghosh A, Hayes MA, McClure EC, Pearson RM, Turschwell MP, Worthington TA, Connolly RM (2020) Indian Sundarbans mangrove forest considered endangered under Red List of Ecosystems, but there is cause for optimism. Biol Conserv 251:108751. https://doi.org/10.1016/j.biocon.2020.108751

Slobodian L, Vidal A (2023) National law and policy to achieve global mangrove goals. Mangrove Law and Policy, Global Mangrove Alliance. https://www.mangrovealliance.org/wp-content/uploads/2023/12/GMA-Policy-Brief_V6.pdf

Spalding M, Kainuma M, Collins L (2010) World Atlas of Mangroves. Earth scan, London, p 305.

Upadhyay VP, Ranjan R, Singh JS (2002) Human-mangrove conflicts: the way out. Curr Sci 83(11):1328–1336

Valiela I, Bowen JL, York JK (2001) Mangrove forests: one of the world's threatened major tropical environments. Bioscience 51(10):807–815

Vasudevan N (2017) Safeguarding the sentinels: the story of mangrove conservation in Maharashtra, Mangal-Van. Mangrove Society of India, Goa, pp 55–65

Waisel Y (1972) The Biology of Halophytes. Academic Press, New York
Worthington T, Spalding M (2019) Mangrove restoration potential: A global map highlighting a critical opportunity. IUCN, University of Cambridge & The Nature Conservancy, p 34
Zimmer M (2018) Ecosystem design: when mangrove ecology meets human needs. In: Makowski C, Finkl C (eds) Threats to mangrove forests. Coastal Research Library, vol 25. Springer, pp 367–376

Post-fire Restoration Plan for Sustainable Forest Management

7

K. Ravichandran

Abstract

The incidence of forest fires in India has increased by an order of magnitude over the past two decades, contributing to land degradation, environmental health deterioration, and greenhouse gas emissions. This study examines the requisite components of a post-fire restoration plan (PFRP) appropriate for India's heterogeneous forest ecosystems. The proposed framework comprises five sequential phases: damage assessment utilizing remote sensing and field surveys; hierarchical planning across temporal scales; implementation of ecological interventions; systematic monitoring; and mechanisms for long-term sustainability. The PFRP is predicated on four guiding principles: preservation of ecological integrity, recognition of fire as an integral ecosystem process, incorporation of landscape-specific contextual factors, and prioritization of ecosystem service continuity. Post-fire restoration in the Bandipur Tiger Reserve case study demonstrates that actively restored plots exhibited recovery of 196 plant species, including 22 endemic taxa, whereas passively restored areas supported only 26 species, with lantana proliferation. Successful restoration necessitates multi-stakeholder collaboration, evidence-based adaptive management protocols, and policy integration across governmental hierarchies. Restoration of fire-impacted areas contributes to the development of resilient forest management strategies in the context of increasing climatic variability and anthropogenic landscape modification in India.

Keywords

Forest fire · Post-fire restoration plan · Resilience · Guiding principles · Active restoration · Passive restoration

---

K. Ravichandran (✉)
Indian Institute of Forest Management (IIFM), Bhopal, Madhya Pradesh, India
e-mail: director@iifm.ac.in

D. Mittal, D. K. Gupta (eds.), *Ecological Restoration in India: Perspective and Practices*, https://doi.org/10.1007/978-981-96-8943-9_7

7.1 Introduction

Forest fires are becoming more frequent and impacting a wide range of forest types and ecosystems worldwide, leading to significant alterations in various environmental factors such as air quality, radiation budget, interactions between land and atmosphere, nutrient cycles, and ecological equilibrium (Giglio et al. 2016). The past decade has seen a significant surge in the incidence of large and uncontrolled wildfires. Nearly 26–29% of global forest loss has been caused by fire. In India alone, there has been a tenfold increase in forest fire incidences in the last two decades (Tyukavina et al. 2022). Fires also affect both the physical and chemical properties of soils, particularly the productive soil layer. Changes in the soil's physical properties lead to loss of productive soil, increased runoff during wet periods, and lower underground water recharge—increasing water deficit over the course of dry periods. Repeated fires deplete seed sources and can affect the regeneration and succession of forest ecosystems. Fire regime shifts in wild and seminatural ecosystems can become a degradation process, with a high impact on net carbon emissions. Climate change and land-use changes are projected to make wildfires more frequent and intense, with a global increase in extreme fires of up to 14% by 2030, 30% by the end of 2050, and 50% by the end of the century (Citaristi 2022). Furthermore, of the land degradation processes, deforestation, increasing wildfires, degradation of peat soils, and permafrost thawing contribute most to climate change through the release of GHGs and the reduction in land carbon sinks following deforestation (Olsson et al. 2024).

Forest fires contribute to increased land degradation, worsen local and regional environmental health problems, emit stored CO_2, and have a major impact on global warming (Flannigan et al. 2000; Henderson et al. 2011). While some forest fires can benefit the ecosystem by clearing obstacles and debris, adding nutrients to the soil, making room in the canopy, and promoting healthier forests in the future, frequent, uncontrollable, and large fires typically harm the ecosystem. Several research studies have also indicated that forest fires can lead to soil erosion and affect water quality (Hewelke et al. 2020; Massman et al. 2003). Therefore, wildfires greatly affect the structure of forests, the operation of ecosystems, and the movement of water and nutrients. In the last two decades, the tree cover loss in G20 countries was 297.40 million ha, of which 111.74 million ha of tree cover loss was due to forest fire (Table 7.1). In India, nearly 34,836 ha of loss was reported during the period.

Forest fires in India are a significant ecological threat. About 36% of forest cover in India falls under extreme, very high, high, and moderate forest fire-prone zones (Forest Survey of India 2019). The country experiences seasonal fires, particularly in tropical dry forests, grasslands, and the temperate zones of the Himalayas. The devastation caused by forest fires not only affects biodiversity but also impacts the livelihoods of millions of people dependent on forest resources. While fire is a natural part of many ecosystems, human activities such as agriculture, deforestation, and poor land-use practices have exacerbated the intensity and frequency of these fires (Fig. 7.1).

Table 7.1 Tree cover loss due to forest fires in G20 countries (2001–2021)

Country	Total tree cover loss (ha)	Tree cover loss from fires (ha)
Argentina	6317526.33	534794.91
Australia	8727950.16	6262638.22
Brazil	62817600.58	9513978.35
Canada	46611577.94	26751110.11
China	10875236.46	893246.31
European Union	*data pending*	*data pending*
France	1333332.74	48209.01
Germany	1143170.60	6417.77
India	2066015.30	34836.44
Indonesia	28562386.82	2843361.80
Italy	415079.35	45809.63
Japan	810806.99	2558.25
Mexico	4481513.78	655931.35
Russia	76008405.95	52788981.99
Saudi Arabia	*data unavailable*	*data unavailable*
South Africa	1530042.50	115730.98
South Korea	288488.43	0.00
Turkey	624593.27	87748.30
United Kingdom	506737.20	23653.49
USA	44274580.82	11131936.29
TOTAL	297395045.22	111740943.20
Total (mha)	297.40	111.74

Source: Global Forest Watch; UNEP

The India State of Forest Report (ISFR) 2023 shows that the country's forest and tree cover now spans 827,357 sq.km, covering 25.17% of the nation's total land area. This includes 715,343 sq.km (21.76%) of forest cover and 112,014 sq.km (3.41%) of tree cover. However, higher forest cover means high fuel loading, which may lead to more intense and larger forest fires (Law and Dickman 1998). Fire incidents have slightly decreased, with 203,544 fire hotspots recorded in 2023–2024, down from 223,333 in 2021–2022 (Fig. 7.2).

FSI has also assessed the forest areas affected by forest fires for the first time for the period November 2023 to June 2024, and it is estimated that 34,562.33 sq.km have been affected due to forest fires (IFSR 2023).

7.2 Restoration of Degraded Lands

The Bonn Challenge Declaration and the UN Declaration of 2021–2030 as the "Decade of Ecosystem Restoration" have placed ecological restoration at the forefront of the world's biodiversity and climate change agendas. Restoring degraded natural areas is a flagship target of the post-2020 global biodiversity framework of the CBD and was put in the spotlight by the UN Decade on Ecosystem Restoration (2021–2030). Restoration is particularly critical where natural systems are so

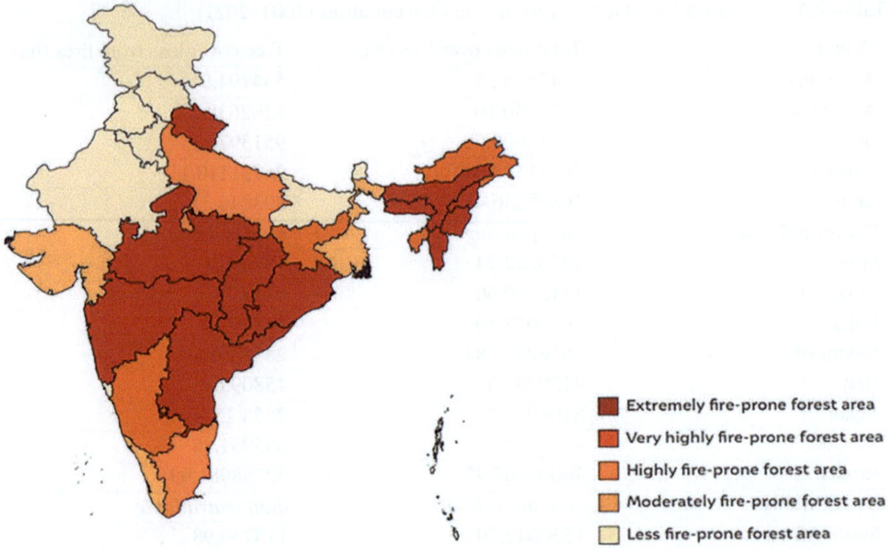

Fig. 7.1 Fire-prone Forest area of India (Mohanty and Mithal 2022)

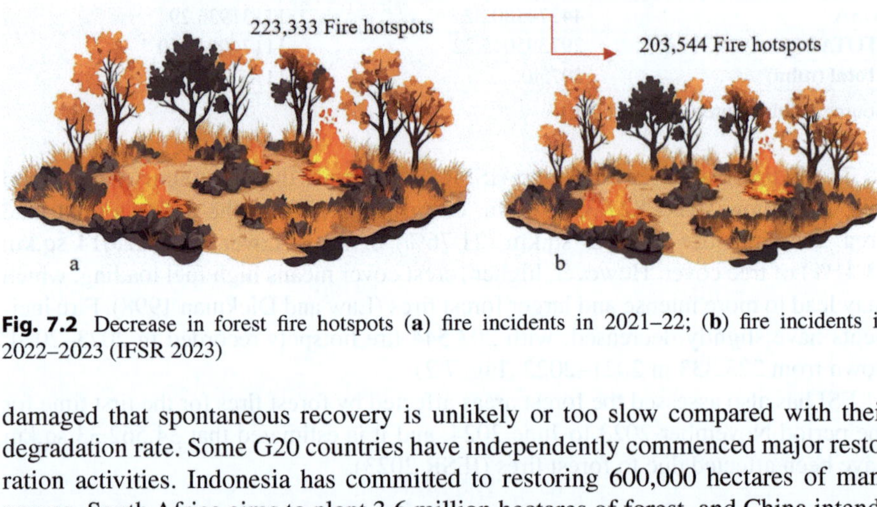

Fig. 7.2 Decrease in forest fire hotspots (**a**) fire incidents in 2021–22; (**b**) fire incidents in 2022–2023 (IFSR 2023)

damaged that spontaneous recovery is unlikely or too slow compared with their degradation rate. Some G20 countries have independently commenced major restoration activities. Indonesia has committed to restoring 600,000 hectares of mangroves, South Africa aims to plant 3.6 million hectares of forest, and China intends to achieve 30% forest coverage by 2050 (Wilson and Lasco 2018).

A robust post-fire restoration plan (PFRP) is essential for mitigating the long-term impacts of wildfires, restoring ecosystems, and enhancing forest resilience. This paper examines the need for a comprehensive PFRP within the context of India's diverse forest ecosystems, the socio-economic significance of forests, and the challenges posed by post-fire recovery. It proposes a strategic approach to post-fire restoration that aligns with sustainable forest management goals, focusing on ecological restoration, community involvement, policy integration, and monitoring systems.

7.3 Post-fire Forest Restoration Planning

Burned forests should have a restoration strategy that includes erosion control to minimize disasters if burned regions are susceptible to secondary damage such as soil erosion or landslides, which are highly typical post-fire natural hazards in India after heavy rain. According to earlier research, a log strip terrace barrier is the best method for stopping soil erosion in burned regions and is an excellent way to make use of dead trees after a fire (Ryu et al. 2006). By optimizing the economic, ecological, and social worth, burned forests can be restored.

A restoration plan must be finished before beginning any restoration. When creating a restoration plan, the following six essential functions should be considered: (i) recreation, (ii) natural environment conservation, (iii) ecosystem conservation, (iv) disaster prevention, (v) water conservation, and (vi) timber production (Ryu et al. 2006).

After a major fire, a restoration plan should also adhere to the relevant land-use regulations. A comprehensive strategy that is appropriate for the new land-use classification should be created if a fire makes it impossible to maintain the old land-use designation. On the other hand, a post-small fire plan should adhere to the principles of sustainable forest management to optimize forest functions.

7.4 Guiding Principles for Post-fire Restoration

Forest fires, while often viewed as destructive events, play a complex and essential role in ecosystem dynamics. The restoration of fire-affected landscapes requires a nuanced approach that balances ecological recovery with an understanding of fire's natural function in forest systems. This restoration plan is guided by fundamental principles that honor the integrity of forest ecosystems, recognize fire as an integral ecological process, consider the broader landscape context, and prioritize the recovery of vital ecosystem services. Rather than attempting to erase the effects of fire, these principles embrace an adaptive perspective that works with natural processes to foster resilient, functional forest ecosystems that can sustain biodiversity, support watershed health, and continue to provide ecological benefits to both wildlife and human communities.

7.5 Principle 1: Ecological Integrity

Forest fire restoration must prioritize ecosystem function over aesthetic or purely commercial considerations. This principle acknowledges that the goal is not merely to replant trees but to restore complex ecological processes and relationships. Restoration success should be measured through indicators of ecosystem health, including biodiversity indices, nutrient cycling efficacy, and functional redundancy of species (Dale et al. 2016). The concept of novel ecosystems (Hobbs et al. 2014) may need to be embraced in cases where a return to historical reference conditions is neither feasible nor desirable due to changing climate conditions.

7.6 Principle 2: Acknowledge Fire as a Natural Process

Restoration strategies must be grounded in the understanding that fire is an integral ecological process rather than simply a disturbance to be suppressed or remediated. Many forest ecosystems have co-evolved with periodic fire, developing adaptations and dependencies on specific fire regimes (Pausas and Keeley 2019). Restoration planning should incorporate knowledge of historical fire return intervals, severity patterns, and seasonal timing specific to the forest type being restored while recognizing that these parameters may need adjustment for future conditions.

7.7 Principle 3: Consider Landscape Context

Each restoration site presents unique challenges requiring tailored interventions (Kotliar et al. 2007). Generic restoration templates must yield approaches developed from specific site conditions and fire characteristics. This includes consideration of:

Pre-fire forest composition and structure
Fire severity mosaic patterns
Topographic position and aspect
Soil properties and burn severity
Proximity to unburned seed sources
Hydrological context
Post-fire climate projections

7.8 Principle 4: Sustaining Ecological Services

Immediate post-fire interventions should prioritize soil conservation to prevent erosion, sedimentation of waterways, and loss of soil organic matter. Methods may include contour log erosion barriers, mulching with native materials, establishing nurse crops for temporary soil stabilization, channel and drainage structure protection, and mycorrhizal inoculation to accelerate soil microbial recovery. Soil protection measures should be implemented within the first year post-fire, with special attention to steep slopes and areas of high burn severity (Robichaud et al. 2010).

These guiding principles can be used to create a restoration framework, which is crucial for rehabilitating ecosystems damaged by wildfires. This framework typically involves a multistep process that includes assessing the affected area, identifying restoration goals, and implementing strategies to restore ecological integrity, project planning, and implementation.

7.9 Rapid Ecological Assessment Framework

A standardized yet adaptable post-fire assessment protocol should guide restoration planning (Fig. 7.3). This assessment should classify burn severity using both remote sensing and ground-truthing, identify natural regeneration potential, map ecological vulnerabilities and opportunities, evaluate erosion risk and watershed impacts, document surviving biological legacies, and forecast secondary disturbance risks (e.g., invasive species, insect outbreaks). This information should form the basis of spatially explicit restoration planning (Morgan et al. 2014).

7.10 Restoration Strategy Development

Restoration planning must determine where natural recovery processes are sufficient and where active intervention is necessary. This restoration framework should consider (i) natural regeneration potential based on surviving trees and soil seed bank, (ii) distance to seed sources and dispersal mechanisms, (iii) risk of competitive exclusion by undesirable species, (iv) presence of biological legacies (surviving organisms, coarse woody debris), and (v) budget constraints and logistical feasibility.

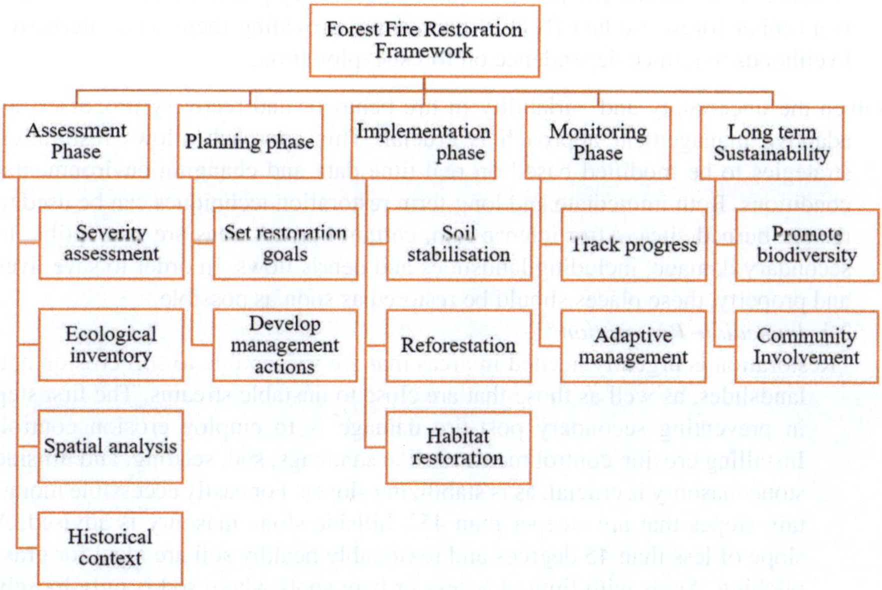

Fig. 7.3 Post-fire development and execution of a plan to restore burned areas

7.11 Five Steps for Forest Fire Restoration Framework

1. *Assessment Phase*

The first step in the restoration process is to conduct an immediate damage assessment using remote sensing, field surveys, and aerial mapping to understand the extent of the damage. Identifying the ecological zones most affected by the fire (e.g., biodiversity hotspots, watershed areas, and critical habitats). Early assessment allows for quick prioritization of restoration efforts and helps direct resources where they are needed most.

This includes evaluating soil erosion risks, water resource depletion, and the displacement of wildlife. Emergency measures such as soil stabilization (through mulching and planting ground cover) and the creation of firebreaks to prevent further spread are implemented. During this phase, relief efforts are also prioritized, providing local communities with immediate food.

2. *Planning Phase (Short Term and Long Term)*

In the short term, the restoration effort focuses on ecological interventions and community empowerment. Replantation of native species—those suited to the specific ecosystem and fire-prone areas—is critical for restoring forest cover and preventing soil erosion. Assisted natural regeneration, alongside the establishment of community-run nurseries and seed banks, supports the replantation efforts. Wildlife recovery programs are initiated to ensure displaced species have access to appropriate habitats. Simultaneously, local communities are trained in sustainable fire management techniques, agroforestry practices, and sustainable non-timber forest product (NTFP) harvesting, providing them with alternative livelihoods to reduce dependence on forest exploitation.

Given the uncertainty and variability in fire behavior and recovery processes, an adaptive management approach is crucial. This approach allows restoration strategies to be modified based on real-time data and changing environmental conditions. Both immediate and long-term restoration techniques can be used to restore burned sites. After intense rain, certain burned areas are susceptible to secondary damage, including landslides and debris flows. In order to save lives and property, these places should be restored as soon as possible.

2.1 *Immediate Restoration*

Restoration is urgently needed in areas that are susceptible to soil erosion and landslides, as well as those that are close to unstable streams. The first step in preventing secondary post-fire damage is to employ erosion control. Installing erosion control measures like sandbags, sod, seeding, and hillside stone masonry is crucial, as is stabilizing slopes. For easily accessible mountain slopes that are steeper than 45°, hillside stone masonry is advised. A slope of less than 45 degrees and reasonably healthy soil are ideal for grass pitching. Areas with limited access or bare spots where sod is only loosely anchored to the ground can benefit from the usage of sandbag pilings. After finishing sandbag piling or stone masonry construction on the slope, areas that need rapid vegetation establishment might be targeted for seed spraying. Erosion control structures stabilize slopes and add vegetation, according to field observations. In burned areas, it has been found that immediate restoration techniques, like hillside stone masonry, sandbags, grass, and seed-spraying treatments, are effective ways to stop soil erosion and landslides.

2.2 *Medium-Term Plan*

As recovery progresses, the focus shifts to landscape-level restoration, promoting agroforestry and mixed land-use practices to integrate forests with agriculture. This not only enhances forest resilience but also generates sustainable livelihoods for local communities. Efforts are also made to diversify income sources, such as through ecotourism and the sustainable harvest of NTFPs like medicinal plants, mushrooms, and bamboo. Fire resilience strategies are further strengthened by implementing practices like thinning dense vegetation, controlled burns, and firebreaks to minimize future fire risks. Additionally, water resource management programs such as rainwater harvesting, watershed management, and riverbank restoration are introduced to improve forest resilience to fires and droughts.

2.3 *Long-Term Plan*

In the long term, the restoration process transitions into continuous monitoring and adaptive management. Monitoring the ecological health of restored forests, including biodiversity levels, soil quality, and vegetation growth, ensures that restoration strategies remain effective and can be adjusted as necessary. A major component of this phase is the integration of fire management strategies into national and state policies, ensuring that future forest management efforts prioritize fire prevention, restoration, and climate change adaptation. Forest governance is strengthened by involving local communities in decision-making processes, establishing legal protections for restored areas, and promoting sustainable forest management practices. The goal of post-fire restoration is not only to repair the damage caused by fire but also to ensure that the recovery process contributes to the sustainability and resilience of the forest ecosystem.

To optimize the economic, ecological, scenic, and environmental aspects, long-term restoration techniques can be employed. The long-term restoration effort should take locals' perspectives into account (Williams and Abatzoglou 2016). To optimize the economic, ecological, scenic, and environmental qualities, it can be separated into two categories: passive restoration and active restoration. Remaining woods and regions that are only slightly disturbed by forest fires can benefit from passive restoration. These regions can naturally recover from the intact canopy and are formally designated as environmental conservation forests. The forest structure should not undergo significant alterations as a result of passive restoration. The social and economic needs of the local population may be overlooked by passive restoration. To meet the needs of the national economy, they can encourage the sustainable production of timber and non-timber forest produce by planting more native tree species.

Firebreaks are necessary in forests to lessen the chance of a fire spreading quickly. Tree species that are resistant to fire, such as those with thick bark and moist leaves, can be used as firebreaks. These trees also need to grow quickly after a fire. Although forest fires burn biomass in forests, they do not harm plant seeds or underground roots. Numerous species can naturally recover from intact canopies. When there are insufficient sprouts and high mortality, planting fire-resistant tree species is a feasible solution. Group planting with mixed species in patches in the form of aided regeneration, as a cost-effective and environmentally friendly substitute for conventional row planting for reforestation of dispersed regions, is the suggested restoration technique.

However, due to neglect in weeding and tending operations, including the maintenance of fire lines and firebreaks in private areas, FRA areas, and, to some extent, government-owned forest areas, forest areas are lost to fire. To prevent this, continuous management measures such as weeding, pruning, thinning, climber cutting, and other tending operations are necessary to suppress forest fires.

Fire burns biomass, and burned areas increase surface flow during rainfall events (Williams and Abatzoglou 2016). Following fires, this may lead to the collapse of soil aggregation and an increase in soil water repellency, which lowers soil permeability and, ultimately, forests' ability to store water. Because they may form deep root systems, oak species thrive on barren slopes with hydromorphic soil conditions that are vulnerable to storm disturbance. Deep-rooted tree species should be planted in forests to preserve the water supply. In forests, planting is quite important, particularly when it comes to reforesting fire-affected degraded forest areas in addition to natural regeneration to speed up the recovery of biodiversity.

A multilayered mixed forest with native species should be created by planting endemic tree species, and planting seedlings of locally existing tree species can help restore native tree species. By restoring forest functions, long-term restoration aims to optimize the ecological, economic, and environmental values. Multilayered, mixed forest establishment and the return of key forest functions, including timber production, water conservation, ecosystem conservation, and recreation to meet the needs of locals, are indicators of restoration success. Ecological services and forest products are examples of economic success indicators. By creating multilayered root structures, additional shallow- and intermediate-rooted species can be planted to expand the soil's capacity to hold water. Therefore, in broad-leaved forests, a greater range of tree species may offer more opportunities for the development of multilayered root structures.

Community involvement may be strongly encouraged to assist restoration efforts. Therefore, post-restoration management will be implemented for a successful restoration site. To assess the effectiveness of management strategies and offer guidance for future management, the site should be periodically observed and monitored so as to take timely interventions, including any additional restoration activities necessary in locations that have not been adequately restored. Cooperation among various stakeholders, including local people, line departments, and research and academic institutions, is also crucial for successful restoration.

3. *Implementation Phase (Ecological Restoration)*

Ecological restoration aims to recover the structure, function, and biodiversity of forest ecosystems following fire damage. Restoration goals may range from restoring preexisting forest structure, functioning, and species composition to rehabilitating desired ecosystem services, such as maximizing carbon sequestration or enhancing biodiversity, to creating forest landscapes that serve additional socioeconomic purposes (Holl 2012; FAO 2019), as shown in Fig. 7.4.

All the above implementation strategies require: (a) soil rehabilitation: after a fire, soil can be left highly vulnerable to erosion. Immediate soil restoration measures like mulching, replanting ground cover, and implementing erosion control structures should be prioritized; (b) reforestation and afforestation: planting native species suited to the local ecosystem should be prioritized over nonnative species to ensure long-term ecological sustainability. Indigenous species help restore biodiversity and promote the recovery of local wildlife populations; (c) water and nutrient management: restoring forest water cycles and addressing nutrient depletion is critical for the success of reforestation efforts. Techniques such as contour bunding, water harvesting, and organic mulching can improve soil quality and water retention; and (d) wildlife recovery: after fires, wildlife populations may experience significant displacement. Programs focused on the recovery of wildlife habitats and corridors should be integrated into restoration plans, ensuring species can re-establish themselves.

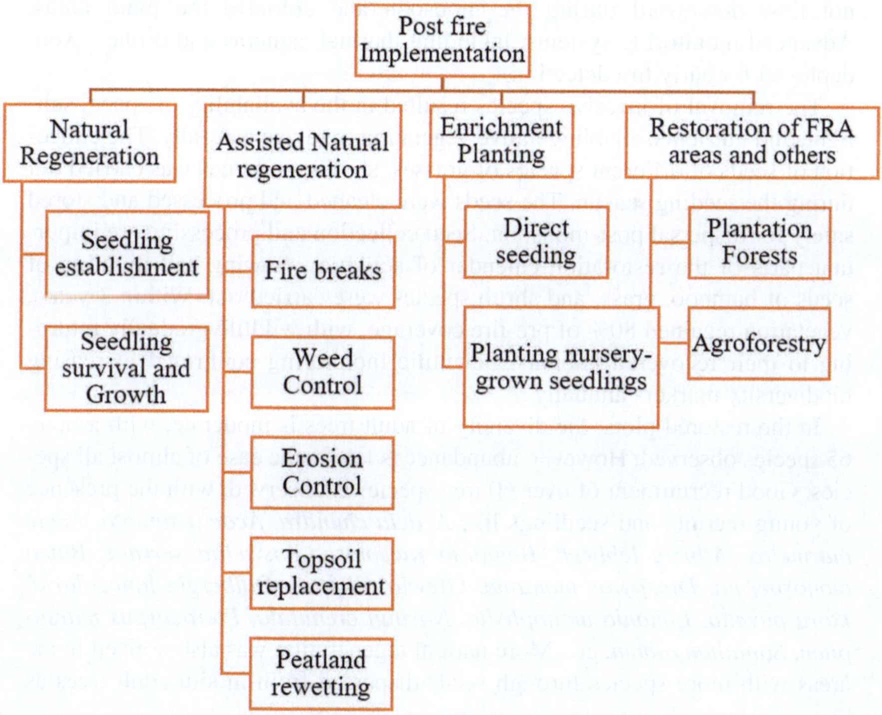

Fig. 7.4 Post-fire implementation strategy

Restoration of Fire-Impacted Areas in Bandipur Tiger Reserve: A Case Study
Following the 2019 forest fire that devastated 2150 ha of Bandipur Tiger Reserve, a remarkable restoration in fire-impacted areas was carried out through coordinated efforts involving local communities. Out of 2150 ha of forest area affected by the 2019 forest fire incident, 340 ha in GS Betta Range was selected for active restoration, and the remaining areas were left for passive restoration. As part of active restoration, 14 restoration plots, each having 25 ha, covering 340 ha, have been taken up for restoration. The main focus was on the removal of *Lantana camara* and other invasive species like *Senna spectabilis, Chromolena odorata, Parthenium hysterophorous*, etc. in forest fire affected areas. Establishment of native species through positive intervention in the forest fire-affected areas though removal of invasive alien species was followed by creation of grasslands through seed sowing and dibbling native endemic species and planting of suitable native species.

The Forest Department partnered with local communities to implement soil stabilization, native reseeding, and the creation of firebreaks. The *Lantana camara* removal was carried out using the cut root-stock method. The main advantage of this adapted method was that there was no disturbance to the soil, and there is no collateral damage to native vegetation, which facilitates faster regeneration. The removal of *Lantana* and other weeds was initiated from the top of the hills and gradually continued downward so that seeds do not flow downward during the monsoon and colonize the plain areas. Advanced monitoring systems, including thermal cameras and drones, were deployed for early fire detection.

The removal of invasive species resulted in the availability of space, sunlight, and nutrients, enabling native vegetation to revive naturally. The collection of seeds of different species of grasses, shrubs, and trees was carried out during the seeding season. The seeds were cleaned and processed and stored safely for dispersal post-monsoon. Seed collection and processing are important parts of the restoration calendar of activities. Sowing and dibbling of seeds of bamboo, grass, and shrub species were carried out. Within 3 years, vegetation regained 80% of pre-fire coverage, with wildlife gradually returning to their recovered habitats. Scientific monitoring confirmed increasing biodiversity markers annually.

In the restored plots, the diversity of adult trees is moderate, with around 65 species observed. However, abundance is low in the case of almost all species. Good recruitment of over 60 tree species is observed, with the presence of young recruits and seedlings like *Acacia chundra, Acacia sinuata, Aegle marmelos, Albizia lebbeck, Bauhinia racemosa, Boswellia serrata, Butea monosperma, Diospyros montana, Grewia tillifolia, Dalbergia lanceolaria, Ixora pavetta, Limonia monophylla, Naringi crenulata, Pterocarpus marsupium, Santalum album*, etc. More natural regeneration was also noticed in the areas with more species through seeds dispersed from in situ adult trees as

(continued)

well as seed migration from other areas aided by dispersal agents like elephants and other ungulates. Further, there is good diversity and abundance of understory plant habits of shrubs, herbs, climbers, etc. The re-emergence of *Dendrocalamus strictus* in some of the restored plots has been noticed, but the presence of *Bambusa bambos* was low. In 3 years' time, 196 species of trees, shrubs, herbs, and grasses, with 22 endemic species and 6 threatened species, were recorded in the restoration plots. The areas left with passive restoration without any intervention except for protection resulted in the recovery of only 26 species. The main reason for very low floral biodiversity in the remaining areas left for passive restoration was the proliferation of lantana and other weeds, which made the habitat impenetrable for other native grasses, herbs, shrubs, and tree species, resulting in poor biodiversity.

This clearly demonstrates that restoration of fire-affected areas through aided regeneration is a must, and it should be carried out properly with a pre-assessment survey, followed by broadcasting and dibbling of native grasses, bamboos, and woody perennials, including tree species. Planting seedlings of preferred species native to the area and specific to the particular forest types will help in restoring the fire-impacted areas.

(Source: Restoration of forest fire-impacted areas, Bandipur Tiger Reserve, Bandipur (2022). Conservator of Forests and Field Director, Bandipur, Karnataka Forest Department)

4. *Monitoring Phase*

A robust institutional and policy framework is needed to guide post-fire restoration efforts. The integration of PFRP with India's National Forest Policy and the National Action Plan on Climate Change, which emphasizes the importance of sustainable forest management and adaptation to climate change, can be effective in handling fire. Fire management and restoration require coordination across various government departments, including the Ministry of Environment, Forests and Climate Change, the Ministry of Rural Development, and the State Forest Departments. Collaboration with local authorities, environmental NGOs, and the private sector can enhance the efficiency of restoration efforts. The plan's funding can come from a combination of sources, including government allocations, international climate finance mechanisms like the Green Climate Fund (GCF), and private sector investments. Additionally, community-based funding mechanisms, such as small-scale contributions and labor from local residents, can ensure ownership of restoration efforts, leading to more successful and sustainable outcomes.

5. *Long-Term Sustainability*

Restoration is the long-term process of regaining ecological integrity and enhancing human well-being, including a range of interventions such as natural regeneration, mixed-species plantation, and agroforestry, among others (Chazdon et al. 2017). Post-fire restoration plan (PFRP) that supports sustainable forest management in India, drawing from global best practices, local needs, and the principles of adaptive management. The following are the keys to the successful restoration of forest fire-impacted areas.

1. *Integrate Ecological and Social Factors*: Restoration plans should be tailored to the specific ecosystem, incorporating local community needs and ecological requirements to ensure sustainability and acceptance.
2. *Implement Adaptive Management*: Regular monitoring and evaluation should be conducted to adjust restoration strategies based on ecosystem responses and changing environmental conditions.
3. *Promote Sustainable Forest Practices*: The plan should adopt precision silvicultural practices, including tending, weeding, and prescribed burning, to reduce fuel loads and enhance forest resilience against wildfires.
4. *Develop Emergency Stabilization Plans*: Establish landscape emergency stabilization plans to reduce sediment transport into water sources and protect water quality.
5. *Enhance Community Engagement*: Engage local communities in restoration efforts to foster support and ensure long-term success of restoration projects.
6. *Invest in Research and Development*: Continuously update restoration strategies based on new research findings to improve effectiveness and adaptability in diverse ecosystems.

Successful restoration largely depends on the plan and relies on the active involvement of multiple stakeholders, including local communities, who are crucial for both implementing restoration activities and ensuring long-term forest sustainability. Empowering local communities and engaging them in post-fire restoration activities not only improves the success of restoration but also strengthens the social fabric. Training communities in fire prevention, restoration techniques, and sustainable forest management practices can increase resilience. Providing alternative livelihoods to communities that depend on forest resources, such as firewood, non-timber forest products, and agriculture, is essential. This can include promoting agroforestry, ecotourism, and sustainable harvesting practices.

Post-fire restoration is a crucial component of sustainable forest management. A well-thought-out post-fire restoration plan and its implementation can improve the resilience of forest ecosystems, restore ecological functions, and lessen the detrimental effects of forest fires. India can create a more resilient and sustainable forest landscape that benefits both the environment and the people who depend on it by emphasizing ecological restoration, community involvement, policy integration, and efficient monitoring. Successful restoration of fire-impacted areas will ensure the long-term health of India's forests and the ecosystems that are linked to them, especially in light of the increasing risks posed by land-use change and climate change.

References

Chazdon RL, Brancalion PHS, Lamb D, Laestadius L, Calmon M, Kumar C (2017) A Policy-Driven Knowledge Agenda for Global Forest and Landscape Restoration. Conserv Lett 10:125–132. https://doi.org/10.1111/conl.12220

Citaristi I (2022) United Nations Environment Programme—UNEP. In: The Europa directory of international organizations 2022. Routledge, pp 193–199

Dale VH, Hughes MJ, Hayes DJ (2016) Climate change and the future of natural disturbances in the central hardwood region. In: Natural disturbances and historic range of variation: type, frequency, severity, and post-disturbance structure in central hardwood forests USA. Springer, pp 355–369

FAO (2019) Restoring forest landscapes through assisted natural regeneration (ANR). A practical manual. FAO

Flannigan MD, Stocks BJ, Wotton BM (2000) Climate change and forest fires. Sci Total Environ 262(3):221–229

Forest Survey of India, (FSI) (2019) India state of forest report 2019, vol 1, 16th edn. Ministry of Environment, Forest and Climate Change, Govt. of India, pp i–183

Giglio L, Schroeder W, Justice CO (2016) The collection 6 MODIS active fire detection algorithm and fire products. Remote Sens Environ 178:31–41

Henderson SB, Brauer M, MacNab YC, Kennedy SM (2011) Three measures of forest fire smoke exposure and their associations with respiratory and cardiovascular health outcomes in a population-based cohort. Environ Health Perspect 119(9):1266–1271

Hewelke E, Górska EB, Gozdowski D, Korc M, Olejniczak I, Prędecka A (2020) Soil functional responses to natural ecosystem restoration of a pine forest peucedano-pinetum after a fire. Forests 11(3):286

Hobbs RJ, Higgs E, Hall CM, Bridgewater P, Chapin FS III, Ellis EC et al (2014) Managing the whole landscape: historical, hybrid, and novel ecosystems. Front Ecol Environ 12(10):557–564

Holl KD (2012) Restoration of tropical forests. In: van Andal J, Aronson J (eds) Restoration ecology: the new frontier, 2nd edn. Wiley-Blackwell, Hoboken, pp 103–114

ISFR (2023) India state of forest report 2023. Forest Survey of India, Ministry of Environment Forest and Climate Change, Dehradun

Kotliar NB, Kennedy PL, Ferree K (2007) Avifaunal responses to fire in southwestern montane forests along a burn severity gradient. Ecol Appl 17(2):491–507

Law BS, Dickman CR (1998) The use of habitat mosaics by terrestrial vertebrate fauna: implications for conservation and management. Biodivers Conserv 7:323–333

Massman WJ, Frank JM, Shepperd WD, Platten MJ (2003) In situ soil temperature and heat flux measurements during controlled surface burns at a southern Colorado forest site. In: Omi PN, Joyce LA (eds) Fire, fuel treatments, and ecological restoration: conference proceedings. Proceedings RMRS-P-29, vol 29. US Department of Agriculture, Forest Service, Rocky Mountain Research Station, pp 69–88

Mohanty A, Mithal V (2022) Managing forest fires in a changing climate. Council on Energy, Environment and Water, New Delhi

Morgan P, Keane RE, Dillon GK, Jain TB, Hudak AT, Karau EC et al (2014) Challenges of assessing fire and burn severity using field measures, remote sensing and modelling. Int J Wildland Fire 23(8):1045–1060

Olsson A, Lefvert A, Grönkvist S (2024) An alternative pathway to negative emissions–revitalising the wood-based panel production in Sweden. Environ Dev 50:100973

Pausas JG, Keeley JE (2019) Wildfires as an ecosystem service. Front Ecol Environ 17(5):289–295

Restoration of forest fire impacted areas Bandiour Tiger Reserve, Bandipur (2022) (Unpublished report). Conservator of Forests and Field Director, Bandipur, Karnataka Forest Department

Robichaud PR, MacDonald LH, Foltz RB (2010) Fuel management and erosion. In: Elliot WJ, Miller IS, Audin L (eds) Cumulative watershed effects of fuel management in the western

United States. Gen. Tech. Rep. RMRS-GTR-231, vol 231. US Department of Agriculture, Forest Service, Rocky Mountain Research Station, Fort Collins, pp 79–100
Ryu SR et al (2006) Simulating the effects of prescribed burning on fuel loading and timber production (EcoFL) in managed northern Wisconsin forests. Ecol Model 196:395–406. [CrossRef].
Tyukavina A, Potapov P, Hansen MC, Pickens AH, Stehman SV, Turubanova S et al (2022) Global trends of forest loss due to fire from 2001 to 2019. Front Remote Sens 3:825190
Williams AP, Abatzoglou JT (2016) Recent advances and remaining uncertainties in resolving past and future climate effects on global fire activity. Curr Clim Chang Rep 2:1–14
Wilson D, Lasco R (2018) The context of land cover changes in agriculture and forestry. In: The future of Philippine agriculture under a changing climate: Policies, investments and scenarios. ISEAS-Yusof Ishak Institute

Brave New Wild: Restoring Biodiversity and Values Amid Invaded Landscapes of India

8

Ninad Avinash Mungi, Rajat Rastogi, Yadvendradev V. Jhala, Qamar Qureshi, and Jens-Christian Svenning

Abstract

Human-led species introduction and movement across the world has led to biotic globalization. Such biotic globalization diminishes the natural biodiversity pattern and propels ecosystems into uncertain novel trajectories. A few among the introduced species, aka invasive species, can threaten local biodiversity and human welfare, setting an undesirable trajectory. As invasive species become predominant and, likely, a long-term reality, our conventional conservation models might need a paradigm shift to manage ecosystems with mixed communities of native and alien species. This chapter takes an unconventional detour to refresh the perspectives on emerging ecosystems and their position in modern conservation. It provides a synthesis of alien plant monitoring in India, conducted along

N. A. Mungi (✉)
Center for Ecological Dynamics in a Novel Biosphere Section of Ecoinformatics (ECONOVO); Department of Biology, Aarhus University, Aarhus, Denmark

Wildlife Institute of India, Dehradun, India

R. Rastogi
Wildlife Institute of India, Dehradun, India

German Center for Integrative Biodiversity Research (iDiv) Halle-Jena-Leipzig, Leipzig, Germany

Y. V. Jhala
Wildlife Institute of India, Dehradun, India

Indian National Science Academy, National Centre for Biological Sciences, Bangalore, India

Q. Qureshi
Wildlife Institute of India, Dehradun, India

J.-C. Svenning
Center for Ecological Dynamics in a Novel Biosphere Section of Ecoinformatics (ECONOVO); Department of Biology, Aarhus University, Aarhus, Denmark

D. Mittal, D. K. Gupta (eds.), *Ecological Restoration in India: Perspective and Practices*, https://doi.org/10.1007/978-981-96-8943-9_8

with the national tiger population assessment, quadrennially in the past two decades. This large-scale, long-term monitoring revealed that >66% of natural areas were invaded by a variety of alien plants. Ecosystem processes such as the diversity of plants and herbivores were seemingly reinforcing the resistance to invasion dominance. Nonetheless, expansive invasions compelled local people to adapt to emerging ecosystems by mitigating negative impacts and discovering novel ecosystem services. Managing expansive invasion using conventional methods is costly, with uncertain outcomes and unintended socio-ecological impacts. This chapter highlights an alternative spatiotemporal framework to *resist* invasion during the establishment phase or in areas with guaranteed biodiversity returns, *accept and direct* mixed-species ecosystems toward biodiverse trajectories, and *accept and adapt* to mitigate negative impacts where invasion control is unfeasible. This chapter provides a case for contextualizing the resist-accept-direct (RAD) typology for the management of novel ecosystems in India. Transcending the conventional binaries prevalent in conservation, this chapter highlights the potential for managing novel mixed-species ecosystems for a nature-positive future.

Keywords

Biodiversity · Ecosystem services · Invasion ecology · Nature-based solutions · Novel ecosystems · Restoration

… we have to correct and direct nature, and but for that, we should drown in an ocean of prejudice. But for that, there would never have been a single great man. Fyodor Dostoevsky, Crime and Punishment, 1866

8.1 Background

Globalization has accelerated the movement of life-forms beyond their biogeographic barriers at an unprecedented rate (IPBES 2023). The magnitude of such a movement has led to amalgamations of native and alien species interacting in novel ways and producing what appear to be novel ecosystems (Hobbs et al. 2009). Novel biotic communities have become a hallmark of the Anthropocene (Svenning et al. 2024a). Emerging biotic novelty is distorting known natural patterns of biodiversity and the conventional conservation models based on such patterns (Fricke and Svenning 2020; Svenning et al. 2024b). Subsets of introduced species have pervasively proliferated in the introduced regions, reducing biodiversity and important ecosystem services, and are termed invasive species (Roy et al. 2024) (Table 8.1). Such biological invasions precipitate cascading impacts on the functioning of the biosphere, societies, and economies, thereby endangering our collective welfare (Diagne et al. 2021; Pyšek et al. 2020). There is a growing global consensus on the severity of these biological invasions and the need to reduce/mitigate negative impact via management (i.e., prevention, control, and/or eradication of invasive species) (Blackburn et al. 2014; IPBES 2023). However, despite the resources invested

Table 8.1 Key terms and their meanings, as used in this study

Concept	Meaning
Alien species	Species introduced to a new locality by human agency that are able to establish populations in the new locality. When an introduced species starts to have negative impacts, it is referred to as an "invasive alien species," i.e., the concepts of introduced and invasive are hierarchical (IPBES 2023). Synonymous with "introduced species"
Biological invasion	Process involving the intentional or unintentional transport or movement of a species outside its natural range by human activities and its introduction to new regions, where it can establish and spread (IPBES 2023; Roy et al. 2024)
Ecological restoration	The process of assisting the recovery of ecosystems that have been degraded or destroyed, as well as conserving the ecosystems that are still intact (UNEP 2022)
Invasive alien species	A subset of introduced species that have negative impacts (Roy et al. 2024)
Mixed-species community	Used here to refer to a community of species consisting of a mix of populations of both native and introduced species, and where the number of introduced species present can range from one to several (McGeoch et al. 2024a)
Native species	A species within its natural range, including shifting its range, without human involvement (IPBES 2023)
Nature-based solutions (NbS)	Measures building on ecological processes and associated human adaptations to control biological invasions and their undesirable impacts, while contributing to long-term ecological autonomy, resilience, and good quality of life. These measures must have the potential to naturally upscale via spontaneous ecological and social processes (IUCN 2020; Mungi et al. 2025)
Novel ecosystems	Self-developing systems that have a biotic or abiotic structure different from the historical baseline, as a result of direct or indirect human agency (Hobbs et al. 2006; Kerr et al. 2024)

in invasion management (USD 120.5 billion in the past 50 years (Diagne et al. 2021)), retaining uninvaded ecosystems is increasingly becoming infeasible (Mormul et al. 2022). With mixed-species novel ecosystems as the new normal comes the uncertainty in the functioning of these ecosystems, their sustenance, and adaptive management to restore biodiversity with equitable social benefits.

While numerous studies have documented the negative consequences of alien species, a few but increasingly studies are documenting their positive impacts (Sax et al. 2022). Alien species invasions on islands have particularly negative impacts on biodiversity (Bacher et al. 2024), but the pattern elsewhere is more complex. For example, introduced herbivores impact vegetation similarly to the comparable native herbivores (Lundgren et al. 2024). Sometimes, the impacts of introduced megafauna are comparable to the ecological functions lost due to historical megafauna persecutions (Lundgren et al. 2020), questioning the negative connotation of such impacts. Increasing evidence is challenging the conventional binaries of positive and negative impacts and suggesting a more nuanced stance on the complexities of emerging mixed-species novel ecosystems (Lemoine and Svenning 2022). To comprehend these complexities and conceive suitable solutions for elevating biodiversity in mixed-species ecosystems, it is critical to monitor changing ecosystems and assess the functional role of native and alien species in emerging novel ecosystems. This necessitates large-scale, long-term ecosystem monitoring, which is currently limited due to resource constraints, undermining our knowledge and actions for managing the negative aspects of the emerging novel ecosystems.

India has ingeniously linked alien plant monitoring with tiger population assessment, which is an umbrella project by the government of India to quadrennially monitor large carnivores, herbivores, and their habitat at a national scale (Jhala et al. 2021). Alien plant assessment is integrated in this sampling, without incurring any additional sampling costs, apart from training the sampling teams on recording plants. This exercise has been conducted for the past two decades and offers an opportunity to link alien plants with contemporary ecological changes, generating a tool to monitor trophic-level dynamics. This chapter provides a synthesis of the research conducted on alien plants in India and its implications for the adaptive management of invaded ecosystems, biodiversity restoration, and strengthening of community benefits. It highlights patterns of alien plant distribution, their interaction with native biodiversity, the emergence of novel mixed-species ecosystems, novel ecosystem services, and how they shape our collective future. This research from India has global implications, particularly for the tropics and Tiger Range Countries with similar threats from plant invasions, seeking sustainable solutions for stewarding biodiversity amid invaded and mixed-species ecosystems. Key terms and their meanings, as used in this study, are provided in Table 8.1.

8.2 Expanse of Alien Plants in Tropical India

Around ~360,000 km^2 of natural areas have been sampled every four years across India since the year 2006 for assessing tiger habitats. Vegetation of these areas is assessed through systematically placed inventory plots (~160,000 plots) at a 25-km^2 sampling scale (Fig. 8.1a) (Jhala et al. 2021). Sampling revealed at least one alien plant in every 25 km^2 of the sampled areas. Among more than 2000 alien plants that were recorded, 11 were identified as high-concern invasive plants that threaten biodiversity, ecosystem services, and good quality of life (Mathur et al. 2015; Mungi et al. 2019). These invasive plants were found to be potentially occurring in 66% of natural areas in India, with open dry systems and semi-open deciduous systems being the most invaded (Fig. 8.1b) (Mungi et al. 2023b). Species like *Lantana camara* had singularly invaded more than 50% of natural areas. The success of every invasion could be traced to the history of its introduction and propagation by humans. For example, introduction and hybridization between diverse varieties of *Lantana camara* have yielded present species that are complex with high adaptive potential to suit local conditions (Mungi et al. 2020).

Over the past two decades of sampling, alien plants were found to be invading at an average rate of 729 (±155) km^2/year, with the highest rate by *Chromolaena odorata* at 1987 km^2/year (Mungi et al. 2023c). Establishment of invasive plants in new areas was facilitated by ecosystem fragmentation due to linear infrastructure, proximity to human land use, tree felling, altered soil moisture in dry biomes, higher

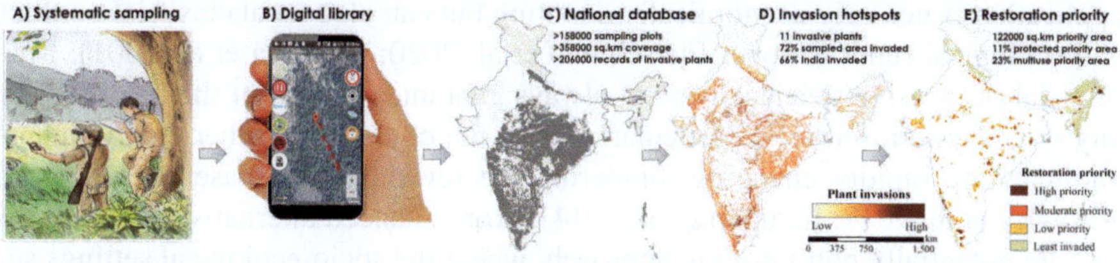

Fig. 8.1 India conducts a four-yearly large-scale assessment of plant invasions (**a**) using mobile applications and digital evidence base (**b**). This has generated a national database on the occurrence of alien plants across 358,000 km^2 (**c**). This assessment revealed that >66% of natural systems were threatened by the invasion of high-concern alien plants (**d**). The data was further used to estimate the rate of invasion, species invasiveness, and availability of native biodiversity to identify restoration priority areas, where minimum investment can guarantee maximum biodiversity returns (**e**). Thus, a large-scale assessment is objectively used to derive state variables of plant invasions for informing invasive plants' control and restoration of invaded ecosystems in India

frequency of fires in wet biomes, and proximity to invaded areas (Mungi et al. 2020, 2021). As these drivers accelerate across India, modifying natural systems (Nayak et al. 2020), invasions seem to be unceasingly expanding in the foreseeable future. Unlike the pace of invasions (hundreds of km^2 per year), measures to contain them operate at conventional scales (a few hectares per year). Invasions are thus transforming ecosystems at a scale that is beyond the traditional management scale to operate (Jackson 2021), which highlights the necessity and urgency of transformative changes in the management paradigm.

8.3 Management Prioritization

Invasive plant management has been practiced in India for more than a century (Bhagwat et al. 2012). The choice of management strategies used in India for controlling invasive plants has largely been constrained by experimentation and is primarily limited to manual control (Babu et al. 2009). There have been experiments on using biocontrol (Sankaran et al. 2017), chemical control, and burning (Mathur et al. 2015), but in the absence of their experimental utility across diverse biomes, the efficacy of such approaches remains unknown. In a few rare experimental frameworks, when multiple techniques were assessed for their efficacy in controlling invasions, manual control has been identified as relatively effective (Nerlekar et al. 2022; Sharma et al. 2023; Sinha et al. 2022). To upscale the manual control technique for managing high-concern invasive plants at their current expanse in India, even a conservative estimate required would need more than USD 13.5 billion, just for one-time management (Mungi et al. 2023b). Moreover, the large-scale management interventions to remove invasive plants from ecosystems are usually met with heavy usage of machinery and/or chemicals (e.g., using bulldozers for removing invasive plants from open ecosystems) and regular eradication drives. Though such interventions might achieve contextual and local success, their large-scale blanket application is not only economically daunting but can also escalate social conflicts and ecological ramifications (Fleischman et al. 2020; Zavaleta et al. 2001). Mere removal of invasive species does not always guarantee revival of the lost biodiversity or ecosystem services, particularly when the climate and other environmental variables are rapidly changing, hindering the revival of the baseline ecosystem (Crowley et al. 2017; Lampert et al. 2014). More nuanced alternatives are required that are essentially bottom-up in approach, where the socio-ecological settings surrounding an ecosystem are considered and subsequently upscaled by adaptively integrating them at various scales and contexts (Eggermont et al. 2015).

As managing large-scale pervasive invasions seems infeasible due to limited resources and often ineffective strategies, it is essential that management areas are prioritized for guaranteed outcomes with optimal investments. There is an inevitable necessity for adaptive management approaches like the Resist-Accept-Direct (RAD) typology that could provide a strategic and socio-ecologically sensitive framework for delineating the management of biological invasions and biodiversity restoration amid invaded landscapes (Aplet and Mckinley 2017; Schuurman et al. 2022). The RAD typology can help align the invasive species management with adaptive objectivity (Mungi et al. 2025). By being scale-sensitive to the area under invasions, their impacts, ecological interactions, and resource availability, management interventions can (1) "resist" the establishment of invasive plants, or eradicate them in the early stage, with an objective to maintain baseline ecosystem; (2) "accept and direct" ecosystem trajectories toward a desirable state by accommodating certain benign alien species with nonnegative impacts or via the introduction of lost ecological functions/interactions; and (3) "accept and adapt" the ecosystem transformation and facilitate community adaptations to a novel ecosystem state caused by invasion, such that the adaptation reduces negative impacts of invasions on human well-being and facilitates biodiversity (McGeoch et al. 2024a). These categorical interventions are not a rigid ternary but a flexible continuum, where different typologies and ecological status of the ecosystem and invasion can simultaneously have a bearing on any selected intervention (Fig. 8.2).

The RAD typology was mobilized contextually in India to assign management categories based on the prevalence of invasive plants, their long-term invasion trends, the availability of native plant diversity and cover, the conservation value of the areas, and resource constraints. Key restoration sites where biodiversity restoration could be guaranteed were identified using spatial simulation of the invasion and management tradeoffs (Fig. 8.2a). During the initial phase of invasion or in case only a small area is invaded, management can control existing invasions and *resist* future invasions. In case of expansive invasions, management can actively *accept and direct* the coexistence of uninvaded areas along with mixed-species areas by reducing the dominance of invasive species. In case of pervasive long-term invasions, management can *accept and adapt* to a novel mixed-species ecosystem by mitigating the negative impacts of invasive species.

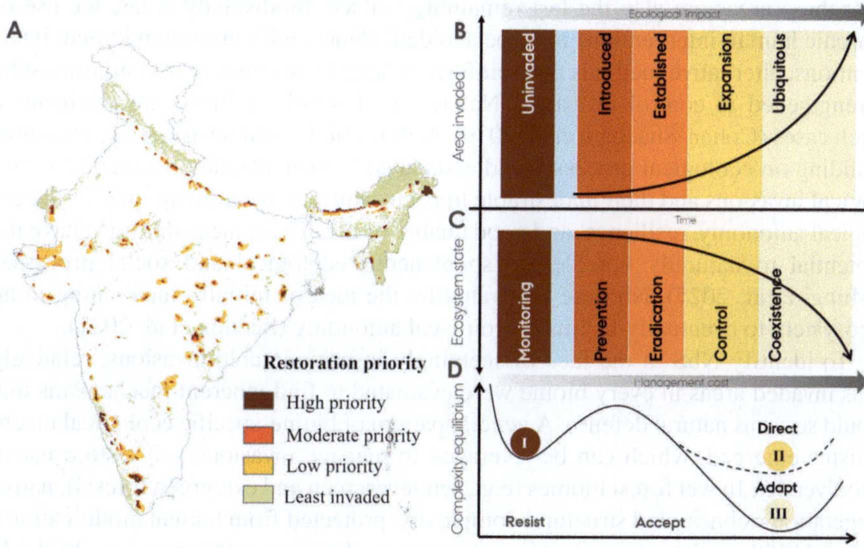

Fig. 8.2 Linkage between restoration priorities (**a**), invasion process (**b**), management regime (**c**), and Resist-Accept-Direct (RAD) typology (**d**). (**b**) An alien species expands in an area over time, eventually pervasively integrating/shifting the system. (**c**) The management cost and ecological impacts of the alien species increase with this quantum, and the potential of restoring the baseline system becomes more uncertain. (**d**) Invasions and management options could be compared with the alternate stable-state scenario, where prior to alien species introduction, a baseline state exists (I), comprising native species. Invasions can lead to higher interactions between native and alien species, with the baseline state emerging into an alternate invaded state (II) and, eventually, to a novel mixed-species state (III). Management interventions for these states can be aligned with the Resist-Accept-Direct typology, where in the initial phase, actions could *resist* and eradicate invasion; in the mixed-species stage, actions could *accept and direct* the system toward native species predominance; and once invasion predominates, people might *accept* the change and *adapt* to reduce the negative outcomes. The color scheme provides a comparison between the restoration priorities (**a**), the invasion process (**b**), and management options (**c**)

8.3.1 Resisting Invasion Dominance

With more than two-thirds of natural systems in India under plant invasions, it becomes pertinent to identify the last of the least invaded ("native") systems, where priority investments could guarantee maximum biodiversity returns. The least invaded native areas were the protected evergreen forests in the Western Ghats and Northeastern hills, the Thar desert, the rocky grasslands of peninsular India, and the least modified core regions of tiger reserves. These areas are threatened by neighboring invaded regions that act as a source for future invasions (Mungi et al. 2020, 2021). In total, a 122,532-km^2 area across India was found to be least invaded and of high priority for safeguarding its biodiversity and functions from imminent invasions (Fig. 8.1e) (Mungi et al. 2023b). Once these regions are safeguarded from invasions, the surrounding invaded areas could be subsequently managed to reduce invasions and/or their impacts, setting off a strategic chain reaction. Considering that these areas are also the last remaining "intact" biodiversity areas, the use of chronic human interventions must be avoided. Hence, with minimum human interventions, alternative methods that reinforce inherent resistance to invasions must be strengthened to control invasions. Nature-based solutions (NbS) are pertinent in such cases (Cohen-Shacham et al. 2019). NbS for biological invasions are measures building on ecological processes and associated human adaptations to control biological invasions and their undesirable impacts while contributing to long-term ecological autonomy, resilience, and good quality of life. These measures must have the potential to naturally upscale via spontaneous ecological and social processes (Mungi et al. 2025). NbS are motivated by the idea of initially intervening in an ecosystem to eventually facilitate ecological autonomy (Seddon et al. 2021).

To identify NbS in the face of seemingly insurmountable invasions, relatively less invaded areas in every biome were evaluated to find inherent mechanisms that could serve as natural defense. A general pattern of biome-specific ecological mechanisms emerged, which can be leveraged to manage invasions and restore native biodiversity. In wet forest biomes (e.g., semievergreen and evergreen forests), native vegetation richness and structural complexity, protected from human modifications, offered biotic resistance against invasive species dominance (Mungi et al. 2021). In humid, semi-open biomes (e.g., floodplain savannas and moist deciduous forests), restoring optimal densities of megaherbivores reduced invasions (Mungi et al. 2023b). In dry open and semi-open ecosystems (e.g., semiarid grasslands and dry savannas), restoring herbivore diversity and controlled fire reduced the impacts of invasions. Thus, the protection of old-growth forests, conservation and recovery of natural densities of megaherbivores, and maintenance of seasonal soil aridity and fire regime can act as potential NbS to reduce and/or mitigate contemporary and future invasions. However, the effectiveness of these mechanisms diminished in the presence of human modifications that fragment wet forests, reduce the abundance of megaherbivores, and alter soil moisture or fire regimes. The role of these ecological mechanisms as NbS for biological invasions was notably observed in core regions of the tiger reserve, where, owing to limited human modification in the past few decades, natural functions are largely restored (Jhala et al. 2021). Thus, the protection of ecosystems for tigers is augmenting the reinforcement of NbS within systems, making them resistant and resilient to the impact of biological invasions (Fig. 8.3).

8.3.2 Directing Mixed-Species Ecosystems Toward Native Species Predominance

Large parts of India's protected ecosystem are mixed communities of native and invasive plants, where invasions are rapidly increasing (Mungi et al. 2023c). These mixed communities often occur as native vegetation interspersed with homogeneous, dense patches of invasive plants, along with large areas containing a mix of sparse invasive and native plants. Increasing dominance of multiple dense patches of invasive plants can increase cumulative negative impacts on native species

Fig. 8.3 Invasion control mechanisms. The use of controlled burning in dry savannas reduces the prevalence of herbaceous invasive plants (**a**) and promotes the growth of native forage plants used by herbivores (**b**). Moderately productive ecosystems devoid of burning and megaherbivores were invaded by invasive plants (**c**) and were controlled in the presence of these consumers (**d**). Fragmentation of forests promoted plant invasions (**e**) that were resisted if the structural and functional diversity of old-growth forests was protected (**f**). Woody invasive plants are removed from dry forests (**g**), and the extracted biomass is used by local communities for commercial and household purposes (**h**). (Photo credits: A: Ninad Mungi, B: Parul Sen, C: Saurabh Verma, D: Y.V. Jhala, E: Ashwin Warudkar, F: Abir Jain, G: Rajat Rastogi, H: (Kannan et al. 2014))

(Rastogi et al. 2023). Hence, in such mixed communities, management can redirect the trajectory by reducing the dominance of invasive plants while restoring the dominance of native plants. Methods like cut rootstock (Babu et al. 2009) and girdling have been widely used for removing populations of woody invasive plants like *Lantana camara*, *Prosopis juliflora*, and *Senna spectabilis* from such mixed communities. Generally, these plants are removed from an area, expecting the recovery of native plants due to reduced competition from invasions. However, due to the existing seed bank, dispersal from proximate patches, or vegetative spread, these plants often reinvade the restored area (Kannan et al. 2013). Hence, the fate of such control actions must be evaluated to strategically restore native systems by active supplementation of lost ecological processes.

Broadly, restoration could be categorized into active and passive categories. Active restoration encapsulates strategic and chronological processes from the identification of priority sites up to the recovery of native biodiversity (Box 8.1). It also entails developing guidelines (removal, disposal, and utilization of invasive plants), generating resources (e.g., human resource and business models), and building infrastructure (e.g., herbivory control fence and nursery) for use at various stages in restoration. Owing to its resource- and time-intensive nature, active restoration is a limited enterprise, despite its successful outcome. On the other hand, passive restoration entails all those methods where interventions are limited to removing invasive plants, expecting that native plants will naturally regenerate after removing invasive plants. Sometimes, initial seeding and plantations are conducted without controlling for fire or herbivory or without guaranteed growth, hence aligning more to passive models of restoration. In India, passive restoration is more widely practiced, while active restoration is confined to a few important regions. Key examples of active restoration arise from protected areas (Babu et al. 2009; Sinha et al. 2022) and a few multiuse areas (Nerlekar et al. 2022; Osuri et al. 2024).

A comparison was explored between the outcomes of "active restoration," where native plants are actively regenerated and consumer regimes (grazing and fire) are maintained at prescribed levels, and "passive restoration," where invasive plants are removed with the expectation of natural regeneration of native plants. For such a comparison, vegetation changes following invasive plant control in seasonally dry grassy ecosystems across 16 protected areas in tropical India were monitored (Bandhavgarh, Bandipur, Corbett, Kanha, Manas, Panna, Pench MH, Pench MP, Rajaji, Ranthambhore, Sariska, Sathyamangalam, Satpuda, Similipal, Tadoba, and Valmiki). Sites where invasive plants were removed during 2004–2011 were identified, along with the type of restoration actions. Cover of alien and native plants (grasses, legumes, forbs, and shrubs) was measured every fourth year to observe long-term vegetation changes over a 16-year period (Fig. 8.4). Active restoration sites witnessed a substantial increase in native plants across the years, whereas alien plants increased till 8 years and did not change significantly thereafter. Passive restoration sites witnessed a consistent increase in alien plants (including those that were initially removed), whereas native plants did not change significantly and remained at a low cover. These patterns are likely influenced by restoration activities, herbivory, propagule pressure, and fire and water regime, which must be

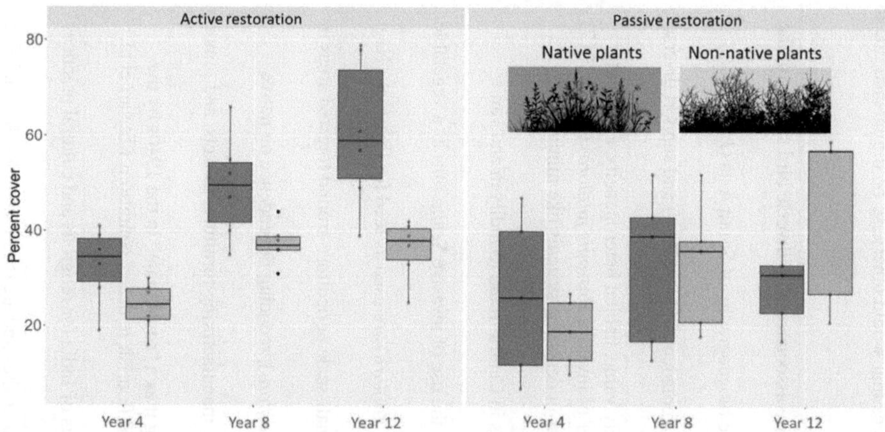

Fig. 8.4 Vegetation changes across dry grassy ecosystems in protected areas, following invasive plant removal. Sites with "active restoration" observed a consistent increase in native plant cover (native grasses, legumes, forbs, and shrubs), whereas alien plant cover increased in the first four years and remained similar after that. Sites with "passive restoration" observed an increase in native plants over the first eight years, and then, it declined, while alien plants continued increasing and eventually predominated

experimentally investigated to guide future restoration. Such vegetation recovery patterns highlight the necessity of considering restoration as an integral part of invasive species management for achieving long-term success in ecosystem recovery. It also questions the efficacy of one-time invasive plant removal used widely across India. It highlights the necessity of deploying strategic long-term resources for effective restoration using a scientific framework. Considering the importance of these systems in supporting a rich assemblage of herbivores and dependent carnivores, the resources utilized can help direct these mixed communities toward a biodiverse trajectory.

8.3.3 Adapting to the Novelty of Invaded Ecosystems

A large part of India's natural ecosystem is outside protected areas and is invaded by multiple invasive plants. The expansive and old-growth invasions have changed the functioning of these multiuse spaces, particularly considering native communities altered by traditional and commercial human use. Here, a legacy of overharvesting, unsustainable land use change, global changes, and commercial use has depleted the structure and functions of native ecosystems and allied services. In such scenarios, alien species and their interactions with native species can provide novel services and opportunities for a nature-positive future (Table 8.2). The cover provided by invasive plants acts as a habitat refugium for native species in heavily modified anthropogenic landscapes, where native plants are dwindling. A few examples include lions in the multiuse area using *Prosopis juliflora* patches as a

Table 8.2 Select examples of alien species from mixed-species ecosystems, providing novel ecosystem services that would otherwise be diminished due to ongoing anthropogenic overharvesting, biodiversity decline, and global changes

Services and overview	Examples
Regulating: In novel ecosystems with mixed species, alien species can assist ecological functions that are lost due to biodiversity decline, human disturbance, global changes, or erosion of baseline ecological interactions	Mixed communities of native and alien plants (e.g., *Lantana camara*) are associated with diverse bird species in the Western Ghats (Aravind et al. 2010)
	Alien plants can add to the food resources for wild mammals across the biogeography of South Asia (Jadeja et al. 2013; Rastogi et al. 2023; Subedi 2012)
	Alien plants add to habitat heterogeneity of the novel ecosystem (e.g., Keoladeo National Park) and support diverse migratory and resident species
	Alien species that do not decline native biodiversity can elevate diversity value, habitat heterogeneity, and habitat cover important for many endangered species (e.g., lion using the cover of invasive *Prosopis juliflora*) (Singh 2017)
	Alien plants assist in the recovery of soil carbon, biomass, and nutrients in degraded systems like mining sites (Ahirwal et al. 2017; Negi et al. 2019)
Provisioning: People, particularly the local communities, have adapted to utilizing novel ecosystems with mixed species. Alien species offer alternatives for local communities, who would otherwise be dependent on already-depleting native species, due to overharvesting or global changes	Alien plants are utilized for their medicinal value by local communities in Central India (Sandilyan and van't Klooster 2016; Wagh and Jain 2018)
	Parts of several alien plants are used as food resources by people (e.g., the use of invasive *Senna tora* as a vegetable by underprivileged communities) (Srivastava et al. 2018)
	Parts of several alien plants (e.g., pods and leaves of invasive *Prosopis juliflora*) are used as feed for livestock (Negi et al. 2019; Patnaik et al. 2017)
	Parts of alien plants are used for mulching, manure, and compost in small-scale agriculture around forested areas in India (Negi et al. 2019)
	Alien species can provide timber for use as fuelwood (e.g., *Prosopis juliflora*), reducing dependence on native plants (Baka and Bailis 2014; Patel 1985)
	Alien plants (e.g., *Lantana camara*) are used by local communities for manufacturing furniture, baskets, and minor timber products (Kannan et al. 2014)
Recreational: Mixed-species ecosystems are used by people for their intrinsic, esthetic, and cultural values, which are usually diminishing due to ongoing biodiversity crises and global changes	The alien plant *Prosopis juliflora* was accorded protection by declaring it as a "royal plant" in the 1940s by the princely state of Jodhpur in present-day India for its ability to grow and flourish in desert conditions (Tewari et al. 2011)
	Flowers, fruits, and leaves from many alien plants are now used in parts of India for religious and cultural practices (Singh et al. 2022)
	Alien plants are perceived as "green cover" in degraded areas like mining sites and areas of landslides, highlighting the intrinsic quality of nature to revive itself

refuge and dispersing tigers often using dense thickets of *Lantana camara*. It is true that had native plant cover not been anthropogenically reduced, it could have been used by these species in multiuse areas, and a mere refugium does not qualify as an ecological habitat (as sometimes, even anthropic infrastructure is used by these species). Nonetheless, naturalist records account for alien plants provisioning food resources for diverse taxa (e.g., fruits for birds, flowers for pollinators, and forage for herbivores), thus becoming a source of novel regulatory services (Table 8.2). Thus, identifying and leveraging novel ecosystem services arising from novel ecosystems of alien and native mixed communities can be seen as nature-based adaptations to minimize the negative impacts of invasions (Mungi et al. 2025). However, when typical populations of invasive species are considered, their disservice may far surpass their services, necessitating relevant mitigation strategies. As most of these areas are owned by private individuals, communities, and corporations, it becomes pertinent that human interventions reduce the negative impact by the sustainable management of invasive species populations. Such social initiatives for managing invasions are seen to be more effective if they generate monetary gains.

The utility of alien species to people has been considerable, given that a majority of food crops, agricultural crops, and domestic animals are alien (Ewel et al. 1999). Local communities across India are also adapting to live with an overwhelming abundance of alien species (Fig. 8.3h). For example, local communities in the Western Ghats traditionally used bamboo for the production of baskets, fences, and furniture. But due to overharvesting and the decline in bamboo in the region, they have started to use invasive *Lantana camara* plants (Kannan et al. 2014). This provided reprieve to the depleting populations of native plants while also resulting in the decline of invasive plants. Many tiger reserves are now training local communities to utilize invasive plant biomass to manufacture products on a commercial scale, which will provide livelihood alternatives to the people as well as support management of invasions in the area. If implemented and upscaled with necessary subsidies and capacity building, this could offer a viable economic model for the utilization of invasive plants and their management through public participation. Thus, alien and invasive plants are adding to nature's contribution to people through novel ways and supporting a good quality of life. However, it is important to prevent such economic models from converting natural regions into commercial plantations of invasive plants for production purposes or from overharvesting natural areas. It is also important to recognize that the social value of an invasive species could sometimes conceal its larger environmental cost, hindering its management and resulting in severe biodiversity decline. For example, control of alien tree plantations from desert systems is hindered due to their associated social and emotional values. In such cases, where bottom-up approaches are lacking, policies could assist in reducing the negative ecological consequences until public values are adapted. Thus, there is a necessity to dovetail management adaptations and inculcate evidence-based appreciation of ecosystems among people. These adaptations should direct future investments toward ecosystem management and education.

8.4 Novel Ecosystems

As the Anthropocene unfolds, anthropogenic influence is visible across natural systems and functions. Invasion of alien species is one such sign of the Anthropocene, which also exacerbates ongoing global changes, biodiversity crises, and erosion of ecosystem services (Malhi 2017). As species are being introduced at an unceasing rate, the quantum of future invasions would only add to the present levels (Mormul et al. 2022). Their cascading consequences on biodiversity and ecological interactions would trigger rapid shifts within ecosystems and their states (Svenning and Sandel 2013). The scale and pace of current management approaches are inadequate to manage the complexities brought about by seemingly insurmountable invasions. It necessitates transformative changes in conservation science, beginning with the reformation of ideas on alien species and ecosystem management.

The idea of alien species being "a problem" assumes that ecosystems consist solely of coevolved species and the introduction of alien life-forms can only "destabilize" the system as they are not coevolved. Recent studies have challenged these claims, as alien species are interacting with native species, following mechanistic ecological principles at par with native species. For example, introduced megaherbivores exert top-down effects on native plants, similar to the effects of extinct megaherbivores that once existed in the ecosystems (Lundgren et al. 2020). Our studies on novel interactions between alien plants and native species in India are suggestive of interactions following mechanistic ecological principles (e.g., resistance of biodiversity-rich regions to new species, megaherbivores controlling dominance of plants, and natural disturbance regime controlling plant diversity). Novel interactions between mixed species and the environment can also trigger expression of novel plastic responses in invasive species. For example, varying habitat characters in India trigger the expression of varying ecotypes of the invasive *Lantana camara* (Mungi et al. 2020). Given sufficient time, evolutionary mechanisms can subsequently enact novel ecological interactions within mixed-species ecosystems (Prentis et al. 2008), coming close to our contemporary idea of coevolution. Thus, in the context of the mechanistic definition of ecosystems, novel interactions between native and alien species can be viewed as a sign of emerging novel ecosystems. Globally, such novel ecosystems are being investigated for their role in biodiversity recovery and reinstation of lost ecological interactions and ecosystem services (Hobbs et al. 2009). A similar scope of inquiry can be made for the emerging ecosystems across India, where alien and native species coexist. Emerging novel ecosystems become more relevant when viewed along with ongoing global changes, biodiversity crises, and our aspirations for nature recovery (Hobbs et al. 2013).

Conventional binaries in conservation science have largely hindered progress in the management of mixed-species ecosystems (Lemoine and Svenning 2022). Binaries, such as native vs. alien, eradicate vs. accept, positive impacts vs. negative impacts, invasive vs. naturalized, and manage vs. not manage, are largely considered outdated approaches to contemporary ecosystem management. They undermine both nature's complexities and the human resoluteness to understand and conserve it. Transcending these binaries, it is urgent to rethink approaches for stewarding biodiversity and nature's contribution to people through novel mixed-species systems (Mungi and Qureshi 2018). It is important to acknowledge that species cannot be defined by mere political borders (Davis et al. 2011), the addition of new species cannot always be detrimental (Sax et al. 2022), a species cannot always be "declared" as invasive/naturalized across a political unit (Lemoine and Svenning 2022), ecosystem restoration does not imply pushing the ecosystem back in time (Higgs et al. 2014), and changes in species composition and interactions are not very uncommon across ecosystems (Hobbs et al. 2018). As conservation becomes more interventionist, contextuality becomes the key (Corlett 2016). Conventional approaches, such as in India, where the efficacy of a management intervention in one area triggers its blanket application over arguably diverse systems, are seemingly inadequate. Blanket application of a strategy developed to manage invasive species, without considering ecological contextuality and conservation value, can have negative socio-ecological ramifications that must be prevented (Kopf et al. 2017). Thus, science and management pertaining to novel mixed-species ecosystems need transformative reforms and adaptive conservation goals.

Despite the global tumult of categorically perceiving the negative impacts of alien species, the bias has not captured public perception (at large) in India, mostly due to socio-cultural reasons. Hence, valuing novel mixed-species ecosystems could be socially suited. It can help catalyze adaptive interventions for stewarding biodiversity and services in emerging novel ecosystems. A scientific approach for streamlining these perceptions, interventions, and conservation goals can ensure nature recovery in our rapidly changing and globalized world.

8.5 Conclusion and Future Directions

In India, management experimentation and research on novel mixed-species ecosystems are still at a nascent stage. Lessons from global research can guide the development of an approach based on the invasion process and management constraints. Adopting a unifying framework on the invasion–management spectrum (Blackburn et al. 2011), stages of invasion can assist optimal management decisions. During the initial stage, when the expanse of invasion is limited, eradication could be attempted with a vision to revive preinvasion biodiversity. However, with increasing expanse and time, the cost of eradication can be massive, and management may focus on containment of a species to reduce its documented negative impact on native biodiversity. With further spread, the interaction between native and alien species could result in novel mixed-species ecosystems (but see invasional meltdown (Simberloff and Von Holle 1999)). Here, adaptive management is used to mitigate the negative impacts of invasions and steward biodiversity and services in the unfolding system. These three management approaches are not definitive but adaptive and contextual and must be developed using robust scientific evidence in the milieu of social values through a bottom-up approach. It is important to understand the stages of invasion and their relationship with native biodiversity to inform adaptive management, which necessitates long-term monitoring and scientific experimentation. There have been studies and reviews on the necessity of a dedicated policy framework for the management of invasive species in India (Kannan et al. 2013; Mathur et al. 2015). Considering adaptive management described in this study, the points below highlight areas for future improvements to overcome the existing lacuna in research and management in India.

1. Management experimentations across diverse ecosystems are necessary to identify scale-sensitive and ecosystem-specific effective restoration models for mixed-species ecosystems. In situ restoration experiments must be supported at large scales.
2. Early detection of a new species is critical to eradicate it in the early stages and prevent future invasions. Smart technologies and approaches must be employed at high-risk pathways to detect and eradicate species. (It is important to recognize that the stage of invasion is not defined by political boundaries but ecological areas. A species that is invasive in one part of India can be at an early stage of invasion in another eco-region.)
3. Industrial research must focus on the commercial utilization of invasive plant biomass that can generate revenue necessary for large-scale, long-term management. It can also reduce the dependence on native plant biomass.
4. Participation of people, particularly local communities, must be achieved in the management of alien species in multiuse areas to steward an inclusive model of ecosystem restoration.
5. Biological invasions must be a critical aspect of "environmental impact assessment." Activities that obliterate the inherent resistance of natural ecosystems to invasions (e.g., mining, infrastructure development, etc.) must be identified for to provide ex gratia for arising management actions and future biodiversity restoration.
6. Policies, guidelines, and protocols must be developed across administrative scales to scientifically streamline the restoration and management of mixed-species ecosystems.

Restoration remains a vital facet for the future of India's ecosystems. Growing connectivity also introduces more alien species, which are further dispersed via increasing public movement. Simultaneously, increasing human population, people's dependencies on natural resources, intensifying land use changes, changing climate and soil moisture regimes, and many allied aspects of regional and global changes are degrading native ecosystems. Alien plants tend to exploit such conditions and dominate. Thus, the conservation of native ecosystems, management of alien plants, and restoration of managed areas emerge as inherently complementary. Treating these aspects in silos has globally failed and must be avoided (McGeoch et al. 2024b). India's conservation policies must integrate invasive species management and restoration to align with global and regional necessities. Placing restoration and invasive species management in policies will generate cascading impacts on strengthening finances, social models, scientific experiments, and success stories on restoration. Rare glimmers of successful restoration would show the potential to upscale and contribute to a nature-positive future. This chapter outlines ideas that could help in imagining and navigating decisions about the emerging new wild, fit for our new world.

Box 8.1 Key Dimensions for the Successful Restoration of Invaded Areas in India

Here is an outline of key aspects to consider while restoring invaded areas in protected areas like national parks and tiger reserve cores in India (Fig. 8.5). Other areas where local people utilize the ecosystem are complex, and restoration actions must strengthen inherent people–nature relationships. Considering that limited studies have scientifically experimented on this topic, it is important to experiment with long-term restoration across ecosystems. Nonetheless, based on primary field experience, below are the key points to consider during restoration:

1. *Priority zonation*: Considering the expansive levels of invasions across protected areas, this step assigns priority to those areas where concerted efforts could guarantee benefits to native biodiversity and ecosystem restoration. Using the national prioritization model and data (Mungi et al. 2023a, Fig. 8.1), high-resolution zonation could be achieved at a compartment scale. Following the compartments of highest priority in every range, buffering compartments should be subsequently ranked such that the cumulative restoration widens.
2. *Invasion removal*: Methods like cutting rootstock (Babu et al. 2009) and girdling are used for removing woody plants, while manual uprooting (Sinha et al. 2022) and burning (Sharma et al. 2023) are typically used for herbaceous plants in fire-adapted ecosystems (Fig. 8.5c, d). Control methods must essentially remove regenerative plant parts from the control sites. In case of manual control, removed invasive plant biomass must be taken away (without dispersing) from the site, as it can act as a source of vegetative reproduction.
3. *Active restoration using scientific methods*: The control sites are vulnerable to many invasions following the control and can set a trajectory for more invaded ecosystems (Fig. 8.4). Hence, active restoration of native vegetation and ecological processes must be done by using reference uninvaded sites from the same ecosystems (Fig. 8.5e, f). As soil characteristics are altered by invasions (e.g., increased nitrogen), initial restoration must target restoring natural edaphic characteristics using nature-based interventions (e.g., restoring nitrogen-dependent native plants from reference sites). During the initial period, supplementation of native plants, soil manipulation (e.g., dung of wild herbivores), herbivory control, and fire

(continued)

Box 8.1 (continued)

control could be considered to provide respite for regenerating native vegetation.

4. *Active rewilding of restored sites*: Following the initial period to achieve native vegetation dominance, active rewilding must be practiced, enhancing the resilience of restored sites for future invasions (Fig. 8.5g, h). For example, ecological processes such as controlled fire and herbivory can control the overabundance of ruderals and common species in restored grassy systems. A diverse herbivory guild can control the dominance of future invasions (Mungi et al. 2023a). Ecological processes, including below-ground biota, herbivory, fire, seasonal soil moisture regimes, and other trophic complexities, can enhance the diversity and resistance and resilience of restored sites to future invasions and other undesirable disturbances.
5. *Early detection*: Restored sites cannot always be resistant to future invasions; hence, they must be actively monitored to detect invasions in early stages and arrest their dominance. Active detection can be systematically integrated into patrolling (without compromising the objective of protection) using specialized modules in patrol devices, such as MSTrIPES (Jhala et al. 2019). Such monitoring techniques will enhance strategic restoration efforts, generate long-term evidence on reinvasions and secondary invasions, and strengthen invasion control with the least investment.
6. *Economic utilization of invasive biomass*: Given the expansive levels of invasions, the biomass generated by removing invasive plants could offer a valuable asset for economic gains. Many agencies have launched community-based businesses to produce crafts, furniture, and other products (e.g., Lantana elephants) utilizing invasive plant biomass. This has multifaceted benefits, including reducing dependence on native plant extractions and generating income for strengthening restoration activities. Considering the limited market for these products, other commercial uses, such as the production of paper, composite wood, ethanol, and other chemicals, must be explored.
7. *Social model*: As restoration is a social enterprise, strengthening social participation during various stages of restoration is vital. It can not only help restoration but also ensure inclusive and equitable benefits to local communities. Key interfaces for participation could include community-based nurseries of native plants and seed banks, removal of invasive plants from parks, establishment of biomass-processing units to compress plants into deliverable raw products, and establishment of a participatory end-to-end market chain of invasive produce. Key initiatives by the government and local administration that enhance participatory initiatives could enhance resources for launching such models.

Fig. 8.5 A case of restoring grassy areas dominated by invasive alien plants and native woody ruderal plants in central India (**a** and **b**). Herbaceous plants were removed by manual uprooting (**c**), and woody plants were removed by the cut rootstock method (**d**). Control sites were actively restored using quarantined seeds of native plants (**e**), employing active seeding repeated seasonally (**f**). Restored sites were controlled for fire and herbivory during the initial period (**g**) and subsequently rewilded by allowing herbivores and periodic controlled burning. These systematic processes led to native plant–dominated areas within 20–24 months of initial management. (Credits: Ravikant Mishra)

Acknowledgments We thank the National Tiger Conservation Authority, Wildlife Institute of India, State Forest Departments, sampling teams and coordinators, researchers, scientists, and students who participated in the All India Tiger Monitoring projects. We thank Dr. Vishnupriya Kolipakam for the support and initiative on this chapter. NAM and J-CS were funded by the Danish National Research Foundation Center for Ecological Dynamics in a Novel Biosphere (ECONOVO) (grant DNRF173 to J-CS) and the VILLUM Investigator project, "Biodiversity Dynamics in a Changing World" (grant 16549 to J-CS).

References

Ahirwal J, Maiti SK, Satyanarayana Reddy M (2017) Development of carbon, nitrogen and phosphate stocks of reclaimed coal mine soil within 8years after forestation with Prosopis juliflora (Sw.) Dc. Catena 156:42–50

Aplet GH, Mckinley PS (2017) A portfolio approach to managing ecological risks of global change. Ecosyst Health Sustain 3:e01261

Aravind NA, Rao D, Ganeshaiah KN, Uma Shaanker R, Poulsen JG (2010) Impact of the invasive plant, Lantana camara, on bird assemblages at Malé Mahadeshwara Reserve Forest, South India. Trop Ecol 51:325–338

Babu S, Love A, Babu CR (2009) Ecological restoration of lantana-invaded landscapes in Corbett Tiger Reserve, India. Ecol Restor 27:467–477

Bacher S, Galil BS, Nuñez MA, Ansong M, Cassey P, Dehnen-Schmutz K et al (2024) IPBES invasive alien species assessment: Chapter 4. Impacts of invasive alien species on nature, nature's contributions to people, and good quality of life. Zenodo

Baka J, Bailis R (2014) Wasteland energy-scapes: a comparative energy flow analysis of India's biofuel and biomass economies. Ecol Econ 108:8–17

Bhagwat SA, Breman E, Thekaekara T, Thornton TF, Willis KJ (2012) A Battle lost? Report on two centuries of invasion and Management of Lantana camara L. in Australia, India and South Africa. PLoS One 7:e32407

Blackburn TM, Pyšek P, Bacher S, Carlton JT, Duncan RP, Jarošík V et al (2011) A proposed unified framework for biological invasions. Trends Ecol Evol 26:333–339

Blackburn TM, Essl F, Evans T, Hulme PE, Jeschke JM, Kühn I et al (2014) A unified classification of alien species based on the magnitude of their environmental impacts. PLoS Biol 12:e1001850

Cohen-Shacham E, Andrade A, Dalton J, Dudley N, Jones M, Kumar C et al (2019) Core principles for successfully implementing and upscaling nature-based solutions. Environ Sci Pol 98:20–29

Corlett RT (2016) Restoration, reintroduction, and rewilding in a changing world. Trends Ecol Evol 31:453–462

Crowley SL, Hinchliffe S, McDonald RA (2017) Conflict in invasive species management. Front Ecol Environ 15:133–141

Davis MA, Chew MK, Hobbs RJ, Lugo AE, Ewel JJ, Vermeij GJ et al (2011) Don't judge species on their origins. Nature 474:153–154

Diagne C, Leroy B, Vaissière A-C, Gozlan RE, Roiz D, Jarić I et al (2021) High and rising economic costs of biological invasions worldwide. Nature 592:571–576

Eggermont H, Balian E, Azevedo JMN, Beumer V, Brodin T, Claudet J et al (2015) Nature-based solutions: new influence for environmental management and research in Europe. GAIA-Ecol Perspect Sci Soc 24:243–248

Ewel JJ, O'Dowd DJ, Bergelson J, Daehler CC, D'Antonio CM, Gómez LD et al (1999) Deliberate introductions of species: research needs: benefits can be reaped, but risks are high. Bioscience 49:619–630

Fleischman F, Basant S, Chhatre A, Coleman EA, Fischer HW, Gupta D et al (2020) Pitfalls of tree planting show why we need people-centered natural climate solutions. Bioscience 70:947–950

Fricke EC, Svenning J-C (2020) Accelerating homogenization of the global plant–frugivore meta-network. Nature 585:74–78

Higgs E, Falk DA, Guerrini A, Hall M, Harris J, Hobbs RJ et al (2014) The changing role of history in restoration ecology. Front Ecol Environ 12:499–506

Hobbs RJ, Arico S, Aronson J, Baron JS, Bridgewater P, Cramer VA et al (2006) Novel ecosystems: theoretical and management aspects of the new ecological world order: novel ecosystems. Glob Ecol Biogeogr 15:1–7

Hobbs RJ, Higgs E, Harris JA (2009) Novel ecosystems: implications for conservation and restoration. Trends Ecol Evol 24:599–605

Hobbs RJ, Higgs ES, Hall C (2013) Novel ecosystems: intervening in the new ecological world order. Wiley

Hobbs RJ, Valentine LE, Standish RJ, Jackson ST (2018) Movers and stayers: novel assemblages in changing environments. Trends Ecol Evol 33:116–128

IPBES (2023) IPBES invasive alien species assessment: full report. Zenodo

IUCN (2020) Global standard for nature-based solutions. A user-friendly framework for the verification, design and scaling up of NbS

Jackson ST (2021) Transformational ecology and climate change. Science 373:1085–1086

Jadeja S, Prasad S, Quader S, Isvaran K (2013) Antelope mating strategies facilitate invasion of grasslands by a woody weed. Oikos 122:1441–1452

Jhala YV, Qureshi Q, Gopal R, Bonal BS, Swain D, Nayak AK et al (2019) Development and implementation of MSTrIPES in Tiger Reserves. National Tiger Conservation Authority, Government of India, and Wildlife Institute of India, Dehradun

Jhala Y, Gopal R, Mathur V, Ghosh P, Negi HS, Narain S et al (2021) Recovery of tigers in India: critical introspection and potential lessons. People Nat 3:281–293

Kannan R, Shackleton CM, Shaanker RU (2013) Playing with the forest: invasive alien plants, policy and protected areas in India. Curr Sci 104:1159–1165

Kannan R, Shackleton CM, Shaanker RU (2014) Invasive alien species as drivers in socio-ecological systems: local adaptations towards use of Lantana in Southern India. Environ Dev Sustain 16:649–669

Kerr MR, Ordonez A, Riede F, Svenning J-C (2024) A biogeographic–macroecological perspective on the rising novelty of the biosphere in the Anthropocene. J Biogeogr 51:575–587

Kopf RK, Nimmo DG, Humphries P, Baumgartner LJ, Bode M, Bond NR et al (2017) Confronting the risks of large-scale invasive species control. Nat Ecol Evol 1:0172

Lampert A, Hastings A, Grosholz ED, Jardine SL, Sanchirico JN (2014) Optimal approaches for balancing invasive species eradication and endangered species management. Science 344:1028–1031

Lemoine RT, Svenning J-C (2022) Nativeness is not binary – a graduated terminology for native and non-native species in the Anthropocene. Restor Ecol e13636

Lundgren EJ, Ramp D, Rowan J, Middleton O, Schowanek SD, Sanisidro O et al (2020) Introduced herbivores restore Late Pleistocene ecological functions. Proc Natl Acad Sci 117:201915769–201915769

Lundgren EJ, Bergman J, Trepel J, le Roux E, Monsarrat S, Kristensen JA et al (2024) Functional traits – not nativeness – shape the effects of large mammalian herbivores on plant communities. Science 383:531–537

Malhi Y (2017) The concept of the Anthropocene. Annu Rev Environ Resour 42:77–104

Mathur VB, Kaushik M, Bist SS, Mungi NA, Qureshi Q (2015) Management of human-wildlife interaction and invasive alien species in India (No. TR 2015/004). Wildlife Institute of India, Dehradun

McGeoch MA, Clarke DA, Mungi NA, Ordonez A (2024a) A nature-positive future with biological invasions: theory, decision support and research needs. Philos Trans R Soc B Biol Sci 379:20230014

McGeoch MA, Ordonez A, Howard PL, Groom QJ, Shrestha BB, Fernandez Trigoso M et al (2024b) IPBES invasive alien species assessment: Chapter 6. Governance and policy options for the management of biological invasions

Mormul RP, Vieira DS, Bailly D, Fidanza K, da Silva VFB, da Graça WJ et al (2022) Invasive alien species records are exponentially rising across the Earth. Biol Invasions 24:1–13

Mungi NA, Qureshi Q (2018) On the history, politics and science of invasion ecology. DIALOGUE: Sci Sci Soc 1:1–16

Mungi NA, Kaushik M, Mohanty NP, Rastogi R, Johnson JA, Qureshi Q (2019) Identifying knowledge gaps in the research and management of invasive species in India. Biologia 74:623–629

Mungi NA, Qureshi Q, Jhala YV (2020) Expanding niche and degrading forests: key to the successful global invasion of Lantana camara (sensu lato). Glob Ecol Conserv 23:e01080

Mungi NA, Qureshi Q, Jhala YV (2021) Role of species richness and human impacts in resisting invasive species in tropical forests. J Ecol 109:3308–3321

Mungi NA, Jhala YV, Qureshi Q, le Roux E, Svenning J-C (2023a) Megaherbivores provide biotic resistance against alien plant dominance. Nat Ecol Evol 7:1645–1653

Mungi NA, Qureshi Q, Jhala YV (2023b) Distribution, drivers and restoration priorities of plant invasions in India. J Appl Ecol 60:2400–2412

Mungi NA, Rastogi R, Qureshi Q, Jhala YV (2023c) Plant invasions and restoration priorities in India. In: Status of tigers, co-predators and prey in India, 2022. National Tiger Conservation Authority, Government of India, New Delhi, and Wildlife Institute of India, Dehradun

Mungi NA, Gloria AO, Rastogi R, Svenning J.-C (2025) Expanding the Resist–Accept–Direct framework for developing nature-based solutions and societal adaptations to biological invasions. People and Nature. https://doi.org/10.1002/pan3.70073

Nayak R, Karanth KK, Dutta T, Defries R, Karanth KU, Vaidyanathan S (2020) Bits and pieces: forest fragmentation by linear intrusions in India. Land Use Policy:104619

Negi GC, Sharma S, Vishvakarma SC, Samant SS, Maikhuri RK, Prasad RC et al (2019) Ecology and use of Lantana camara in India. Bot Rev 85:109–130

Nerlekar AN, Mehta N, Pokar R, Bhagwat M, Misher C, Joshi P et al (2022) Removal or utilization? Testing alternative approaches to the management of an invasive woody legume in an arid Indian grassland. Restor Ecol 30:e13477

Osuri AM, Kasinathan S, Raman TRS, Mudappa D (2024) Restoration opportunities beyond highly degraded tropical forests: insights from India's Western Ghats. Biol Conserv 291:110519

Patel VJ (1985) Rational approach towards fuelwood crisis in rural India. Econ Polit Wkly 20:1366–1368

Patnaik P, Abbasi T, Abbasi SA (2017) Prosopis (Prosopis juliflora): blessing and bane. Trop Ecol 58:455–483

Prentis PJ, Wilson JRU, Dormontt EE, Richardson DM, Lowe AJ (2008) Adaptive evolution in invasive species. Trends Plant Sci 13:288–294

Pyšek P, Hulme PE, Simberloff D, Bacher S, Blackburn TM, Carlton JT et al (2020) Scientists' warning on invasive alien species. Biol Rev 95:1511–1534

Rastogi R, Qureshi Q, Shrivastava A, Jhala YV (2023) Multiple invasions exert combined magnified effects on native plants, soil nutrients and alters the plant-herbivore interaction in dry tropical forest. For Ecol Manag 531:120781

Roy HE, Pauchard A, Stoett PJ, Renard Truong T, Meyerson LA, Bacher S et al (2024) Curbing the major and growing threats from invasive alien species is urgent and achievable. Nat Ecol Evol:1–8

Sandilyan S, van't Klooster CIEA (2016) The other sides of invasive alien plants of India – with special reference to medicinal values. J Nat Conserv 31:16–21

Sankaran KV, Tjitrosemito S, Sastroutomo SS (2017) Impacts and management options for Mikania micrantha in plantations. In: Invasive alien plants: impacts on development and options for management. CAB International, Oxfordshire, pp 39–58

Sax DF, Schlaepfer MA, Olden JD (2022) Valuing the contributions of non-native species to people and nature. Trends Ecol Evol 37:1058–1066

Schuurman GW, Cole DN, Cravens AE, Covington S, Crausbay SD, Hoffman CH et al (2022) Navigating ecological transformation: resist–accept–direct as a path to a new resource management paradigm. Bioscience 72:16–29

Seddon N, Smith A, Smith P, Key I, Chausson A, Girardin C et al (2021) Getting the message right on nature-based solutions to climate change. Glob Chang Biol 27:1518–1546
Sharma M, Devi A, Badola R, Sharma RK, Hussain SA (2023) Impact of management practices on the tropical riverine grasslands of Brahmaputra floodplains: implications for conservation. Ecol Indic 151:110265
Simberloff D, Von Holle B (1999) Positive interactions of nonindigenous species: Invasional meltdown? Biol Invasions 1:21–32
Singh HS (2017) Dispersion of the Asiatic lion Panthera leo persica and its survival in human-dominated landscape outside the Gir forest, Gujarat, India. Curr Sci 112:933–940
Singh R, Tiwari AK, Singh GS (2022) Perceptions of the impacts of invasive alien plants in the riparian zone of the Ganga River: insights from Varanasi, India. River Res Appl 38:1495–1509
Sinha A, Nath A, Lahkar BP, Brahma N, Sarma HK, Swargowari A (2022) Understanding the efficacy of different techniques to manage Chromolaena odorata L., an invasive alien plant in the sub-Himalayan tall grasslands: toward grassland recovery. Ecol Eng 179:106618
Srivastava, Pan RS, Bhatt BP (2018) Antioxidant and nutritional potential of some underutilized leafy vegetables consumed by tribals of Jharkhand, India. Curr Sci 114:1222–1233
Subedi N (2012) Effect of Mikania micrantha on the demography, habitat use, and nutrition of Greater One-horned Rhinoceros in Chitwan National Park, Nepal. Forest Research Institute University, Dehradun
Svenning J-C, Sandel B (2013) Disequilibrium vegetation dynamics under future climate change. Am J Bot 100:1266–1286
Svenning J-C, Kerr MR, Mungi NA, Ordonez A, Riede F (2024a) Defining the Anthropocene as a geological epoch captures human impacts' triphasic nature to empower science and action. One Earth 0
Svenning J-C, McGeoch MA, Normand S, Ordonez A, Riede F (2024b) Navigating ecological novelty towards planetary stewardship: challenges and opportunities in biodiversity dynamics in a transforming biosphere. Philos Trans R Soc B Biol Sci 379:20230008
Tewari JC, Krishnan PR, Harsha SL, Bohra HC (2011) Prosopis juliflora: past, present and future
United Nations Environment Programme, & Food and Agriculture Organization of the United Nations (2022) The UN decade on ecosystem restoration (2021–2030): flagship initiatives
Wagh VV, Jain AK (2018) Status of ethnobotanical invasive plants in western Madhya Pradesh, India. S Afr J Bot 114:171–180
Zavaleta ES, Hobbs RJ, Mooney HA (2001) Viewing invasive species removal in a whole-ecosystem context. Trends Ecol Evol 16:454–459

Yamuna Biodiversity Park: Restoring Natural History of Yamuna Floodplain in Delhi Stretch

9

Faiyaz A. Khudsar, Mohan Singh, Umendra Sirohi, Harmeek Singh, Ekta Khurana, M. Faisal, and A. K. Singh

Abstract

Rivers are critical determinants for the future sustenance of human civilisations. However, the majority of rivers are facing a series of problems altering their ecological processes. There is a need to find out a new strategy for the restoration of degraded landscapes and riverscapes, especially in urban centres, and now Delhi Biodiversity Parks have emerged as a third major conservation strategy after in situ and ex situ, where the initiative was taken at the ex situ level but finally achieved in situ.

Yamuna Biodiversity Park spreads over an area of 185 ha in the upstream of the Delhi stretch of the River Yamuna, situated in Jagatpur Village, near Wazirabad. This is a successful model to restore characteristic floodplain wetlands, grasslands and floodplain forests, adopting some of the principles of ecological restoration through assemblage of species in the form of biological communities belonging to the Yamuna River basin.

Once a barren, degraded riverscape, the Yamuna Biodiversity Park has now transformed into a dynamic and complex ecosystem with restored floodplain wetlands, grasslands and forest communities harbouring organisms of various trophic structures which interact with each other. The established functional ecosystems are now homes to many plants, insects, birds, reptiles and mammalian species. Some of the species, such as Hog deer (*Axis porcinus*) and Siebold water snake (*Ferania seiboldii*), have been rediscovered after several decades, while the establishment of heronry dominated by Black-crowned night heron (*Nycticorax nycticorax*) and the arrival of many migratory birds during winter, along with Nilgai (*Boselaphus tragocamelus*) and Wild pigs (*Sus scrofa cristatus*), are now residents of the park and have marked the testimony for the suc-

F. A. Khudsar (✉) · M. Singh · U. Sirohi · H. Singh · E. Khurana · M. Faisal · A. K. Singh
Biodiversity Parks Programme, Centre for Environmental Management of Degraded Ecosystems, University of Delhi, Delhi, India

D. Mittal, D. K. Gupta (eds.), *Ecological Restoration in India: Perspective and Practices*, https://doi.org/10.1007/978-981-96-8943-9_9

cessful restoration of the floodplain with its characteristic natural ecosystems. Yamuna Biodiversity Park is ready to be replicated elsewhere for the restoration of floodplains, which will help riverine ecosystems to become functional ecosystems in a span of 7–10 years and support the rejuvenation of the river.

Keywords

Wetland restoration · Yamuna floodplains · Biodiversity conservation · Biodiversity park

9.1 Introduction

Without water, there is no life on this planet, and the rivers are a major source of water for humans. The great human civilisations flourished along the rivers and also perished when the rivers were dead or degraded. Therefore, rivers are critical determinants for the future sustenance of human civilisations.

Riverscapes are complex, interconnected, dynamic socio-ecological systems, and healthy rivers provide the majority of requirements for the survival and developmental needs of mankind (Hand et al. 2018; Dunham et al. 2018). In fact, rivers are complex gross landforms that drain the earth's surplus surface water and discharge it into oceans/lakes and form an important component of the hydrological cycle that regulates climate and weather patterns. During their course across the landmass, they give rise to a wide range of landforms that support rich biodiversity and generate a wide range of ecological services.

However, in recent decades, the majority of rivers have faced a series of problems because of large-scale alterations to satisfy human needs. Multiple human-induced drivers, such as land use change, affect the capacity of a catchment to divide rainfall into evaporation and runoff (Kanade and John 2018); water diversion leads to aquifer exploitation; wastewater contaminants allow the nutrient enrichment of floodplain wetlands and weed invasion, and rivers modify the natural flow (Girolamo De et al. 2015); rampant sand mining and stream channelisation cause geomorphic and hydrological changes; and all of these ultimately change the dynamics and complexity of a river ecosystem.

Manipulating rivers beyond their carrying capacity lost their ability of the ecosystem processing, which includes self-cleansing and healing naturally; this leads to severe floods and droughts. All of these are matters of increasing concern for the ecology, health and well-being of all life forms, including humans, and the economies sustained by rivers. In fact, the river stretch that passes through the urban areas is highly compromised due to unplanned urbanisation that led to encroachment and increasing pollution, and ultimately, in the face of escalating extreme weather patterns, multidimensional issues cannot be addressed within the river channel alone. It is imperative to look at the channel-floodplain connectivity, considering other complexities such as tributaries, rain, floodplain forest, wetlands, subsoil and groundwater aquifers.

Since all pollution and degradation of rivers are due to human actions, the key to restoration is for all stakeholders, including communities along the rivers, to come together to stop the sources of degradation and support the restoration efforts, as these can take many years.

River Yamuna, one of the major tributaries of the Ganges, is not an exception. The River Yamuna between Prayagraj and Agra was used for navigation in 1940. The extensive channelisation of water through the upper Yamuna canal since 1789 and the western and eastern canals in 1810, and the discharge of sewage, especially between Delhi and Agra, converted it into a sewer (Gopal and Sah 1993). The River gets its characteristics partially only after its confluence with the Chambal River. In fact, all of these led to degradation of the riverscape and loss of biodiversity, ultimately affecting the River Yamuna.

River Yamuna is one of the important rivers, and the Yamuna River basin is considered the second-largest basin after the Ganga River in northern India, which covers a distance of 1376 km and passes through four major states: Uttarakhand, Haryana, Delhi and Uttar Pradesh. The River goes through the main townships of Haryana, such as Sonipat, Panipat and Delhi state, then through Ghaziabad, Mathura, Agra, Etawah and Prayagraj of Uttar Pradesh, and contributes about 85% of pollution in the Yamuna, and Delhi tops the list of cities that contribute the maximum pollution load to the River Yamuna (Gopal and Sah 1993).

River Yamuna passes through Delhi and some stretches of Uttar Pradesh, covering a distance of about 50 km. The River Yamuna in Delhi's stretch begins from village Palla, located at an altitude of 210.3 m in the northeast, and further flows southwards to a distance of about 50 km and goes out at Jaitpur village, located at an altitude of 198.12 m in the south. It can be categorised ecologically into two stretches: the upstream from Palla to Wazirabad barrage and the downstream from Wazirabad to Jaitpur Village near Kalindi.

About a 25 km stretch in the upstream of Wazirabad reservoir receives water from the Western Yamuna canal, which joins the river at Palla, and the Eastern Yamuna canal joins it at Wazirabad barrage. Both the canals originate from the Hathnikund barrage, and downstream, there is no flow from the barrage during the lean period, and the only flow is from the canal. This stretch of river has no source of sewage. However, the 22–25 km stretch of the river between Wazirabad barrage and Okhla barrage receives only sewage from 22 drains and clean water from Wazirabad barrage, largely during the monsoon. The upstream of the Okhla barrage receives polluted water from the Hindon River through the Hindon cut canal. The Shahadra drain, carrying sewage from the Trans Yamuna area, discharges its content into the river after the Okhla barrage. Hence, the most polluted stretch in Delhi is between Wazirabad and Kalindi.

River Yamuna in Delhi's stretch does not have its natural banks but is largely surrounded by roads and bunds and intercepted by multiple water breaks. The width of the floodplain on either side of the main channel is about 2 km, and the waterways of the river vary from 100 m to 500 m, particularly downstream during the dry season for a period of about 8–10 months. There are islands of various shapes and sizes throughout the river.

There is no discharge of sewage into the river between Palla and Wazirabad; however, the release of some pollutants is reported from upstream sometimes. The river is bounded on both the eastern and western sides by the embankment or embankment cum road and also has a large floodplain along with the oxbow lakes and floodplain wetlands. The Bhalswa Lake is a large oxbow water body, also highly degraded and in need of attention. Hence, identification of network wetlands in both the eastern and western banks is crucial, followed by their ecological restoration. The downstream of the river from Wazirabad to Jaitpur Village is highly polluted, with the drains discharging large quantities of sewage, and still the Okhla Bird Sanctuary acts as an important area for bird stopover (Manral and Khudsar 2013). Therefore, it is critical to restore the floodplain for the rejuvenation of the river.

It is imperative to understand the river and its importance before planning for any restoration. A river is not a pipeline which carries water from one place to the other, nor a place for sand and boulders only being used as construction material, nor only a marketable commodity to bottle and sell in the market. In fact, a river is a living system and is not restricted to the main river channel only. Rivers can be seen as a complex ecological system, comprising landscape and waterscape features which include waterways or floodways, followed by the riparian zone (including marshy areas,), active floodplain, embankment (natural/manmade) and upland outside the embankment. There is strong ecological integrity among all the components.

9.1.1 Ecological Restoration

Continuing degradation of river ecosystems and loss of aquatic biodiversity are widespread. Restoration of the floodplain can only help rejuvenate the river. The objective of river restoration should be the reestablishment of river habitat and biodiversity attributes. Restoration is mainly about enhancing ecological integrity (Baron et al. 2002), which is, in fact, the ability to self-sustain an ecological unit such as a species, population, community, ecosystem and ecological processes such as nutrient cycling and sediment movement. The principles of ecological restoration involve the designation of a site for restoration, ascertaining the ecological history of the site, identification of a reference ecosystem, preparation of site prescriptions and finally, systematic interventions based on available resources.

There is a need to come up with a comprehensive conservation strategy for restoring riverscapes/landscapes, which includes floodplains. *The Concept of Biodiversity Parks*: a new conservation strategy (Babu C.R. *per. Communication*, 2002) evolved and was implemented for the first time in India and perhaps in the world by the Delhi Development Authority in joint collaboration with CEMDE, Delhi University. 'Biodiversity Park is first of its kind and a unique blue green infrastructure for restoring degraded landscapes/riverscapes by re-establishing the characteristic ecosystems through assemblage of species in the form of biological communities belonging to specific ecological ranges with high biodiversity values and rendering several ecosystem services having conservation, education and recreational importance' (Niangthianhoi and Khudsar 2015).

A floodplain restoration programme has been initiated in a stretch of the Yamuna River in Delhi to identify and restore the floodplain wetlands and re-establish the characteristic grasslands and floodplain forest using a biodiversity park model and designated as *Yamuna Biodiversity Park*, the first biodiversity park of India. The Hon'ble National Green Tribunal, a green judicial institution of India, has given direction in a matter that the setting up of biodiversity parks on floodplains shall also be considered as an important component for river rejuvenation (NGT 2019).

9.1.2 Yamuna Biodiversity Park

The Yamuna Biodiversity Park, situated at 28°44′N and 77°12′E on the west bank of floodplains in the upstream of the Yamuna River in Delhi, spreads over an area of 184 ha near Jagatpur Village, Wazirabad, on the flat alluvial floodplains of the River Yamuna. The park is divided into two phases: Phase I, consisting of 63 ha located on the inactive floodplain, which was originally a part of the Yamuna floodplains (about 50 years ago) but was cut off from the existing floodplains due to the construction of a permanent embankment in 1978, and Phase II, which consists of the rest of the 121 ha and is located on the active floodplain of the River Yamuna. Phase I does not receive floods, whereas Phase II receives floods periodically. A 200 m wide ecological corridor has been established connecting both the phases of Yamuna Biodiversity Park, facilitating the movement of animals, as shown in Fig. 9.1.

Ecological restoration initiative was taken up in 2002, designating a site in Phase I with an objective to identify and restore the wetlands and re-create historically representative forest communities and grasslands within a defined space to achieve ecological integrity for establishing functional ecosystems. It was envisaged in the beginning itself that characteristic floodplain wetlands, grasslands and forest communities would be re-established through restoration, which would enable many species of insects, birds, reptiles and mammals to reclaim their historical geographical distribution ranges. In other words, ecological restoration of floodplain ecosystems will facilitate natural rewilding.

9.1.3 Initial Intervention

The site designated for the Biodiversity Park was highly alkaline with a saline subsoil aquifer. The site is represented largely by a few species of salt-loving bushes, such as *Suaeda fruticosa*, *Cressa cretica* and *Heliotropium curassavicum*, and some common weeds. In such circumstances, how to restore the pristine ecological glory of the Yamuna floodplain was a challenge. Initial interventions for habitat improvement were critical, and therefore, both site- and time-specific interventions were planned and scheduled. Basic principles of ecological restoration were used to re-establish floodplain ecosystems comprising characteristic wetlands, marshes, grasslands and floodplain forests.

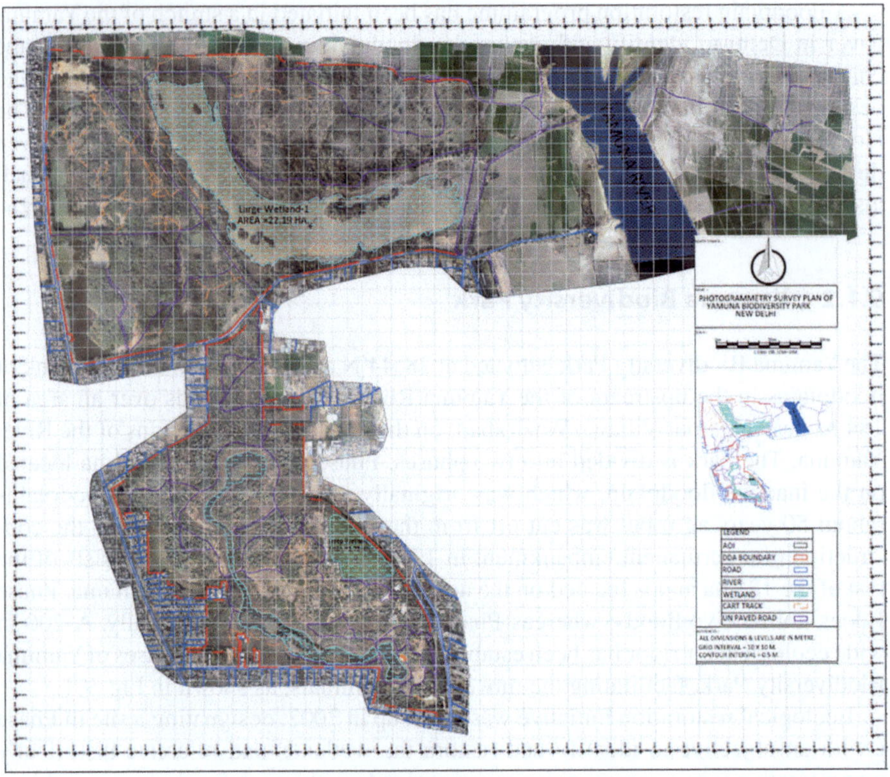

Fig. 9.1 Map of Yamuna Biodiversity Park

To address the issues related to soil alkalinity, it was decided to raise the land level to facilitate leaching by precipitation. To procure soil on-site for raising land and also to avoid ecological disturbances to other sites due to soil excavation and transportation costs, a systematic survey was carried out on the given site. Some of the indicator species, such as *Typha* and *Phragmites*, were observed, suggesting a high subsoil aquifer. In other words, they were historically wetlands which got silted or filled up over a period of time.

The identified wetlands were desilted, and excavated silt was used to create mounds of different sizes and shapes. These mounds were developed with the objective to raise the level, which allowed leaching of salts, thereby permitting the habitat to reclaim and regain some fertility, besides contributing to habitat diversity, influencing microclimatic variables and carving conducive niches.

Some areas were ploughed before any floral interventions. Ploughing helped break the hard clay in soils, which facilitated aeration for the establishment of plant root systems and also helped proper mixing of biological inputs given initially, such as farmyard manure and leaf litter. Initially, *Leptochloa fusca*, a grass species, was introduced for its known characteristics of alkalinity reduction, followed by other grass species such as *Chrysopogon zizanioides*, *Cenchrus ciliaris*, *Bothriochloa and*

Desmostachya to initiate ecosystem processing. Further, some of the legume species, such as *Crotalaria* sp., *Sesbania* sp., *Indigofera tinctoria*, *Rhynchosia* sp. and *Leucaena leucocephala*, were used for improving the nitrogen content of soil, and many of them either eliminated their own in succession or were removed, especially *Leucaena leucocephala.* Since the assigned site was in a human-dominated landscape, a three-layer biological fencing was established along the human settlement comprising *Dendrocalamus stictus*, *Terminalia arjuna*, *Albizia* sp., etc., and some of the tall grass species.

9.1.4 Restoration of Floodplain Wetlands

Wetlands are critical within a larger ecosystem, particularly in floodplains, for initiating ecosystem processing and ultimately helping in the establishment of functional ecosystems. These floodplain wetlands also serve as a nursery for the river and surrounding riparian ecosystems by sustaining insects, fishes and plant propagules which ultimately reach the river in the next flood, besides providing water to the river during lean periods through lateral flow. These wetlands also act as kidneys by containing silt and other pollutants and not allowing them to go to the river. Therefore, identification and restoration of both the catchment and treatment wetlands in the floodplain are critical for rejuvenating rivers.

Initially, two wetlands were identified and restored in Phase I: a narrow, shallow wetland of more than 1.5 km in length with an average width of 20–25 m and average depth varying from 0.25 m to 1m and a deep wetland of about 03 ha with the average depth varying from 0.25 m to 3.5 m. Both the wetlands were identified and restored, fulfilling different objectives. The narrow, shallow wetland was restored to simulate the River Yamuna and also helped in the creation of habitats by waders for providing roosting, nesting and foraging sites, whereas the deep wetland was restored to attract dabbling-diving resident and migratory birds. These wetlands were restored considering the physical and biological attributes such as depth and slope gradients, presence of vegetation with patchiness and openness, and physiochemical parameters for creating foraging guilds—presence of invertebrates and vertebrates in submerged and emergent aquatic vegetation such as *Paspalum, Polygonum, Potamogeton* sps., *Hydrilla*, *Vallisneria* and *Ipomoea aquatica*, and surrounding *Tamarix* sps., *Typha* and *Phragmites* sps. Fingerlings of many native species of fishes, such as Catla (*Catla catla*), Rohu (*Labeo rohita*), Mrigal (*Cirrhinus cirrhosis*), *Puntius* spp., *Channa* spp., Asian stinging catfish (*Heteropneustes fossilis*), *Clarias* spp., *Chitala chitala*, *Wallago attu* and *Sperata seenghala*, were introduced periodically along with the macrophytes, which initiated ecosystem processing, carving a variety of niches. An existing heavily silted paleochannel was identified in the active floodplain, i.e. Phase II, and restored through desiltation with varying depth (0.25 m–5 to 7 m) and slope gradient. Embankment was made from silt in a way to contain a substantial amount of floodwater. Grass species such as *Paspalum*, *Chrysopogon zizanioides*, *Saccharum spontaneum and Saccharum bengalense* were planted above the riparian zone of the desilted wetland, followed by the establishment of floodplain forest communities in the upland along the embankment.

9.1.5 Establishment of Biological Communities

Now, the challenge was to design and decide on biological communities belonging to the Yamuna River basin. A rapid survey was carried out in the Yamuna River basin, including some of the major tributaries such as the Kuno River and Chambal to ascertain the diversity, form and structure, succession and trophic structure of existing biological communities, grassland and wetlands with the objectives to simulate them at the site through assemblage of species. Based on the survey, different plant species were assembled in the form of biological communities, looking at the life history traits such as seed size (large seeds influenced by survival rather than growth), growth rates, patterns of senescence, dormancy, dispersal, shade dependence and tolerance to moisture stress.

Initially, about 25–30 species of grass were identified and introduced to begin the successional process, comprising *Leptochloa fusca*, *Sporobolus*, *Cenchrus ciliaris and Cenchrus setigerus*, *Bothriochloa*, *Saccharum spontaneum*, *Saccharum bengalense*, *Chrysopogon zizanioides* and *Imperata cylindrica*. Seeds of various plant species belonging to the Yamuna River basin were collected from the forest floor and germinated.

Based on the field survey, it was planned to establish various forest communities representing the Yamuna River basin, comprising tropical dry deciduous forest, riparian forest communities and sprawling floodplain grassland. Hence, the assemblage of different plant species representing top canopy, middle storey, understorey and ground vegetation was carried out during monsoon season in a way also to establish an interface or ecotone for facilitating energy and material exchange between terrestrial and aquatic ecosystems, which ultimately maintained dynamic environments critical for riverine ecosystems and associated various organisms (Chen et al. 2019; Stutter et al. 2012). Such an arrangement of establishing communities not only provided complexity of structure but also their functions that offered microhabitats for many faunal species, facilitating the reclaiming of their former historical geographical ranges and their successful survival.

The established forest communities largely consist of *Mitragyna parvifolia*, *Terminalia arjuna*, *Dalbergia sisoo*, *Syzygium cumini*, *Feronia limonia*, *Terminalia bellirica*, *Terminalia chebula*, *Terminalia tomentosa*, *Albizia* spp., *Bombax ceiba*, *Haldina cordifolia*, *Bridelia retusa*, *Holoptelia integrifolia*, *Madhuca latifolia*, a few *Prosopis cineraria*, *Lannea coromandelica*, *Soymida febrifuga*, etc., assembled as the top canopy. *Senegalia catechu*, *Aegle marmelos*, *Ziziphus mauritiana*, *Trewia nudiflora*, *Phyllanthus emblica*, *Cordia myxa and Cordia gharaf*, *Mimusops hexandra*, *Cassia fistula*, *Syzygium heyneanum*, *Kydia calycina*, *Bauhinia* spp., *Butea monosperma*, *Diospyros montana*, *a few Vachellia nilotica*, *etc*., assembled as the middle storey. *Vitex negundo*, *Grewia* spp., *Carissa carandas*, *Woodfordia fruticosa*, *Dendrocalamus strictus*, *a few Capparis sepiaria*, *Salvadora oleoides and Salvadora persica*, etc., assembled as understorey. *Justicia adhatoda*, *Abutilon theophrasti*, along with many other annual and perennial grass species, represented ground vegetation. Keystone species such as *Ficus religiosa*, *Ficus benghalensis* and *Ficus racemosa* were also considered as a part of forest communities for facilitating diversity and interactions. They were assembled in the form of various forest communities.

9.1.6 Establishment of Floodplain Grasslands

It is a well-known fact that rivers carry seeds of many woody shrubs and herbaceous weedy species during the flood, and many of them get settled in the floodplain grasslands. Due to low-intensity floods because of barrages and damming of rivers and also followed by extreme weather patterns, which affect the intensity of rainfall. Now these seeds are not able to move further and get established as woodland, replacing grasslands. Therefore, it is desirable to identify such weed invasions, followed by eradication and establishment of characteristic grassland.

Sites were identified, weeds were eradicated, and characteristic grassland comprising short grasses such as *Sporobolus*, *Imperata cylindrica*, *Dichanthium annulatum* and *Cenchrus* spp. (on upland); medium-size grasses such as *Leptochloa fusca* and *Bothriochloa;* and tall grasses such as *Chrysopogon zizanioides*, *Saccharum spontaneum* and *Saccharum bengalense* were planted to re-establish floodplain grasslands.

The restored wetlands, grasslands and floodplain forest communities increased the complexity, diversity and abundance of floral components, which ultimately led to enhanced faunal diversity.

9.1.7 Present Status

The restored wetlands store substantial amounts of water throughout the year and become perennial wetlands, having different types of sediments and substrates colonised by characteristic submerged and floating macrophytes such as *Nelumbo* spp., *Hydrilla*, *Potamogeton* spp., *Vallisneria* and *Chara*, besides zooplankton and phytoplankton and many characteristic micro- and macroinvertebrates. Noteworthy species are the rare and small snails like the freshwater limpet (*Ferrissia baconi*), the rams' horn (*Segmentina calatha*), the large *Lymnaea* (Radix) *ovalis* and also the recorded Apple snail (*Pila globosa*) in an active floodplain. All these are associated with the undisturbed submerged or floating macrophyte. The habitat processing with the specific floral interventions enabled many insect and bird species to find their suitable niches. Aquatic macrophytes are primary producers and therefore act as a base for herbivorous and detritivorous food chains, becoming food for invertebrates, fish and birds, and organic carbon for bacteria, besides providing substrate for many invertebrates and different stages of fish, amphibians and reptiles (Timms and Moss 1984; Dvořák 1996).

Restored wetlands and marshes became homes for many resident and migratory birds, which have been identified using a field guide (Kazmierczak 2000). The narrow, shallow wetland with its restored *Tamarix-Phragmites* forest community became the lonely heronry in Delhi, which facilitated the nesting of Black-crowned night herons, Darter (Niangthoi and Khudsar 2015), Little egrets (*Egretta garzetta*), Little cormorant (*Microcarbo niger*), etc. Deep wetlands became homes for many migratory and residents birds, such as Gadwall (*Mareca strepera*), Northern Shoveller (*Spatula clypeata*), Northern pintail (*Anas acuta*), Eurasian wegion (*Mareca penelope*), Garganey (*Anas querquedula*), Common teal (*Anas crecca*), Bar-headed goose (*Anser indicus*), Greylag goose (*Anser anser*), Red-crested pochard (*Rhodonessa rufina*), Common pochard (*Aythya ferina*), Ferruginous pochard (*Aythyna nyroca*), Tufted duck (*Aythyna fuligula*), Spot-billed duck (*Anas poecilobyncha*), Lesser whistling duck (*Dendrocygna javanica*) and Ruddy Shelduck (*Tadorna ferruginea*). Predators like the Marsh harrier (*Circus aeruginosus*) also made an appearance in the wetlands of the Yamuna Biodiversity Park.

The restored forest communities represented by about 1000 plant species now established as three storeyed communities, which provided multiple niches and supported nesting and foraging guilds, attracted many resident and migratory birds such as three species of Parakeets (*Psittacula krameri, P. eupatria P. cynocephala*), three species of Bulbul (*Pycnonotus cafer, P. jacosus, P. leucotis*), Indian peafowl (*Pavo cristatus*), two species of babbler (*Argya striata, A. malcolmi*), Yellow footed green pegion (*Treron phoenicoptera*), Grey hornbill (*Ocyceros birostris*), Golden oriole (*Oriolus kundoo*), three species of Owls (*Tyto alba, Ottus lettia, Athene brama*), Orange headed thrush (*Zoothra citrina*), Red-breasted flycatcher (*Ficedula parva*), Taiga flycatcher (*Ficedula albicilla*), Grey-headed canary flycatcher (*Culicicapa ceylonensis*) and Paradise flycatcher (*Terpsiphora paradise*). The forest communities also attracted birds of prey such as the Shikra (*Accipiter badius*), White-eyed buzzard (*Butastur teesa*) and Crested serpent eagle (*Spilornis cheela*). Predation is known to be important for establishing ecological balance, and therefore, the arrival of owls and other birds of prey, which are typical woodland birds, helped in scavenging and disposing of carrion, controlling rodent populations and offering various natural pest control. They are known as top predators, effectively maintaining prey–predator dynamics and ultimately supporting the functionality of ecosystems.

The established grasslands of Yamuna Biodiversity Park are unique ecosystems, not only providing a wide range of ecological goods and services but also forming a complex ecosystem with the associated animals. The beginning of grass herbivory was witnessed from the grasshopper and reaches to small Cervidae such as Hog deer (*Axis porcinus*) and large Bovidae such as Nilgai (*Boselaphus tragocamelus*) and often other ungulates like Wild pigs (*Sus scrofa cristatus*), including other small mammals like Indian Hare (*Lepus nigricollis*), short-tailed bandicoot rat (*Nosokia indica*) and Wroughton's free-tailed bat (*Otomops wroughtoni*) (single record in 2024). These grasslands led to the conservation of soil and water resources besides supporting many birds such as *Prinia* species, Red Avadavat (*Amandava amandava*), Scaly-breasted munia (*Lonchura punctulata*), Indian silverbill (*Euodice malabarica*), Tricoloured munia (*Lonchura malacca*), Bay-backed shrike (*Lanius vittatus*), Long-tailed shrike (*Lanius schach*), Indian roller (*Coracias benghalensis*) and European roller (*Coracias garrulus*) and also birds of prey like Black-shouldered kite (*Elanus caeruleus*).

Restored biological communities through the assemblage of plant species belonging to the Yamuna River basin, along with the floodplain wetlands and characteristic grasslands, facilitated many animals to reclaim their former historical geographical ranges. With the establishment of floodplain forest communities and characteristic grasslands, a barren, degraded, nutrient-deficient riverscape of Yamuna Biodiversity Park provided a complex and diverse habitat that facilitated faunal diversity. In fact, habitat diversity comprises spatial heterogeneity within and between habitat patches of large landscapes and is often regarded as a driver for species diversity (Lengyel et al. 2016). Many butterflies, dragonflies and damselflies are now residents of Yamuna Biodiversity Park.

The elusive Hog deer (*Axis porcinus*) (Singh et al. 2023), Leopard (*Panthera pardus*) (Singh et al. 2021), Seibold snake (*Ferania sieboldii*) (Srivastava et al. 2020) and other fauna, such as the breeding population of Darter (*Anhinga melanogaster*) (Niangthianhoi and Khudsar 2015) and Streaked weaver (*Ploceus manyar*) (Diwakar Prakash et al. 2020), reclaimed their home after a long gap. Many other mammals, such as Wild pigs, two species of Mongoose (*Herpestes auropunctatus*, *H. edwardsii*), Indian crested porcupine (*Hystrix indica*), Small Indian civet (*Viverricula indica*) and Golden Jackal (*Canis aureus*), are commonly seen. Out of which there are many species which are listed in the IUCN Red Data, including Hog deer as endangered, Sand boa (*Eryx johnii*) as near threatened, Leopard as vulnerable and Darter as near threatened species.

The biologically rich forest, grassland and marshy communities consisting of a variety of plants, animals and microbes in interaction with the chemical and physical features of their environment make a complete ecosystem. The restored biological communities of Yamuna Biodiversity Park became home for >1000 plant species along with >200 avifauna, about 85 species of butterflies, 20 herpetofauna and >20 species of mammals—as small as rats and moles and as big as Nilgai (Table 9.1). Important butterfly species such as Danaid eggfly (*Hypolimnas missipus*), Great

Table 9.1 Changes in species diversity in a span of 22 years at Yamuna Biodiversity Park

	Year					
Species group	2002	2007	2012	2017	2022	2024
Terrestrial plants	90	656	856	871	915	915
Aquatic plants	0	76	86	98	101	101
Avifauna	37	168	192	200	204	210
Butterflies	11	25	71	75	82	86
Herpetofauna	3	16	18	18	18	20
Mammals	4	17	18	18	23	23
Fishes	0	12	18	18	18	18

eggfly (*Hypolimnas bolina*), Common gull (*Cepora nerissa*), Gram blue (*Euchrysops cnejus*) and Indian tortoiseshell (*Aglais caschmirencis*) (Biswas et al. 2017) found a home in the park and are protected under the Wildlife Protection Act 1972. Plants and animals living in these ecosystems are interdependent on one another for survival and make a complex trophic structure for maintaining the stability of ecosystem processes. A minimum number of species is essential for ecosystem functioning under constant conditions; however, a large number of species are essential for maintaining the stability of ecosystem processes in a changing environment (Loreau et al. 2001).

Restored wetlands, grasslands and forests communities of Yamuna Biodiversity Park have diverse ecological niches that harbour a wide range of plant and animal species, forming a complex food web.

During the dry season in Yamuna Biodiversity Park, a thick litter layer of leaves on the forest floor attracts a variety of insects and reptiles, which ultimately become an important component of the food chain. Reptiles and insects play an important role in the ecosystem, both as predator and prey species; they also serve as biomonitors in controlling many pests and as excellent ecological indicators due to their high degree of sensitivity to even a minor change in the environment.

Nature Education plays a very important role in the long-term sustenance of any conservation initiative. Therefore, nature education has been one of the major objectives of the biodiversity parks in Delhi, particularly the Yamuna Biodiversity Park. The park acts as a living laboratory for students of all age groups. The park became a knowledge centre for extending training to a variety of national and international professionals in the field of floodplain restoration, besides imparting conservation education to society. The park became a place to witness the birth of a lost ecosystem.

The major societal function rendered by the Biodiversity Park is to enrich the human microbiome of thousands of citizens who use the nature trails of Biodiversity Parks for morning and evening walks through different forest communities.

Yamuna Biodiversity Park was initiated on a barren/usar land with a high pH, where the application of some of the principles of ecological restoration resulted in the establishment of a mosaic of habitats with rich floral and faunal diversity, which functions as a dynamic complex ecosystem. The ecosystem processing has already been initiated, which operates in a way that the organic matter deposition and decomposition, plant regeneration, and pollination have been taking place along with the breeding and nesting of birds, insects, reptiles and some of the mammals. Now the Yamuna Biodiversity Park is considered a wildlife corridor and offers a variety of ecological services such as storage of floodwater, groundwater recharge, mitigating air pollution and climate change, regulating temperature, preserving vanishing flora and fauna and promoting environmental education, inculcating cultural values. Pictorial representation of the restoration trajectory at the Yamuna Biodiversity Park is shown in Figs. 9.2, 9.3, 9.4, 9.5, 9.6, 9.7, 9.8 and 9.9.

Yamuna Biodiversity Park is ready to be replicated elsewhere for floodplain restoration and the rejuvenation of river-extending conservation and education and recreational values, also showcasing probably the first successful model collaborating a government development agency, i.e. Delhi Development Authority, and an academic institution, i.e. University of Delhi.

9.2 Pictorial Representation of Restoration Trajectory at the Yamuna Biodiversity Park

Fig. 9.2 Restoration of narrow shallow wetland in the inactive floodplain

Fig. 9.3 Restoration of deep wetland in the inactive floodplain

Fig. 9.4 Restoration of wetland in the active floodplain

Fig. 9.5 Restored wetlands, grassland and forest communities in inactive floodplain

Fig. 9.6 Restored wetlands, grasslands and forest communities in the active floodplain

Fig. 9.7 (**a**) Initial intervention near Nature Interpretation Centre in inactive floodplain. (**b**) Restored forest communities at the same site

Fig. 9.8 Forest community: Before and after restoration

Fig. 9.9 Hog deer and Leopard arrived in the restored floodplain grassland and forest communities

Acknowledgements Our sincere thanks to Professor C. R. Babu of Delhi University for introducing the concept of Biodiversity Parks to the world as one of the new conservation strategies and for the constant sharing of knowledge. We acknowledge the Delhi Development Authority for providing land and funding support and Delhi University for extending institutional support.

References

Baron JS, Poff NL, Angermeier PL, Dahm CN, Gleick PH, Hairston NG Jr, Jackson RB, Johnston CA, Richter BG, Steinman AD (2002) Balancing human and ecological needs for freshwater: the case for equity. Ecol Appl 12:1247–1260

Biswas J, Dookia S, Faisal M (2017) Butterflies of Delhi with new additions and an annotated checklist from Delhi India. Int J Zool Stud 2(6):04–10

Chen FX, Lu SY, Hu XZ, He Q, Feng CP, Xu QJ, Chen N, Ngo HH, Guo HM (2019) Multi-dimensional habitat vegetation restoration mode for lake riparian zone, Taihu, China. Ecol Eng 134:56–64

Diwakar Prakash, Niangthianhoi SL, Ram Kumar, Khudsar FA (2020) Nest construction process in Streaked Weaver *Ploceus manyar*. Indian Bird. 15(5)

Dunham JB, Angermeier PL, Crausbay SD, Cravens AE, Gosnell H, McEvoy J, Moritz MA, Raheem N, Sanford T (2018) Rivers are social–ecological systems: time to integrate human dimensions into riverscape ecology and management. Wiley Interdiscip Rev Water 5:e1291

Dvořák J (1996) An example of relationships between macrophytes, macroinvertebrates and their food resources in a shallow eutrophic lake. Hydrobiologia 339:27–36

Girolamo De A, Lo Porto A, Pappagallo G, Tzoraki O, Gallart F (2015) The hydrological status concept: application at a temporary river (Candelaro, Italy). River Res Appl. 31:892–903

Gopal B, Sah M (1993) Conservation and management of rivers in India: case study of the River Yamuna. Environ Conserv 20:243–254

Hand BK, Flint CG, Frissell CA, Muhlfeld CC, Devlin SP, Kennedy BP, Crabtree RL, McKee WA, Luikart G, Stanford JA (2018) A social–ecological perspective for riverscape management in the Columbia River Basin. Front Ecol Environ 16:S23–S33

Kanade R, John R (2018) Topographical influence on recent deforestation and degradation in the Sikkim Himalaya in India; Implications for conservation of East Himalayan broadleaf forest. Appl. Geogr. 92:85–93

Kazmierczak K (2000) A field guide to the birds of India, Sri Lanka, Pakistan, Nepal, Bhutan, Bangladesh and the Maldives, 1st edn. Om Book Service, New Delhi

Lengyel S, Déri E, Magura T (2016) Species richness responses to structural or compositional habitat diversity between and within grassland patches: a multi-taxon approach. PLoS One 11(2):e0149662. https://doi.org/10.1371/journal.pone.0149662

Loreau M, Naeem S, Inchausti P, Bengtsson J, Grime JP, Hector A, Hooper DU, Huston MA, Raffaelli D, Schmid B, Tilman D, Wardle DA (2001) Biodiversity and ecosystem functioning: current knowledge and future challenges. Science. 294(5543):804–808

Manral U, Khudsar FA (2013) Assessment of wetland water quality and avian diversity of a human-modified floodplain wetland of River Yamuna. Notulae Scientia Biologicae. 5(1)

NGT (2019) Before the National Green Tribunal (Principal Bench, New Delhi). Original Application No.673/2018 (M.A. No. 1777/2018) date April 8 2019.

Niangthianhoi SL, Khudsar FA (2015) The nesting success of Oriental Darter *Anhinga melanogaster* in Yamuna Biodiversity Park, New Delhi India. J Threat Taxa. 7(14)

Singh M, Sameer Gautam DP, Srivastava PJ, Sirohi U, Singh H, Khudsar AFA (2021) First leopard picture in the Yamuna Biodiversity Park, India. Cats News. No.72

Singh M, Gautam S, Srivastava DP, Jain P, Sirohi U, Singh H, Arjun KE, Khudsar FA (2023) Hog deer *Axis porcinus* reclaiming historical geographical distribution range due to ecological restoration in the Yamuna Biodiversity Park, Delhi, India. Species 24:84s1596. https://doi.org/10.54905/disssi.v24i74.e84s1596

Srivastava DP, Gautam S, Singh M, Sirohi U, Mukherji D, Khudsar FA (2020) First resighting since the1940s of Siebold's water snake *Ferania sieboldii* (Schlegel 1837), in Delhi, India. Reptil Amphib 27(3):525–527
Stutter MI, Chardon WJ, Kronvang BK (2012) Buffer strips as a multifunctional management tool in agricultural landscapes: introduction. J Environ Qual 41:297–303
Timms RM, Moss B (1984) Prevention of growth of potentially dense phytoplankton populations by zooplankton grazing, in the presence of zooplanktivorous fish, in a shallow wetland ecosystem. Limnol Oceanogr 29:472–486

10 Ecological Restoration in Delhi: Challenges and Their Mitigation

Rita Rath, Sadhna Gupta, Neeraja Sood, Versha Goel, and Roopa Rani Samal

Abstract

The Delhi Ridge and the floodplains, the two vital ecosystems in the National Capital Territory of India, are facing significant degradation due to urbanization, pollution, encroachment by the migrant population, and invasive plant species. This chapter explores the challenges encountered in the ecological restoration of the Ridge and the floodplain, including habitat loss, fragmentation, and soil degradation. Urban encroachment has severely impacted native flora and fauna, leading to biodiversity loss and disruption of ecosystem services. Furthermore, invasive species have outcompeted indigenous plants, altering the ecological balance.

Mitigation strategies proposed include the implementation of community-based restoration programmes that engage local populations in conservation efforts. This involves reforestation with native species, removal of invasive plants, and restoration of natural water bodies to enhance biodiversity. Additionally, establishing protective legislation and promoting sustainable land use practices are crucial for long-term success. The chapter emphasizes the importance of multidisciplinary and holistic approaches, integrating ecological, social, and economic perspectives to address the complexities of urban ecological restoration. By encouraging a collaborative framework among stakeholders, the restoration of the Delhi Ridge and the floodplains can serve as a model for similar urban ecosystems facing ecological challenges. Ultimately, this effort aims to restore the ecological integrity of the Ridge and the floodplain while enhancing its resilience against future environmental pressures.

R. Rath · S. Gupta · N. Sood · R. R. Samal (✉)
Department of Zoology, Dyal Singh College, University of Delhi, New Delhi, India
e-mail: roopasamal.zoology@dsc.du.ac.in

V. Goel
Department of Zoology, Maitreyi College, University of Delhi, New Delhi, India

D. Mittal, D. K. Gupta (eds.), *Ecological Restoration in India: Perspective and Practices*, https://doi.org/10.1007/978-981-96-8943-9_10

Keywords

Delhi Ridge · Floodplain · Restoration · Challenges · Mitigation · Biodiversity

10.1 Introduction

Natural resources have been degraded world over, leading to a loss of biodiversity, making it a global concern (Edrisi et al. 2023; Singh et al. 2024). *Direct drivers* of land degradation are the actual activities that humans engage in, such as the clearing of land, excessive grazing, forest burning, wildlife extraction, and injudicious use of fertilizers. *Indirect drivers* of land degradation are the underlying factors that influence people's actions and include factors such as access to land, ownership rights, economic inequality, consumer habits, power dynamics, global perspectives, regulatory policies, and information accessibility (Raghurama et al. 2023). The inclusion of urban areas to strengthen biodiversity conservation efforts becomes especially more relevant due to ongoing urbanization and other forms of land uses, which are impacting the natural and semi-natural habitat (Zari 2018; Kowarik et al. 2019). The connection between the biodiversity of the urban ecosystem and its services has not been thoroughly explored and still remains under-researched (Schwarz et al. 2017). Despite this information deficit, several initiatives, such as the "*Decade on Ecosystem Restoration 2021–2030*" launched by the UN, emphasize the urgent need to improve urban ecosystems for biodiversity conservation and their services. A reconsideration of urban greenspace design and management is essential for ensuring adaptation of such ecosystems to climate change. Urban ecosystems differ significantly from natural ecosystems in both biotic and abiotic characteristics and are heavily influenced by societal norms (McCleery et al. 2014). Due to irreversible abiotic changes, such as technogenic soils and pollution, along with biotic factors like the high prevalence of non-native plants, many of these ecosystems are now classified as novel ecosystems (Kowarik 2011). Restoring novel urban ecosystems, which lack historical precedents, demands tailored strategies and cannot fully rely on traditional ecological restoration methods designed for non-urban areas (Klaus and Kiehl 2021).

In India, destruction of the environment and its biodiversity has been a major problem, especially in metropolitan cities like Delhi, due to rapid urbanization (Bawa et al. 2020). The fundamental ecological challenges for India today are the approximately ninety-six million hectares of degraded land, unparalleled loss of biodiversity, and extreme environmental phenomena (Singh et al. 2021). Delhi, being the capital of India, has faced a lot of anthropogenic pressures in the form of industrialization, immigration, modernization, unplanned settlements, construction work, quarrying/mining, overuse of natural resources, etc. (Saxena et al. 2021). Ecosystem degradation is widely recognized as a key driver of recent extreme climate events, including heat waves, floods, droughts, and landslides. While cities cover only about 3% of the Earth's surface (Liu et al. 2014), the ecosystems within urban areas, along with their services and biodiversity, play a crucial role in supporting sustainable development.

Delhi, a historic city in northern India, spans 58.3 km in length and 48 km in width. Positioned along the western bank of the Yamuna River, it lies between 28.12° and 28.53° North latitude and 76.50° and 77.23° East longitude. Surrounded by the Thar Desert to the southeast, the Indo-Gangetic plains to the northeast, and the Aravalli Range (extending 800 km, with elevations up to 1700 metres) to the south, Delhi boasts of diverse topographical features. With an estimated population of over 32 million in 2023, its high population density adds to the complexity and vibrancy of its urban ecosystem.

The hilly spurs in Delhi, known as "The Ridge," once covered nearly 15% of the city's land area. The Ridge is a long, narrow, elevated landform that forms a continuous crest or chain of hills or mountains. The Delhi Ridge is the northernmost extension of India's ancient Aravalli Mountain Range, which stretches from Gujarat through Rajasthan and into Haryana, eventually reaching Delhi. Extending over 35 km, the Delhi Ridge runs from Bhatti Mines in the southeast near Tughlaqabad, branching in various directions before tapering off near Wazirabad along the Yamuna River.

The Delhi Ridge is made of quartzite rocks. The Ridge has been an attraction for settlers for thousands of years. The two natural features of the city, The Ridge and River Yamuna, have been the preferred and favourite places for the rulers of Delhi. Hence, the struggle for protecting Delhi's "green lungs" started long back. Over the years, the over-exploitation of resources, establishment of recreational parks, public housing, disposal of construction debris, etc. have created a substantial hazard and caused shrinkage of the Ridge. Once contiguous, it is now divided into various fragmented zones, namely, Northern Ridge, Central Ridge, South-Central Ridge, and Southern Ridge. Northern Delhi Ridge covers 87 Hectares (ha) near the University of Delhi. The Central Ridge comprises 864 Hectares of forest area, from Sadar Bazaar to Dhaula Kuan. The South-Central Ridge occupies 633 Hectares of forest around the Sanjay Van area, close to the Jawaharlal Nehru University (JNU). The Southern Ridge, covering 6200 hectares, is the largest section of the Delhi Ridge. The Delhi Ridge spans a total of 7784 hectares in fragmented patches across the city (Sinha 2014). Please refer to Table 10.1.

Table 10.1 General data about the four zones of Delhi Ridge

S. no.	Ridge	Managing agencies	Area (ha)	Proportion (%)
1	Northern	DDA, MCD, and Forest Department	87	1.12
2	Central	Forest Department, DDA, Army, CPWD, NDMC, and MCD	864	11.1
3	South central	DDA	633	8.13
4	Southern	DDA, Sports Authority of India, Department of Forest, Revenue Department	6200	79.65

DDA Delhi Development Authority, *MCD* Municipal Corporation of Delhi, *CPWD* Central Public Works Department, and *NDMC* New Delhi Municipal Corporation

Delhi's floodplains are the low-lying areas near the River Yamuna flowing through the city. They are an important part of Delhi's natural heritage. The significance of floodplains in preserving ecological balance, environmental health, and the resilience of urban areas is crucial.

10.2 Ecological Significance of Delhi Ridge

Delhi Ridge serves several ecological functions, protecting Delhi from westerly winds loaded with sand from the deserts of Rajasthan, lowering the ambient temperature, Sequestering pollution, sheltering flora and fauna, purifying and conserving groundwater, enhancing the environment, providing recreational spots and a calming green look, etc. The ecological significance of the Ridge increases manifold as it harbours many species of vegetation, which help in the amelioration of pollution. The dense branching and leaf arrangement of certain trees enhance their ability to collect dust. Notable examples include *Albizia lebbeck* (siris), *Azadirachta indica* (neem), *Alstonia scholaris* (chattaun), *Acacia nilotica* (desi kikar), *Butea monosperma* (dhak), *Cassia fistula* (amaltas), *Dalbergia sissoo* (shisham), *Ficus religiosa* (pipal), *Ficus glomerata*, *Ficus infectoria* (pilkhan), and *Ziziphus jujube* (ber). These species are also helpful in noise pollution mitigation through masking, absorption, reflection, refraction, deflection, etc. The Ridge of Delhi, in addition to having many ecological functions, is also rich in medicinal plants. Out of 31 plant species that are in high demand both in domestic and international markets, 11 species are found in Delhi (Sinha 2014).

10.3 Need for Restoring Ecosystems

- *Protect and Enhance Urban Habitats*: To enhance the environment and reduce the negative effects of pollution by preserving existing vegetation and through plantations of suitable native species as per the site conditions. To increase the green cover of the area.
- *Preserve and Strengthen Existing Habitats and Ecosystems*: To gradually suppress monoculture plantations of exotic species by raising mixed plantations of native species and efficiently managing the invasive species (Sinha et al. 2022).
- *Create New Habitats*: To enhance the aesthetic beauty of the forest, provide shade along the length of strips passing through the city, and create recreational centres.
- *Preserve and Strengthen Habitat Connectivity* (Liu et al. 2022).
- *Protect Species*: To create favourable conditions for the protection and development of wildlife—flora and fauna.
- *Rehabilitate and Restore Damaged or Destroyed Ecosystems*: Restoring ecosystems helps achieve the Sustainable Development Goals (SDGs) through the protection and improvement of the livelihoods of people depending on them, reduction of the risk of natural disasters, conservation of biodiversity, and regulation of diseases.

10.4 Efforts for Restoration

Realizing the need to preserve, conserve, and restore the degrading urban ecosystems and with the birth of the forest wing in 1950, the administration in India took up the work of managing the ridges through norms of forestry. In 1980, the Lieutenant Governor of Delhi designated 25 sites within the Northern, Central, and South-Central Ridges as Protected Forests under the Indian Forest Act of 1927. However, by 1982, the Town and Country Planning Organization (TCPO) reported that 40% of the Ridge had been destroyed. This was largely due to encroachments and the use of the Ridge land for non-forest purposes, degrading the Ridge from a flourishing ecosystem to one with only a few kikar trees. As a consequence, in 1986, 1880 ha. in the Southern Ridge was declared as a wildlife sanctuary, now known as Asola Bhatti Wildlife Sanctuary (ABWS), and given legal protection. Unfortunately, the other Ridge areas did not receive protection under forest laws, and by the 1990s, less than 5% of the city had green cover, and the Yamuna became a sewage canal for industrial effluents. So in the early 1990s, various voluntary organizations, citizens' groups, non-governmental organizations (NGOs) (like WWF, Kalpavriksh, Srishti, Development Alternatives, and Conservation Society), and activists working on the issue joined together under the banner of the "Joint NGO Forum to Save the Delhi Ridge." The Forum helped mobilize school and neighbourhood organizations and publicized the cause in various media outlets.

Forest landscapes and green covers in Delhi are known by various names such as *parks* (Nehru Park and Indraprastha Park), *gardens* (Budha Jayanti Garden and Lodi Gardens), *nurseries* (Sunder Nursery), *city forests* (Sanjay Van and Mangar Bani), *biodiversity parks* (Yamuna Biodiversity Park), and *sanctuary* (Asola Bhatti Wildlife Sanctuary). The central and the state governments, through various agencies, manage these green spaces. Plantation drives are carried out every year in collaboration with 20 other organizations, under the Delhi Green Action Plan, with the Forest Department as the nodal agency. By controlling the temperature and air quality as well as other ecosystem services through implementations and interpretations, the Delhi High Court and the Supreme Court of India play a critical role in the management of Delhi's green spaces.

India pledged to restore 26 million hectares (Mha) of its degraded lands for restoration by 2030 in the conference of the United Nations Convention to Combat Desertification. India has launched various afforestation programmes to enhance forest tree cover in urban and rural areas. The Government of India initiated a National Mission on Biodiversity and Human Well-being to do research, education, and capacity building for biodiversity, ecosystem services, bioeconomy, climate change, health, and agriculture. Information on the status of the restored sites, scientific understanding of ecological restoration, standards of local practice, and knowledge developed for the restoration of differently degraded ecosystems, in a diverse country like India, is of great significance to meet the goals and commitments of restoration. The collection of scientific studies and their implications for ecological restoration in India is well-documented in the special issue of the *Ecological Engineering* journal (2022), which published 22 original research papers on ecological restoration.

10.5 Government Policies for Restoration

On the World Environment Day, June 5, 2021, the UN declared 2021–2030 as a *Decade on Ecosystem Restoration* for the conservation of biodiversity. The Department of Environment of the Delhi Government, in 2000, started an 8-year plan for the rehabilitation of 2100 acres (850 ha) of Bhatti Mines area at the cost of Rs.1316.19 lacs. Due to the Supreme Court Ruling (2011), the government defined eco-sensitive zones (ESZ) to limit human activity around the forest, consequently reducing the adverse effects on the ecosystem, biodiversity, and natural resources. India joined the Bonn Challenge in 2015, a global challenge to restore landscapes, launched in 2011 by the Government of Germany and the International Union for Conservation of Nature (IUCN), to work towards the SDGs. Out of India's total area of 297,319,000 ha, an area of 26,000,000 ha (8.74%) was to be maintained as protected area, of which the area under restoration is 9,810,940 ha (3.3%) (as per the Ministry of Environment, Forest and Climate Change (MoEFCC), Government of India).

We will now discuss a few case studies on the efforts for restoration of the Delhi Ridge and the floodplains by different stakeholders in different zones of Delhi and understand their achievements and challenges in order to assess and consider the way forwards for mitigation.

10.6 Restoration Case Study No. 1: Northern Delhi Ridge

Location: (latitude 28.41°–28.40°N and longitude 77.12°–77.13°E)

Northern Delhi Ridge (NDR), also called *Kamla Nehru Ridge*, is the smallest and rapidly shrinking section of Delhi Ridge. It denotes the hilly area between Civil Lines on its eastern side, the University Campus on its western and south-western side, and Timarpur residential area on its northern side of North Delhi. Northern Delhi Ridge is highly weathered with undulating topography, and consequently, the Ridge has shallow valleys, gentle slopes, and flat hilltops. NDR is fragmented into patches by various public roads that pass through it. Delhi Ridge was declared a "reserve forest" under the Indian Forest Act, 1927.

10.6.1 Present Status of the Northern Delhi Ridge

10.6.1.1 Flora

The vegetation of NDR is of a soft, thorny scrub type. *Prosopis juliflora* (Vilayati kikar), which forms the dominant tree species and is known to deplete groundwater, was introduced by the British colonial government. It eradicated many native plant species from NDR. For restoration, this tree has been replaced by indigenous species, like *Prosopis cineraria, Acacia leucophloea, Acacia nilotica,* and *Salvadora oleoides* (Kumar et al. 2022).

10.6.1.2 Fauna

The fauna includes the pied kingfisher (*Ceryle rudis*), softshell turtle (*Nilssonia gangetica*), and little cormorant (*Microcarbo niger*), besides other common animals of Delhi Ridge, which are *birds* like the peafowl, shikra, Indian grey hornbill, black kite, common myna, rose-ringed parakeet, and drongo; *mammals* like the Indian palm squirrel, Indian hare, and small Indian civet; *reptiles* like the monitor lizard, Indian cobra, and garden lizard; *amphibians* like the common Indian toad and Indian bullfrog; and *insects and butterflies* such as the common tiger butterfly, blue mormon, dragonfly, and damselfly.

10.6.2 Threats and Challenges to the Northern Delhi Ridge

- No clear demarcation of NDR by proper fencing.
- Another challenge is habitat fragmentation and the resultant loss of native flora and fauna. Areas in the vicinity of NDR have retained the Ridge's forest characteristics and, hence, could have helped in enriching the biodiversity, had the Ridge not been fragmented due to urbanization (Singh et al. 2016).
- Encroachments and heavy traffic pose a threat to NDR.
- Littering and waste disposed by neighbouring hospitals add to the woes of the NDR ecosystem.
- Lack of public awareness.

10.6.3 Initiatives for Restoration at the Northern Delhi Ridge

- Opening of the canopies and cut rootstock methods have been used to manage *P. juliflora* and *Lantana.* By creating canopy gaps, native plants get enough air and sunlight for their growth and stimulate ecological succession.
- On the hilltops and slopes, which have some remnants of *Acacia senegal*, other native species like *Acacia catechu, Acacia modesta, Prosopis cineraria, Holoptelea integrifolia,* and *Lannea coromandelica* have been planted (Delhi Biodiversity Foundation, DDA- Kamala Nehru Biodiversity Park).
- Rejuvenation of existing water bodies has been achieved by channelizing all the drainage into the water bodies. These recharged water bodies provide habitats to many insects and birds and some reptiles like turtles and promote recreational holes for many primates (DDA 2021) (Fig. 10.1).

The Kamala Nehru Ridge plays an essential role in preserving urban biodiversity and acts as a wildlife corridor within Delhi. Its semi-forested environment attracts a variety of species, making it an important area for conservation in the city.

Fig. 10.1 Kamala Nehru Ridge. (Source: https://newdelhitoday.in/biodiversity-parks-in-delhi/)

10.7 Restoration Case Study No. 2: Neela Hauz

Location: (latitude 28.52°N and longitude 77.17°E)

Neela Hauz, a mini biodiversity park located near Vasant Kunj in South Delhi, spans 10 acres and includes a lake nestled in the undulating terrain near Sanjay Van Forest. In 2015, the Delhi High Court ordered its restoration. The DDA, in partnership with the Centre for Environmental Management of Degraded Ecosystems (CEMDE) at the University of Delhi, restored the lake using a Constructed Wetland System (CWS) (Sonowal et al. 2019).

10.7.1 Present-Day Status of Neela Hauz

Neela Hauz plays a critical role in maintaining water levels in South Delhi, particularly in the Mehrauli area. Of the 611 identified water bodies, 274 had dried up, while the remaining 337 were in poor condition. Originally surrounded by dense forests, in Sanjay Van, parts of the catchment area were cleared for institutional development, reducing rainwater flow into the lake and blocking natural drainage paths. To address this, the decision was made to use a mix of untreated sewage and water from sewage treatment plants (STPs) to maintain water levels, as the lake relies solely on rainwater (Jacob 2019). The forest was also invaded by non-native species like *P. juliflora*, which led to the decline of native plants. Gradually, the lake became encroached upon, filled with debris, and polluted with untreated sewage. Water hyacinth (*Eichhornia*) overran the lake as silt and sediment accumulated. About a decade ago, during the construction of a flyover, the lake was nearly buried under debris (Chandra et al. 2018). The lake was desilted, and the extracted material was used to landscape the embankments. The wastewater is processed through the CWS (Constructed Wetland System) before entering the lake. To increase the lake's storage capacity and attract aquatic birds, a small run-off river check dam was also built. The CWS uses a combination of physical and biological processes, operating with zero energy input. It includes two oxidation ponds and four physical channels/tanks to filter and treat the water (Seth et al. 2024).

Fig. 10.2 Restoration of the Neela Hauz wetlands. (Source: Delhi Biodiversity foundation, D.D.A.)

10.7.2 Threats and Challenges to Neela Hauz

1. *Encroachment and Urbanization*: The expansion of nearby urban infrastructure, including institutional buildings and roads, has significantly reduced the lake's catchment area.
2. *Water Supply Issues*: Alternative sources like sewage and treated wastewater are now used to maintain water levels. Dependence on these non-natural sources presents a challenge in maintaining the quality and sustainability of the water supply.
3. *Pollution*: The long-term effectiveness of the CWS in managing pollution without significant human intervention is another challenge.
4. *Invasive Species*: The spread of invasive plant species, such as *Prosopis*, threatens the native biodiversity, which can alter the ecosystem and affect wildlife.
5. *Climate Change*: The hydrology of the lake may be disturbed by erratic weather patterns.
6. *Lack of Public Awareness and Engagement*: Sustaining the restoration efforts requires active community involvement and awareness.
7. *Funding and Resources*: Maintaining the lake and the surrounding biodiversity park requires continuous funding, technical ex pertise, and management.

10.7.3 Initiatives for Restoration at Neela Hauz

Through a combination of ecological restoration, sustainable water management, and active community participation, Neela Hauz can continue to serve as a model for forest conservation and urban ecological regeneration in Delhi (Fig. 10.2).

10.8 Restoration Case Study No. 3: Aravalli Biodiversity Park

Location: (latitude 28.48°N and longitude 77.11°E)

The Aravalli Biodiversity Park is situated on the Mehrauli road at the Delhi–Gurugram border, spanning about 392 acres, with elevations ranging from 248 to 277 metres. Environmental degradation resulted from mining activities, as well as illegal logging, cattle grazing, and waste dumping. The park's terrain features rolling hills, small ridges, valleys, and seasonal streams formed by quarry pits (Chhangani et al. 2013).

10.8.1 Present-Day Status of the Aravalli Biodiversity Park

Haryana Chief Minister, Bhupinder Singh Hooda, officially inaugurated the park on World Environment Day, June 5, 2010. Today, the area plays a variety of roles: it provides a sanctuary for native Aravalli flora and fauna, supports water conservation and recharge, acts as a carbon sink, offers an educational space for environmental awareness, and serves as a recreational spot for Gurugram residents. Unlike conventional parks or gardens, the Aravalli Biodiversity Park embodies the spirit of the Aravallis as an urban forest within the rapidly growing city of Gurugram.

10.8.1.1 Flora

This park has 300 plant species, many of which are rare or endangered, according to the IUCN Red List. More than 100,000 saplings have been planted since 2011, with a survival rate of 85–95%. The park is evolving into a sanctuary for endangered plant species from the Northern Aravallis, which are rapidly vanishing due to ongoing development (Bansal et al. 2023).

10.8.1.2 Fauna

A wildlife survey conducted in 2017 recorded ten species of *mammals*, six of which can be seen in this park (the jungle cat, grey mongoose, golden jackal, Indian civet, Indian hare, and nilgai). Other mammals like the house shrew, Indian bush rat, flying fox, Indian bush rat, striped palm squirrel, and various bat species are also present. The park has a variety of *reptiles* (the monitor lizard, common cobra, saw-scaled viper, Russell's viper, common krait, rat snake, common wolf snake, sand boa, skink, gecko, etc.), *amphibians* (eight species), and *invertebrates*. The park has seen a remarkable increase in *invertebrate diversity*, including *pollinators* like bees, wasps, butterflies, and flies; *herbivores* like aphids, grasshoppers, and cicadas; and *predators* like ants, spiders, and centipedes (Dar et al. 2021). Soil enrichers like earthworms and termites support plant growth, and many of these invertebrates serve as food for reptiles, birds, and mammals. As of June 2020, birdwatchers have recorded 201 bird species in the park (Dutta et al. 2019). Uncommon sightings of the Himalayan griffon vulture and Rufous-tailed scrub robin have sparked continued interest in documenting the park's avian diversity.

10.8.2 Threats and Challenges to the Aravalli Biodiversity Park

1. *Urbanization and Encroachment*: With the rapid growth of Gurugram and Delhi, urbanization, illegal encroachments, real estate developments, and infrastructure projects are increasingly infringing on the park's boundaries, causing habitat loss and fragmentation.

2. *Dumping of Construction Waste and Debris*: The park frequently becomes a dumping ground for construction and demolition debris, which not only degrades the landscape but also hinders restoration efforts and pollutes the groundwater.
3. *Invasive Species*: The presence of invasive plant species, particularly *P. juliflora*, remains a major ecological challenge.
4. *Water Scarcity and Groundwater Depletion*: Over-extraction of groundwater, coupled with reduced recharge due to urbanization, threatens the park's ability to support both flora and fauna.
5. *Climate Change*: Altered weather patterns, increased frequency of droughts, and shifting rainfall can affect the delicate ecosystems of the park, leading to species loss and reduced forest cover.
6. *Deforestation and Habitat Degradation*: Deforestation in surrounding areas, combined with ongoing habitat degradation, remains a challenge.
7. *Pollution*: Air and water pollution from nearby industrial and urban areas pose serious risks to the biodiversity of the park.
8. *Lack of Legal Protection*: Although the Aravalli Biodiversity Park has been recognized as an Other Effective Area-based Conservation Measure (OECM), it does not have full legal protection under Indian law. It has the potential to be designated as a Biodiversity Heritage Site under India's Biological Diversity Act of 2002.
9. *Human–Wildlife Conflict*: As urban areas expand closer to the park, the chances of human–wildlife conflict increase. Animals such as jackals, snakes, and nilgai can wander into residential areas.
10. *Limited Public Awareness and Engagement*: While initiatives have been launched to raise awareness and involve the community, there is still a lack of public understanding. This limits the effectiveness of grassroot conservation efforts and reduces public support for protecting the park.

10.8.3 Initiatives for Restoration at the Aravalli Biodiversity Park

- In 2004, the Supreme Court ordered Haryana to stop mining and groundwater extraction near the Delhi–Haryana border. In 2009, the Municipal Corporation of Gurugram (MCG) transferred degraded land to IamGurgaon (IAG), who set up a native plant nursery and conducted research. The park, designed by architect Atal Kapoor, was planted with over 6000 native species in 2010, supporting the Northern Aravallis' flora and fauna (Amir and Shannon 2023) (Fig. 10.3).
- A list of 200 native forest species was compiled, seed collection drives were launched, and drip irrigation systems were installed. In 2011, the MCG commissioner, Sudhir Rajpal, encouraged local corporations to support the park development, leading to 145,000 plants of over 200 species being planted.

Fig. 10.3 Successful restoration efforts at the Aravalli Biodiversity Park. (Source: Gandhiok 2019) http://timesofindia.indiatimes.com/articleshow/68880325.cms?utm_source=contentofinterest&utm_medium=text&utm_campaign=cppst

- MCG and Hero MotoCorp Limited (HMCL) have signed a 10-year collaboration to improve visitor infrastructure at the park, including an interpretation centre and a research station. The park has been recognized as India's first *Other Effective Area-based Conservation Measure* by the Convention on Biological Diversity (CBD) (Panwar and Dhote 2022).

Addressing these threats will require coordinated efforts among government agencies, conservation organizations, and local communities to ensure the park's long-term survival and its role as a vital ecological asset for the region.

10.9 Restoration Case Study No. 4: Sanjay Van

Location: Mehrauli (latitude 28.53°N and longitude 77.17°E)

Sanjay Van is located within the Mehrauli/South-Central Ridge of Delhi, which serves as an extension of the ancient Aravalli Mountain Range. The Mehrauli Ridge has faced immense pressure from urbanization and development. In the 1970s, the DDA designated 784 acres of this area as a green belt, now known as Sanjay Van. This forest is situated near the Qutub Minar and is bordered by the Qutub Institutional Area, Aurobindo Marg, Mehrauli-Kishangarh, and Aruna Asaf Ali Marg. Sanjay Van is designated as a reserve forest under the Indian Forest Act of 1927 (Chandra et al. 2018).

10.9.1 Present-Day Status of Sanjay Van

10.9.1.1 Flora

Native trees include the ronjh, hingot, bistendu, and jal. Desi kikar, khair, and khejdi/jhand are also present but in limited numbers. The forest also contains thorny, bushy growths like the heens, taint/kareel, jungli karonda, and ber. Under the management of the Horticulture Department, *ornamental trees* such as the amaltas, karanj/papdi, and neem have been extensively planted along roadsides. It also has the maharukh, chudel papdi, gular, semal, siris, pipal, and bargad.

10.9.1.2 Fauna

Once a hunting ground for Firozshah Tughlaq, Sanjay Van now serves as a natural habitat for a variety of wildlife, including blue bulls, golden jackals, and small and large lizards. It also hosts diverse bird species such as grey and purple herons and peacocks. Butterflies, the forest's insect ambassadors, flutter throughout the area. Notable bird sightings include Asian paradise flycatchers, golden orioles, crested pied cuckoos, and Eurasian sparrow hawks. Sanjay Van is a paradise for birdwatchers, with the potential to become a bird sanctuary. Recently, a striped hyena was spotted for the first time, and an Asian paradise flycatcher has nested in the forest after 35 years (Chandra et al. 2018).

10.9.2 Threats and Challenges to Sanjay Van

- To extend the green cover, the Forest Department introduced *P. juliflora*, an exotic, fast-growing tree from Mexico. This proved disastrous, as it has become the dominant species, leading to severe forest degradation and lowering groundwater levels, drying up surface soil, and causing the extinction of native flora. The native Aravalli vegetation, along with introduced ornamental trees, now represents only a small fraction of Sanjay Van's original greenery.
- Additionally, the discharge of sewage and effluents into the forest has further harmed its native flora and fauna, disrupting the ecological balance of South Delhi.

10.9.3 Initiatives for Restoration at Sanjay Van

Since 2010, the DDA has launched a significant initiative to restore Sanjay Van (Fig. 10.4).

- *Tree Plantation*: Approximately 100,000 native trees have been planted over the past few years by reintroducing species such as the khejdi (*P. cineraria*), dhak (*Butea monosperma*), kumttha (*A. modesta*), khair (*A. catechu*), desi kikar (*A. indica*), dhau (*Anogeissus pendula*), kulu tree (*Sterculia urens*), and salai (*Boswellia serrata*). To control the spread of vilayati kikar, measures are being taken to remove its seed pods and young saplings regularly (Sivaramakrishnan 2017).
- *Medicinal Forest*: A medicinal forest at Sanjay Van has been maintained since 2012. The primary goal is to plant native medicinal trees such as the aonla, bael, neem, hingot, amaltas, Arjun, bahera, and sahjan. Significantly, gum guggul (*Commiphora wightii*), an essential medicinal tree endorsed by the Central

Before After

Fig. 10.4 Restoration efforts at Sanjay Van

Council for Research in Ayurveda and Siddha (CCRAS), has been procured from Rajasthan and planted in the area (Sivaramakrishnan 2017).

- *Water Management and Development of Bird Sanctuary*: Through simple and innovative techniques, five interconnected water bodies have been revitalized into clean, flowing year-round water sources, leading to improved biological oxygen demand (BOD) levels that sustain aquatic life. This water is now used for watering saplings (attracting more birds), harvesting rainwater, and recharging groundwater. This will also help recharge the adjacent Neela Hauz. This reservoir could potentially benefit Vasant Kunj in the future. In the efforts to establish a bird sanctuary at Sanjay Van, ongoing work includes revitalizing water bodies, planting bird-friendly vegetation, and introducing fish into the ponds (Sivaramakrishnan 2017).
- *Education and Local Involvement*: Worldwide Nature, in coordination with the DDA, regularly organizes activities for school children, including jungle trails, birdwatching, tree planting, and clean-up drives. Since 2012, these programmes have been expanded to include children with special needs, underprivileged youth, and students from economically disadvantaged schools in nearby areas. Worldwide Nature and the DDA also promote tree planting among locals on special occasions, such as birthdays and anniversaries, by offering free pits and plants to encourage participation (Sivaramakrishnan 2017).

10.10 Restoration Case Study No. 5: Asola Bhatti Wildlife Sanctuary (ABWS)

Location: (latitude 28.24°–28.29°N and longitude 77.11°–77.16°E)

It is a semi-forest wildlife sanctuary covering a 32.71-km^2 area in the southern Delhi Ridge of the Aravalli Mountain Range on the Delhi–Haryana border. It is a vital segment of the leopard wildlife corridor that extends from Sariska National Park in Rajasthan to the Delhi Ridge.

10.10.1 Present-Day Status of the Asola Bhatti Wildlife Sanctuary

Legal protected status to this sanctuary was given in 1986. About 2679.29 acres of land were acquired from three villages, namely Asola, Shapur, and Maidan Garhi. An area of 2166.28 acres of land from village Bhatti was notified as the Asola Bhatti Wildlife Sanctuary in 1991, after imposing a ban on mining at Badarpur. The sanctuary is presently managed by the Conservation Education Centre (CEC) and Bombay Natural History Society (BNHS), Delhi.

10.10.1.1 Flora and Fauna

ABWS is home to various flora, namely, *Butea monosperma*, Indian elm, *Neolamarkia*, and *Tinospora;* fauna, namely, the blackbuck, crested porcupine, golden jackal, small Indian civet, etc.; some endangered species such as the Egyptian vulture, Indian leopard, and red-headed vulture; near-threatened species such as the painted stork, white-faced ibis, and European roller; and rare birds, namely, the black francolin and grey-headed fish eagle (Goel and Rath 2022). The Asola Bhatti Wildlife Sanctuary is unique in harbouring flora and fauna indigenous to one of the oldest mountains of the Aravalli Range. Studies, spanning across 8 years (2014–2022), documenting the avifauna of this wildlife sanctuary, revealed the presence of 250 species and two subspecies. The occurrence of the avifauna species confirms 14 species as single-record (rare/very rare species), 46 as occasional species, 58 as uncommon species, and 134 as common species (Madan et al. 2022).

10.10.2 Threats and Challenges to the Asola Bhatti Wildlife Sanctuary

- *Absence of forest boundaries* has also led to encroachments. Significant portions of the Aravallis are neither legally nor physically protected.
- *Cutting of trees* has resulted in deforestation.
- *Urbanization* is causing a continuous reduction in the forest cover and results in more human–wildlife conflict.
- *Lack of Wildlife Corridor*: The Faridabad–Gurugram highways and railways intersecting the Aravalli Range and the wildlife corridor at various points have led to the deaths of more than ten leopards from January 2015 to 2019.
- *Policy Threats*: MCD (Municipal Corporation of Delhi) is seeking possession of the Bhatti Mines area as an alternative landfilling site. Out of an area of 30 acres of Aravallis used as landfill, 14.86 acres is under the Aravalli Plantation Project and was given legal "forest status." The landfill is very close to the groundwater aquifer. The Bandhwari landfill is estimated to have about 35 lakh tonnes of untreated, mixed waste.
- *Freeing Up the Southern Ridge*: Of the 357.07 hectares of encroached forest land in the southern forest division, the Forest Department has cleared just 79.88 hectares (around 22%). Much of this action was taken in 2019–2020, and the reasons cited for the slow pace of removal of encroachments were several court cases;

land awaiting demarcation; orders from the Delhi High Court, asking for status quo to be maintained till fresh demarcation; issues of manpower; absence of a dedicated fund to reclaim land; and the number of people who might be displaced if large-scale demolitions of unauthorized colonies are initiated. Maidan Garhi police station in Asola, the Indira Gandhi National Open University (IGNOU) (out of the 218 *bighas* on which IGNOU stands, 68 *bighas* are forest land), Radha Soami Satsang Beas (RSSB) (around 95 *bighas* of "forest" land have been identified as encroached by RSSB in Bhatti), many farmhouses, and other private properties are among the encroachments that are considered to be existing in the reserved forest land, and the dispute cases are still pending in the court (Harigovind 2024).

10.10.3 Initiatives for Restoration at Asola Bhatti Wildlife Sanctuary

- *Regeneration projects* (2020) to establish ten micro-habitats across 12 hectares of land, along with an additional 2 hectares dedicated to restoring grasslands (Harigovind 2022).
- *Open land in the Aravalli hill forest* supports biodiversity, particularly mammals, rodents, and reptiles (including grazers and browsers) (Dixit 2021).
- *Plantation* drive of 14 varieties of grasses to help in increasing the microbial population (Gandhiok 2022).
- *Emphasis is on the assistive regeneration* of plants. Assistive regeneration aids in enhancing the existing vegetation, such as palash and dhau, while also removing invasive species like *Lantana*, vilayati kikar, subabul, and *Parthenium* grasses that hinder natural growth (Harigovind 2022).
- *Building of a 45-km-long patrol passage* by the Delhi government outside the sanctuary to prevent encroachment and trespassing of forest land for security (Mani 2022).
- *Conservation of rainwater in the monsoon* season will help supply enough water in the summer season.
- *Encouraging Lake Formation* in the Mining Pit: Under the master plan for the rejuvenation and redevelopment of the Asola Bhatti Mines area, 14 mining pits (four large and ten small) in the Asola Sanctuary are to be transformed into reservoirs that will help recharge Delhi's groundwater (Jain 2022).
- *To promote ecotourism*, the creation of butterfly gardens, wildlife trails, walking tracks, cycle tracks, birdwatching spots, ropeways, and waterfalls is recommended in the master plan (Fig. 10.5).
- *Wildlife Protection Awareness Programme*: A wildlife protection awareness programme event titled "Van Jeev Sanrakshan Abhiyan" was launched at the ABWS by the government to sensitize people about the importance of wildlife protection in October 2021.

Fig. 10.5 Artificial waterfalls and floating mountains to boost ecotourism at ABWS. (*Image Source: Hindustan* |*Times*)

While progress has been made in restoring the ecosystem by granting protected status to the Asola Bhatti Wildlife Sanctuary, additional efforts are necessary for ecological restoration that follows a natural progression, supports assisted regeneration, and addresses ongoing threats to prevent further degradation of the forest post-restoration.

10.11 Restoration Case Study No. 6: Mangar Bani Forest

Location: (latitude 28.37°N and longitude 77.17°E)

Mangar is a 700-year-old village, nestled in the rocky Aravalli hills, covered with unique native trees, bounded by ABWS to the north and Gurugram to the west. Mangar Bani is a forested landscape, adjacent to Mangar village, and is located at the intersection of the boundaries of Faridabad, Gurugram, and Delhi (Abhir 2024). Mangar Bani is a sacred grove extending over 274 hectares (677 acres). The present condition is conserved because of the development of the forest management system by the local community, ensuring enough feed for their animals and resources (Verchot and Biswas 2021).

It acts as a corridor for wildlife between the southern Aravallis in Rajasthan and the Delhi Ridge (Sinha 2015; Arora 2018a). Wetlands, such as Lulu Lake wetland, are present on the hills of Mangar Bani. The villages of Gwal Pahari, Baliawas, and Bandhwari lie near the forested parts of Southern Delhi Ridge. Mangar Bani is perhaps the last green patch between a polluted Delhi and a concrete Gurugram, which has been preserved by the local Gujjar community and attracts a lot of birds, both indigenous and migratory (Bani 2015).

10.11.1 Present-Day Status of Mangar Bani

10.11.1.1 Flora

It is an ecologically sensitive biodiversity area that is under threat. It has more than 30 native tree species thriving, some of which have gone extinct elsewhere in Delhi, along with 100 native shrubs and herbs, all of which provide shelter to the wildlife, which includes rare birds and animals, including leopards (Abhir 2024). Ninety-five per cent of the Mangar Bani Forest is covered with the slow-growing native plant species, such as the dhau tree (*Anogeissus pendula*), covering the rocky slopes. *Capparis sepiaria* (heens), *Ziziphus mauritiana* (wild ber), and *Butea monosperma* (palash) occur in the flatter areas (Agarwal 2020). The flora and ecology have degraded and severely need restoration by replacing the invasive vilayati kikar with the three-layer cover of native grass, shrubs, and trees. Even today, the villagers guard the forest zealously, and not a single tree is cut or pruned. As per the Forest Department's vegetation analysis, in 2014, Mangar Bani had around 599,889 trees above a certain girth and many more below this (Arora 2018a).

10.11.1.2 Fauna

Mangar is astonishingly rich in bird, mammal, reptile, and other faunal life. The fauna includes *mammals,* namely, leopards, bears, hyenas, porcupines, civet cats, jackals, four-horned antelopes, deer, nilgais, and langurs; more than 200 species (219) of birds; and 90 species of *butterflies* (Abhir 2024). It is also an important wetland for the migratory birds, as there are several dozen lakes formed in the abandoned open pit mines. Some of the bird and butterfly species are almost endemic to this area and are rarely found elsewhere. The online database e-Bird lists and documents 207 species of birds in just 17 sq. km. Possibly, 20 per cent of these bird species are tropical dry forest specialists, including the yellow-crowned woodpecker, common woodshrike, black-headed cuckooshrike, white-bellied drongo, and the white-browed fantail flycatcher, all relatively rare in Delhi (Fig. 10.6). Other forest-preferring species include the jungle prinia, cinereous tit, yellow-eyed babbler, chestnut-shouldered petronia, small minivet, Tickell's blue flycatcher, and the rare Indian pitta. However, the people who intend to clear the forest land for the purpose of construction work still continue to undertake illegal activities. These activities also hamper the animal movement in this important wildlife corridor (Pati 2023, 2024).

Apart from its biodiversity values, the Mangar Bani and its surroundings form a water catchment, which acts as an important source of groundwater recharge. It also helps regulate the climate and monsoon patterns in the area. Ecological sustainability of a forest area is based on the presence of the entire micro-habitat required by the organisms present there, and Mangar Bani, being a small patch of Aravallis, cannot sustain itself unless its connectivity to the surrounding areas with dense forest cover is protected. Mangar forms the largest contiguous stretch of forest of biological value in Delhi NCR. The entire Gurugram–Faridabad stretch could thus be ecologically secured for the future (Agarwal 2020).

Fig. 10.6 A few examples of avifauna at the Mangar Bani Forest. (Source: https://wildhawk.in/219-species-of-birds-found-near-camp-wildhawk-survey/)

10.11.2 Threats and Challenges to Mangar Bani

- *The geographical boundaries* of Mangar Bani have never been identified. The remaining hills are experiencing varying degrees of degradation due to ongoing mining activities and the harvesting of fuel wood and fodder (Verchot and Biswas 2021).
- *Encroachment, illegal construction, and illegal tree felling* are significant threats to the forest. As increasing numbers of people visit this unsecured and vulnerable site, the deterioration of the area will accelerate (Pati 2023).
- *The waste-to-energy plant* near Bandhwari and the landfill situated in the two sand mine pits, used for disposing of untreated toxic waste, are creating an environmental hazard by leaching toxic pollutants into the forest and contaminating the aquifers in both Delhi and Gurugram (Arora 2018a).
- *The unclear legal status* of the forest poses a persistent threat to its security, as it has never been designated as a protected forest area (Abhir 2024).
- *The changing ownership* of land in Mangar has heightened the risk of deforestation for the sacred grove, as the area has been divided and sold to private parties over the years (Verchot and Biswas 2021).
- *The buffer zone surrounding the forest* underwent significant changes in a short span of time, being reduced from 500 metres to 60 metres and then restored back

to 500 metres within less than 2 years. Such abrupt fluctuations in the status and size of the buffer zone create opportunities for construction activities to begin in close proximity to the forest (Abhir 2024).

- *The Mangar forests* face additional threats from illegal clearing and fragmentation. Large-scale tree felling, in spite of the legal restrictions imposed by the National Green Tribunal (NGT) to maintain the status quo in Mangar, would cause the deterioration of the ecosystem (Sinha 2015; Pati 2024).
- *The proposal for developmental activities* in 23 villages within the Aravalli region of Faridabad, including Mangar, as part of the Mangar Draft Development Plan (DDP) for 2031, aims to kickstart various construction initiatives across an area of 10,484 hectares (Bani 2015).

10.11.3 Initiatives Taken at Mangar Bani

- In 2015, the chief minister of Haryana announced that Mangar Bani would be designated as a *protected forest*, with a 500-metre buffer zone (Shahabuddin 2015).
- In June 2024, IUCN, in collaboration with the Haryana State Forest Department and supported by the MoEFCC, organized a workshop focused on *Forest Landscape Restoration (FLR)*, as part of the Bonn Challenge pledge (IUCN 2024).
- *The Mangar Eco Club* for children was set up in 2015 to bring about a change in their outlook and help them experience the wilderness. The club receives support from teachers, birdwatchers, and organizations like BNHS, Laksh Foundation, and CEDAR (Centre for Ecology Development and Research). This initiative aims to foster a love among students for the rich biodiversity found in the sacred grove.
- *To enhance groundwater recharge*, the Haryana government is constructing check dams in the forest hills. These structures will help prevent waterlogging during rainfall and ensure water supply for wildlife in the summer months (Arora 2018b).
- *NGT, in 2017*, asked the Central Pollution Control Board (CPCB) to test the groundwater sample near the Bandhwari sewage treatment plant (NGT 2017).
- *Involvement of the Local Community*: The preservation of this remnant forest was possible due to the efforts of the local community.

10.12 Restoration Case Study No. 7: Tilpath Valley Biodiversity Park (TVBP)

Location: (latitude: 28.47°N and longitude: 77.23°E)

Tilpath Valley is a historic green expanse covering 172 acres, situated on the Southern Ridge adjacent to the Aravalli Range of Haryana. It lies immediately northwest of the Asola Bhatti Wildlife Sanctuary and south of Sainik Farm. The park, which has been under development since 2015, is linked to the Asola Bhatti Wildlife Sanctuary. It was officially inaugurated as Delhi's third biodiversity park in

2018, with development carried out by the DDA. The park was inaugurated by the Delhi Lieutenant Governor, along with eminent ecologist Prof CR Babu (Correspondent TOI 2018b; Agarwal 2021). The hills surrounding this park are also connected with other valleys of the Asola Bhatti Wildlife Sanctuary and the Aravallis of Haryana. Attempts were made to repopulate the TVBP by translocating animals from Madhya Pradesh, although the park officials are certain that, since Tilpath Valley shares a border with both Aravallis and ABWS, there will be natural migration of the animals from there (Correspondent TOI 2018b).

Before the restoration process, the area showed the presence of hilltops and ridges, various types of slopes (steep/gentle), and even open mine pits. Dried-up springs were found, as there were no recharging zones due to the destruction and clearing of the vegetation. The area was devoid of any greenery and was highly degraded, with only invasive species such as *P. juliflora*. In addition, *Lantana* and *Parthenium* could also be found invading the entire area (Correspondent TOI 2018b). This area is important not only from the point of biodiversity but also mythologically, being one of the villages demanded by the Pandavas from Duryodhana. Initiatives were thus undertaken by the DDA to restore this degraded Aravalli area. Attempts were made to conserve the relicts of flora and fauna of Aravallis and recreate its native biotic communities (Delhi Biodiversity foundation, DDA- Tilpath Valley Biodiversity Park).

The valley acts as a recharging zone of the ground aquifers since it receives the drainage from Chhatarpur and all surrounding hill ranges. It has the potential to store and recharge millions of cubic metres of rainwater harvested annually. Being rich in relict Aravalli vegetation, the DDA declared Tilpath Valley to be developed as a biodiversity park since it has the potential to attract large carnivores from the Haryana range. In 2016, the park witnessed the presence of over 105 tree and shrub species, 32 butterflies, 15 reptiles and amphibians, 103 avifauna, and more than eight mammal species.

10.12.1 Present-Day Status of the Tilpath Valley Biodiversity Park

10.12.1.1 Flora

The park now covers grasslands and hills. The invasive species were replaced with three layers of forest-community plant species, such as mahua, haldu, shisham, and bael. Based on the topography of the terrain, the park has been restored with varied plant species. The plains are covered with grasslands to provide a habitat for herbivores, which, in turn, attract carnivores. Grasslands enhance the trophic level of a particular area and also generate a variety of ecological services for society. For a similar purpose, water bodies were created (Correspondent TOI 2018b; Agarwal 2021). The major grass species are *Cenchrus* (buffelgrass), *Heteropogon, Chrysopogon, Aristida, Desmostachya* (halfa/darbha grass), etc. The gentle slopes and upper reaches of the valleys of the park are mostly covered with the tropical dry deciduous forests dominated by species like the mahua, tendu, haldu, dhak, chilbil, shisham, and bael. The steep slopes are developed into open thorny forests, with the dominant plant species being the bael, guggul, siris, dhawa, sami, etc. The ridges and plateaus are mostly covered by native shrub species such as *Capparis* (karel), *Carissa* (karaunda), *Cordia* (lisoda), *Ehretia* (chamror), *Grewia* (phalsa), *Lycium* (goji berry), and *Ziziphus* (ber). A three-storey canopy has been established in the valley using exclusively native species. It was decided that the variety of the medicinal plant wealth present in the park should be utilized by Ayush healthcare system and pharmaceutical companies for product development (Correspondent HT 2018a).

10.12.1.2 Fauna

Grassland *birds* such as the bee-eater, crested lark, grey francolin, house sparrow, Indian paradise flycatcher, Indian silverbill, nightjar, paddy field pipit, pied bush chat, red avadavat, and sirkeer malkoha cuckoo can be spotted. The presence of many *reptilian species* like the gecko, Indian cobra, Indian rock python, and Pakistani ribbon snake has been recorded. *Mammalian species* such as the hyena, Indian hare, golden jackal, mongooses, nilgai, porcupine, small Indian civet, spotted deer, sambar, and leopard could be spotted. However, no recent marks of leopards have been spotted. There was a proposal to ecologically diversify the park by reintroducing the mammalian *herbivorous* fauna like the chital deer and chinkara gazelle and *Cervidae* like the hog deer. For self-sustainability, the ecosystem should have producers, herbivores, and primary and secondary carnivores (Agarwal 2021, 2022). Wildlife surveys are carried out by tracking pugmarks using pressure impression pads (PIPs) and photographing the animals. TVBP is located near Tughlaqabad and ABWS areas, where leopard pugmarks have previously been observed. It is possible that the leopards and hyenas spotted randomly might have been wandering animals from these adjacent areas. This has led to the successful rehabilitation of mammalian species, marking a significant advancement, since reintroducing mammals can be challenging. Consequently, the plan to translocate herbivorous mammals from other regions has been abandoned (Agarwal 2022). The growing population of jackals is another indication of the advancement in the restoration process. The area has been fully revitalized, attracting a resurgence of birds, butterflies, and reptiles (Fig. 10.7).

10.12.2 Threats and Challenges to the Tilpath Valley Biodiversity Park

The Tilpath Valley Biodiversity Park faces several threats and challenges, including habitat degradation due to urban encroachment, invasive species, pollution, and climate change. Additionally, human–wildlife conflicts and limited resources for conservation efforts hinder the protection of its rich biodiversity, impacting the park's ecological balance and long-term sustainability. Many other challenges faced are similar to those of ABWS discussed earlier.

Fig. 10.7 Fauna at the Tilpath Valley Biodiversity Park after restoration. (Source: Priyangi Agarwal 2022, TOI) http://timesofindia.indiatimes.com/articleshow/93307401.cms?utm_source=contentofinterest&utm_medium=text&utm_campaign=cppst

10.12.3 Initiatives Taken for Restoration at the Tilpath Valley Biodiversity Park

- During the restoration process, invasive foreign species were removed and replaced with a three-tiered forest community consisting of native trees, shrubs, and grasslands.
- A total of 105 tree species belonging to 15 plant communities were planted at Tilpath Valley.
- Low-lying and flat areas were transformed into grasslands, where broadleaved tree species were planted. Additionally, the shrubland was restored to attract reptiles and birds.
- In order to keep track of the species arriving every month, animals were mapped and tracked through PIPs.
- In a series of plantation drives,
  - In 2015, the DDA, in collaboration with the Times of India (TOI), planted over one lakh native saplings as part of the Hero–TOI Green Drive.
  - 20,000 more plants were planted by TOI again in 2016 (Correspondent TOI 2018b).

10.13 Restoration Case Study No. 8: Tughlaqabad Biodiversity Park

Location: (latitude: 28.50°N and longitude: 77.26°E)

The Tughlaqabad Biodiversity Park spans 200 acres and features a gentle eastern slope of the Tughlaqabad Fort area, characterized by its plateau, ridges, and valleys. The park is developed with an objective to restore the forest communities, the surroundings of Tughlaqabad Fort to their unique semi-arid forest ecosystems, and the waterbodies by the development of CWS (Delhi Biodiversity Foundation, DDA-Tughlaqabad Biodiversity Park).

It is located in the Southern Ridge (200 acres) and includes the eastern gentle slope of the Tughlaqabad Fort area. The biodiversity park is made up of peaks, valleys, and a gentle eastern slope of the plateau. The terrain is characterized by undulating ridges and depressions, with most ridges being flat and either barren or with scattered *P. juliflora* bushes. Numerous shallow depressions across the landscape serve as rainwater recharge zones. Four major water bodies (wetlands) collect surface drainage from the Fort area; however, these water bodies are contaminated by untreated sewage discharged from encroachments within the Tughlaqabad Fort area. The storm drains, originally designed for surface water drainage, have been repurposed as sewers, allowing sewage to flow into the water bodies by gravity. In 2015, the Delhi High Court recommended that the DDA should replicate its successful management of wetlands in the Yamuna Biodiversity Park at this site. Consequently, in 2016, the DDA was instructed to provide recommendations for transforming the forested district park and the lake near Tughlaqabad Fort into a biodiversity park (Rajput 2016).

10.13.1 Present-Day Status of the Tughlaqabad Biodiversity Park

10.13.1.1 Flora

The park, at present, showcases the diverse flora typical of the Delhi Ridge and the Aravalli Range. *Trees* like *Acacia* species, *P. juliflora* (slowly being replaced by native species), *Butea monosperma*, *Azadirachta indica*, *Aegle marmelos* (bael), pipal, banyan, and ber; *shrubs* like *Carissa carandas* (karonda), *Capparis decidua* (kair), and *Cassia auriculata*; *grasses and herbs* like *Cynodon dactylon* (Bermuda grass), *Dactyloctenium aegyptium* (crowfoot grass), and *Dichanthium annulatum* (marvel grass); *climbers* like *Abrus precatorius* (rosary pea) and *Cissus quadrangularis* (hadjod); and *medicinal plants* like *Tinospora cordifolia* (giloy), *Aloe vera*, and *Ocimum tenuiflorum* (tulsi) are seen in the park.

10.13.1.2 Fauna

It also supports a rich fauna, much of which is adapted to the semi-arid, rocky terrain of Delhi Ridge. *Mammals*, like the golden jackal, Indian hare, five-striped palm squirrel, small Indian civet, and nilgai; *birds*, like the Indian peafowl, black kite, Indian grey hornbill, shikra, yellow-footed green pigeon, red-vented bulbul, Indian robin, and purple sunbird; *reptiles*, like the Indian monitor lizard, garden lizard, Indian cobra, rat snake, Russel's viper, and Indian star tortoise; *amphibians*, like the Indian bullfrog and Indian toad; and *insects and butterflies*, like the *Papilio polytes*, *Danaus chrysippus*, *Junonia orithya*, honeybee, dragon fly, and damselfly are found in this biodiversity park.

10.13.2 Threats and Challenges to the Tughlaqabad Biodiversity Park

Tughlaqabad Biodiversity Park is confronted with numerous threats and challenges. These include severe pollution from untreated sewage, which degrades water quality in the wetlands. Habitat loss due to encroachments and invasive species disrupts the native ecosystem. Furthermore, urban development pressures and insufficient resources for conservation hinder effective management, jeopardizing biodiversity.

10.13.3 Initiatives Taken at the Tughlaqabad Biodiversity Park

- *Infrastructure Development*: This includes the construction and reinforcement of a boundary wall to deter cattle grazing and prevent illegal dumping of solid waste, establishment of nature trails, and construction of wicket gates for visitors' access and movement (Delhi Biodiversity Foundation, DDA- Tuglaqabad Biodiversity Park).
- *Site Preparation and De-weeding*: To manage exotic and invasive weeds like *Argemone* sp., *Hyptis suaveolens*, and *Parthenium*, continuous uprooting is being carried out. Additionally, *Lantana camara* and *P. juliflora* are being removed using the cut rootstock method. The canopies of *P. juliflora* are being opened to allow sunlight to reach newly planted saplings, and the cut canopies are repurposed as fencing material.
- *Development of Biological Communities*: A proposal has been made to establish 30 biological communities, representative of the Aravalli Mountain Range. So far, 30,000 saplings from six of these communities have already been planted.
- *Fencing Strategies for the Protection of Saplings*: The three-layered fencing consists of thick or unpruned branches of *P. juliflora*, along with twigs from *Capparis sepiaria* (Fig. 10.8).
- *Gully Plugging and Water Conservation*: Check dams are being constructed across gullies using loose boulders and logs from *P. juliflora*. The water collected in these structures will help enhance groundwater levels and decrease run-off in the Kalkaji drain. This initiative will improve soil moisture and mitigate soil erosion.
- *Establishment of Nursery and Compost Pit*: A nursery has been established to cultivate native tree saplings and propagate grass species, currently housing a stock of 15,000 saplings. Plans are also in place to construct a net house and a polyhouse for further development.

Fig. 10.8 Fencing strategies for sapling protection at the Tughlaqabad Biodiversity Park. (Source: https://newdelhitoday.in/biodiversity-parks-in-delhi/)

- *Recreational Park*: Massive plantation of desired plant communities is done each year in the monsoon season.
- *Restoration of Four Waterbodies and the Development of CWS to Treat Raw Sewage*: The action plan for the restoration of four waterbodies and the construction of a wetland system is being implemented.
- *Introduction of Flora and Fauna*: The introduction of aquatic vegetation, including phytoplankton and zooplankton, along with fish species that feed on algae, floating root mats, water lilies, and other macrophytes, is planned. A butterfly park and herbal garden are also planned for development. Peripheral bamboo thicket, comprising ten species, will be planted along Maa Anandmai Marg to act as a dust trap and noise barrier. Green walkways will be developed along the fort wall. In Tughlaqabad Biodiversity Park, the species count of dragonflies and damselflies has increased from seven in 2022 to 14 in 2023, as the CWS has provided more habitats for these odonates to breed locally (Gandhiok 2023).

A case study of the restoration efforts of the floodplain of Delhi is discussed below.

10.14 Restoration Case Study No. 9: Yamuna Biodiversity Park

River Yamuna plays a pivotal role in shaping an impressive array of ecological variations in the National Territory of Delhi. It displays a transition between the semi-arid and arid deserts of the west and the fertile Gangetic plains of the east, thereby showing both the biogeographic zones (Sinha 2014).
Location: (latitude: 28.76°N and longitude: 77.23°E)

The Yamuna Biodiversity Park is located at Jagatpur Khadar, Wazirabad, on the banks of Yamuna River and occupies 184 hectares of biodiversity area in Delhi (Sharma 2012). The development of the park was commenced in 2005 by the DDA, in collaboration with CEMDE, University of Delhi. It is the first biodiversity park in the country. The decision to bring back the lost natural heritage of Delhi through the establishment of biodiversity parks was taken in the year 2003, and the objective was to oversee the development of this park and create more such biological parks in the future and sustain them too. Almost a 22-km stretch of Yamuna River around Wazirabad had become an open sewer, and the floodplains were reduced to channel banks, making it a dead river plain. Even the Air Quality Index (AQI) was quite severe due to increased concentration of PM 2.5 in the winter months. The floodplains were lying barren due to the presence of sodic soil, hampering the growth of the plants (Patel 2019).

10.14.1 Present-Day Status of the Yamuna Biodiversity Park

10.14.1.1 Flora

Forest communities, grassland communities, rich wetlands, and a variety of fruit-bearing plants, as well as a rich flora of medicinal herbs, represent the flora here. Trees like the jamun, Arjun, shisham, babool, pilu, chilbil, anjan, kadamb, and mahua were established based on the community structure and function. There are 900 species of plants, such as the adina, sal, teak, and hardwickia, covering the floodplain area. Many fruit-yielding plants like the anar, ber, khirni, cheeku, guava, amla, citrus fruit plants, grape, and loquat are also planted and attract many frugivorous birds. There are many familiar fruits, along with some unfamiliar fruits, such as *Mimuscops*, which is a locally extinct species (Delhi Biodiversity Foundation, DDA- Yamuna Biodiversity Park).

10.14.1.2 Fauna

The biodiversity park before reforestation had very limited animal diversity: a few mongooses, lizards, and some 31 species of birds, until 2004, but with the restoration efforts since 2014, it has become home to a diversity of fauna. A large number of species of birds are able to locate their suitable niches. The park is now full of bird species like the shikra, muniya, prinia, shrike, bulbul, and parakeet. The forest communities are maturing because of the nesting success of a number of resident birds by ensuring a continuous food supply and shelter. The butterfly conservatory harbours around 82 species of butterflies, the prominent ones being the lemon butterfly, peacock butterfly, pansy, common castor, etc. The conservatory is a home for many moths, aphids, flies, spiders, bees, etc. The park has more than 35 species of dragonflies and damselflies that act as biological control agents and regulate the replication of insect vectors.

By 2014, the park exhibited many migratory birds (coming from Siberia, Europe, and Central Asia), around 200 species of other birds,75 species of butterflies, big mammals like wild boars and porcupines, and a few species of snakes. Since the park harbours a rich variety of floral components from various plant communities, there is enhanced faunal diversity too.

The Yamuna Biodiversity Park has been transformed from a barren and degraded landscape into a luxurious forest grassland and wetland community (Fig. 10.9). This park is different from recreational parks in having a diverse ecosystem that harbours native plants and animals and renders multiple ecological services, such as carbon sequestration, recharge of groundwater, and reduction of pollution load, besides

Fig. 10.9 This figure presents a comparative view of the nature reserve area within Yamuna Biodiversity Park. (Source: Delhi Biodiversity foundation D.D.A.)

contributing to urban environmental sustainability and resilience to climate change. It serves as an effective carbon sins and acts as an excellent filter for a variety of sources of air pollution by trapping dust and helps in mitigating pollution and adapting to climate change. This park forms an integral part of blue–green urban infrastructure.

10.14.2 Threats and Challenges to the Yamuna Biodiversity Park

The Yamuna Biodiversity Park is threatened by urban expansion, encroachments, pollution, and water contamination. Industrial growth nearby puts a strain on its ecosystems, while invasive species and habitat degradation challenge efforts to conserve biodiversity, disrupting the park's ecological balance and native wildlife.

10.14.3 Initiatives Taken for Restoration at the Yamuna Biodiversity Park

- To reduce the salinity in the soil, a large number of native plants were grown. The scientists at Delhi University planted specific species of grass in order to bring the pH level of the soil to an optimum. With the introduction of such grass, the pH level was brought from 10 to neutral. Earlier, the electrical conductivity was somewhere between 1500 and 5000 (S/m).
- To start the restoration work, a large number of surveys of the forest and the wetland areas around the Yamuna River and its tributaries were conducted. The restoration work started with the planting of almost 25 species of grasses, such as *Leptochloa fusca, Bothriochloa,* and *Cenchrus*. Different kinds of wild plants were planted using seeds collected from the natural forest habitat. Around 157 acres of ecosystem were restored in Phase 1 by developing a grassland, two wetlands, and some forest communities.

- Two wetlands that simulate natural water bodies were also created in the park; one was narrow, simulating the river, and was 1.8 km in length, and the other one was open and deep and covered approximately 2 hectares.
- This park was developed in two phases: Phase 1 of the park focused on barren floodplains, which started in 2005 and occupies 63 hectares, and Phase 2, spanning 121 hectares, focused on the active floodplains. The goal was to preserve the wild genetic resources of agricultural crops, ensure the conservation of a wide variety of life forms, recharge groundwater, and enhance the availability of freshwater, making an ideal habitat for the resident and migratory birds (Jain 2009).
- A few wetland systems were constructed and developed for the treatment of sewage water, which would enter directly into River Yamuna (Harjani 2013).
- There are two zones in the Yamuna Biodiversity Park:
  - *The visitor zone* comprises many modules like mounds from the basin of the river, wetlands, etc. Even repositories like that of medicinal plants, groves and conservatories of fruit plants, and a butterfly conservatory have been developed that act as a living museum for educating people about environmental awareness, also promoting ecotourism and educating our civil societies. The visitors' area features miniature representations of various forest ecosystem along with Yamuna River Basin.
  - *Natural conservation zone* indicates the nature conservation zone consisting of a vast forest, grassland, and wetland communities (Delhi Biodiversity Foundation, DDA- Yamuna Biodiversity Park).

After reviewing several urban restoration projects in Delhi, India, let us now explore the key aspects and considerations essential for planning and executing successful restoration initiatives in urban areas.

10.15 Key Threats and Challenges to the Restoration of Delhi Forests

India, covering 2.4% of the world's land area, is home to 8% of global biodiversity. As a mega-diverse country, it supports 18% of the global population, but this immense pressure is causing a rapid loss of its rich biological resources. Although ecosystem restoration targets are well-proposed in various programmes and policies of India, the scientific insights, knowledge, awareness, and willingness to implement policy initiatives for ecosystem restoration are inadequate (Singh et al. 2022). The threats and concerns for a few of the restoration sites of Delhi as individual case studies have been dealt with in detail in the preceding section of this chapter. The key challenges identified from the case studies of these sites can be summarized as follows:

1. *Urbanization Pressure*: Leading to loss of green cover, reduced habitat, pollution, etc.
2. *Encroachment and Illegal Status*: Due to unauthorized settlements and lack of proper legal protection status conferred by governmental policies or their strict implementation, hampering successful restoration processes.

3. *Invasive Species*: *P. julifera* has degraded the flora and fauna of Delhi.
4. *Water Scarcity-*: Availability of water is a critical need for restoration. Declining water table and lack of regular water sources hamper the revival of aquatic ecosystems, such as the wetlands.
5. *Pollution*: The sites get affected by air, soil, and water pollution from nearby urban areas by industrial or vehicular emissions, waste dumping, etc.
6. *Loss of Biodiversity*: A significant loss of native species occurs due to the degradation of land. Re-establishing a balanced ecosystem or reintroducing endangered species requires expertise and careful planning and is time-consuming.
7. *Human Activity*: Land use changes, monoculture, and other agricultural practices; deforestation; over-exploitation of resources; infrastructure development; high-impact tourism; cultural practices; poverty; lack of education; etc. are some of the major human activities impacting restoration efforts.
8. *Lack of Public Awareness and Involvement*: Support from local communities is an essential part of successful restoration efforts.
9. *Institutional and Governance Issues*: Weak governance, inconsistent or unclear policies on land use, lack of proper enforcement of environmental laws, and poor land use practices are some of the challenges that can hinder long-term restoration success.
10. *Climate Change*: Rising temperature and irregular rainfall patterns are additional challenges impacting restoration processes and the maintenance of ecosystems in biodiversity parks.
11. *Funding and Resources*: Sufficient and adequate financial resources are needed to implement restoration projects and maintain biodiversity parks.

Addressing these challenges requires a multidisciplinary approach, involving urban planning, ecological expertise, strong governance, environmental laws, and community participation to effectively restore and conserve Delhi Ridge and the floodplains.

10.16 Mitigation Strategies for the Restoration of Delhi Forests

The environmental, ecological, and social values of forests are now being recognized. Forests are now understood as interconnected with the broader ecosystem, rather than as separate, isolated entities. These considerations should be taken into account when developing the management objectives for Delhi's forests. Thus, the general objectives of management can be summed up as follows:

1. *Safeguarding Forest Ecosystem*: Residents living near the forest often encroach on the area to meet their needs and dispose of their household waste there. Numerous workshops and small industrial units are established near the forest boundary. It is crucial to fence off areas adjacent to these industries, workshops, housing estates, and slums using a combination of live hedges and barbed wire.

The Forest Department has implemented a Ridge Protection Management System to address the problem of encroachment.

2. *Sustainable Urban Greening*: Planting indigenous species in the vacant areas of reserved and protected forests, along with the *Gaon Sabha*, enhances the city's aesthetic appeal and helps mitigate vehicular pollution. Free distribution of high-quality seedlings and imparting of technical knowledge to the public should be promoted. Hardy plants that can withstand the extreme environmental conditions of the urban area should be planted. Some examples of various avenue trees planted in Delhi include *Azadirachta indica* (Safdarjung Road and Ashok Road), *Ficus religiosa* (Mandir Marg), *Ficus virens* (Zakir Hussain Marg), *Tamarindus indica* (Akbar Road), *Putranjiva roxburghii* (Racecourse Road), *Ficus amplissima* (Krishna Menon Marg), *Syzygium cumini* (Rajpath and Tughlak Road), and *Terminalia arjuna* (Janpath and Teen Murti Marg) (Sinha 2014).
3. *Effective Nursery Development Strategies*: The site for a new nursery should be selected on the basis of distance, terrain, water availability, seed quality, and the suitability of the plants to Delhi's climate. Species that are naturally adapted to Delhi's climate are the khair, kikar, neem, pipal, papri, jungli jalebi, jamun, gullar, shisham, etc.
4. *Approaches for the Conservation of Soil and Moisture*: The conservation of soil and moisture is a crucial measure to prevent the degradation of ecosystems. The plan for each site should be site-specific to the soil and moisture parameters. Rainwater harvesting and prevention of run-off by channelizing all the drains to a common reservoir is important. Species like *Dendrocalamus strictus, Acacia nilotica, Azadirachta indica,* and *Albizia lebbeck* are suitable for the afforestation of gully beds, whereas *Dalbergia sissoo* and *Acacia catechu* do well on river beds, gully slopes, and humps.
5. *Conservation and Enhancement of Vegetative Cover*: Delhi had a total of 19.97% of its land under forest and tree cover in 2011 (State of Forest Report 2011), which increased to 21.72% in 2021 (State of Forest Report 2021), but still needs further expansion. The management of existing forest areas, their maintenance, afforestation, and enrichment should be carried out by the Forest Department along with the active participation of the citizens of Delhi. This can be done by planting and development of nurseries; creation of new city forests; monitoring, consolidation, and mapping of forest land; development of water reservoirs; protection through fencing and boundary wall creation; eco-restoration of riverine areas; etc. Rather than relying on barren/open lands to plant trees, we should utilize all three dimensions available in the city, including the development of rooftop gardens (similar to rooftop solar policy) and vertical gardens along the overhead metro lines and flyovers, which is already being done and needs to be replicated across the city.
6. *Tree Relocation Strategies*: The practice of tree relocation is used for various reasons such as landscape redesigning, construction work, or conservation efforts. The health, age, type of tree, timing of relocation, soil condition, and the

environmental factors of the relocation site are all important considerations. Post-relocation care is also very critical.

7. *Rehabilitation Through Tree Replacement*: Dead, damaged, or very old trees should be replaced. It should be done with indigenous species like neem, pipal, jamun, and pilkhan for ensuring better survival.
8. *Removal of Tiled Surfaces and Surrounding Tree Guard*: The growth of trees is adversely affected by concretizing, tiling, and black-topping close to the tree stem, which prevents percolation of groundwater and proper root aeration, and the tree suffocates to death. The department is already engaged in such removal activities with the support of the National Green Tribunal.
9. *Compensatory Reforestation*: Any individual or user agency wishing to cut down a tree must obtain prior permission from the Forest Department. It should be ensured that, for every single tree felled, a total of ten saplings are planted and maintained by the Forest Department as well as the user agencies. This has resulted in an increase of the area under green cover and ensured densification of the existing forest patch. Orders were issued to plant ten times the number of plants felled for the construction of flyover on Vikaspuri -Meerabagh road in Delhi.
10. *Herbal Gardens*: Eleven species of high-demand medicinal plants in the world are found in Delhi. These are *Tinospora cordifolia, Emblica officinalis, Aegle marmelos, Phyllanthus amarus, Withania somnifera, Andrographis paniculata, Saraca asoca, Solanum nigrum, Rauwolfia serpentina, Asparagus racemosus, and Ocimum.* The National Medicinal Plant Board of India has given recommendations for their cultivation since a bulk of the ingredients are used in the preparation of herbal products. A medicinal and aromatic *herbal garden* (*Amrit Udyan*) of around 35 species has been developed at the Rashtrapati Bhavan by the Board. The State Government is also advised to develop such gardens at appropriate places and create awareness among the society.
11. *Revitalization of Degraded Areas*: The ecologically and environmentally sensitive sites need to be specially examined, and suitable provisions need to be made for them in the prospective plans. For degraded areas or areas requiring densification in a short time period, bamboo plantation is a good option. It is the fastest-growing plant among the woody species. India has 120 bamboo species belonging to 23 genera (Tewari et al. 2019), the commonest being *Dendrocalamus strictus and Bambusa bambos.* In Delhi, the introduction of *D. strictus* may be quite useful in restoring the green cover. The Forest Department of Delhi, realizing the importance of this, was engaged in the plantation of bamboo in Garhi Mandu and areas near ITO Chungi in 2009–2011.
12. *Innovative Monitoring Methods*: The advent of network-based technologies and the use of drones are some of the technological advancements for monitoring. Drones are currently being utilized to enhance the monitoring of extensive forest areas in Delhi. Similarly, connected sensors can be calibrated to assess various ecosystem services provided by urban forests and trees.
13. *Creating an Inventory of All Trees*: Understanding the distribution, age, size, quantity, and species of trees within Delhi, along with their spatial arrangement

and their ecosystem services, will facilitate effective management of the biodiversity they encompass (Babu and Singh 2024).

14. *Environmental Legal Framework*: Environmental protection and conservation have been advanced through the implementation of various laws and policy instruments aimed at promoting sustainable development (Gupta 2022; Ali 2023). The Indian Constitution is one of the few in the world that contains specific provisions on the environment. While these laws might be an integral part of our Constitution, environmental degradation is one of the most threatening realities at present. Environmental Courts having civil and criminal jurisdiction must be established to deal with environmental issues in a speedy manner. After years of deliberation, the National Green Tribunal Bill was introduced in the Indian Parliament on July 29, 2009. The bill provided for the establishment of a National Green Tribunal (NGT), meant to offer fast redressal of cases on environmental protection and the conservation of natural resources and forests. NGT is a multifaceted, multi-skilled body that combines the services provided by the existing courts, tribunals, and inspectors in the environmental field. In Delhi, it directed for Ridge consolidation, encroachment prevention, increase of staff strength, settlement of rights, and imposition of a fine to deter debris dumping in the Ridge's forest land. The Forest Department, strengthened by the NGT directives, has developed the Ridge Protection Management System to help locate and generate changed data of the Ridge through satellite imaging. Financial support for this work is provided by the Government of Delhi and the Ridge Management Board. Development of a mosaic of forest areas on geo-portals to help in the identification of the Ridge on the ground, as well as encroachment, is also planned. The NGT takes a practical stand and tries to bring about a balance between sustainable development and conservation, implying that development is also necessary. The Tree Authority constituted under the Delhi Preservation of Trees Act, 1994, includes government representatives, non-governmental organizations, eminent scientists, and environmentalists for preserving and enhancing the forests of Delhi.
15. *Many eco clubs* have been established in schools/colleges of Delhi, which conduct environmental activities throughout the year. At the end of every year, the Annual Eco-Meet and environmental exhibitions are organized, wherein the representatives from all the eco-club schools participate and exhibit various eco-friendly products.

10.17 Future Approach for Forest Restoration and Sustainable Development

The future approach for forest conservation and sustainable development should focus on a combination of ecological restoration, community involvement, and long-term management strategies. Here are the key strategies for ensuring its conservation and sustainability:

1. *Ecological Restoration and Native Species Revival*: Indigenous plants that once thrived in the area should be replanted to help revive the natural ecosystem that supports local wildlife and biodiversity. Periodic removal of the invasive species should be undertaken.
2. *Water Conservation and Groundwater Recharge*: A focused approach on enhancing groundwater recharge is necessary. This could include removing obstructions, re-establishing *nallahs* (small streams), restoring the abandoned mine pits as wetland habitats, and enhancing rainwater harvesting to increase water retention and maintain water levels.
3. *Climate Resilience and Adaptation*: Restoration efforts should integrate climate resilience by planting drought-resistant species to ensure long-term survival and stability of the ecosystem and also enhancing ecosystem services like carbon sequestration to mitigate climate effects.
4. *Biodiversity Monitoring and Research*: Continuous monitoring of water and soil quality, biodiversity health, and plant growth is essential to track the success of restoration efforts and make necessary adjustments. The lake and the surrounding forest areas should be developed to create suitable habitats for local and migratory bird species, reptiles, amphibians, and small mammals.
5. *Utilizing soil microbes to decompose landfill waste* can enhance soil quality and, in turn, support vegetation growth. Bioremediation is an environmentally friendly, cost-effective, and efficient approach for managing environmental pollutants. This biotechnical process cleans up contaminants from polluted sites by employing biological organisms such as bacteria, fungi, and plants (Muhammad et al. 2024).
6. *Protection of the sanctuary by a continuous boundary wall* to keep animals protected and prevent human encroachment.
7. *Establishing safe wildlife corridors* to facilitate animal movement, along with a buffer zone surrounding these areas and promoting wildlife crossings (Sharma 2019).
8. *Proper wildlife census* using pugmarks, trap camera, or radio collars (Madan 2021).
9. *Public Awareness and Education*: Engaging the public through educational programmes and awareness campaigns to sensitize the masses about the importance of wildlife protection (March 3: World Wildlife Day). Encouraging active participation of local residents in restoration activities will help in maintaining long-term sustainability. Companies can be encouraged to contribute through corporate social responsibility (CSR) initiatives, making conservation a collective effort.
10. *Sustainable Ecotourism*: Sustainable tourism can provide economic benefits while promoting conservation. Cultural and recreational activities should be provided at the biodiversity parks to achieve a wider community participation. Facilities like birdwatching areas, nature trails, educational tours, and environmental workshops can be established without harming the ecosystem to encourage responsible tourism. Restoration and maintenance of historical monuments/remains should be considered. Vehicular movement and the number of people

visiting these sites should be controlled. A balance between accessibility and ecological preservation must be maintained to protect the park's natural resources.

11. *Strengthening Legal Protection*: It is critical to secure legal protection for the biodiversity parks and designate the parks as *Biodiversity Heritage Sites* under India's Biological Diversity Act, 2002. Legal safeguards would help prevent illegal mining and encroachments, illegal dumping, and degradation while protecting the park from future urban pressures.
12. *Policy Support and Funding*: Ensuring strong policy backing and financial resources for forest conservation, pollution control, and water management from governmental bodies such as the DDA and the Delhi Jal Board and also efforts from non-governmental organizations (NGOs), corporations, and local communities for their technical expertise and understanding of the local conditions will be crucial for the long-term sustainability of the ongoing restoration and conservation initiatives.

10.18 Key Messages for Policymakers

1. Restoration of degraded ecosystems in India is crucial from economic and ecological viewpoints; it can help reverse the significant losses arising from land degradation and biodiversity loss and contribute to mitigating climate change, thereby improving human well-being.
2. Restoration is a long-term conscious activity that needs continued commitment in terms of both funding and the efforts of all stakeholders.
3. Restoration is not the same as tree plantation, and actions for climate change mitigation do not universally correspond with restoration.
4. Identifying the drivers of land degradation and addressing them sensibly are critical to ensure the success, long-term sustainability, and cost-effectiveness of all restoration efforts.
5. Restoration efforts that aim to maximize biodiversity and ecosystem functioning by reintroducing diverse mixtures of native species can provide greater benefits than approaches that employ monoculture or low-diversity plantations.
6. Effective community partnerships, involving indigenous and local communities, especially ensuring women's participation and encouraging participation of NGOs, businesses, and private sectors from the time of planning to implementation and monitoring, can greatly enhance the success of restoration efforts.
7. Plans for multi-use and multifunctional landscapes for restoration programmes can optimize cost-effectiveness, address the needs of different interest groups, and provide greater social, economic, and ecological benefits, thereby ensuring long-term sustainability of restoration efforts.
8. Restoration efforts can be successfully scaled up through the generation of awareness to change entrenched philosophies that are detrimental to restora-

tion, greater multilateral coordination between implementing nodal agencies, and the removal of institutional and socio-economic barriers that hinder restoration efforts.

9. India's national and international commitments for restoration can be met and scaled up by establishing an effective network of nurseries and seed banks, developing restoration knowledge resource centres and extensive data for native species, creating strong financial and institutional frameworks, and encouraging active engagement of various stakeholders in restoration efforts.
10. Rigorously executed, well-designed, and transparent monitoring efforts that track both the ecological and social outcomes of restoration efforts are vital to assess the progress, dynamically manage restoration methods, and determine tactics to enhance future restoration success.

10.19 Conclusion

In the past decade, the emphasis on urban green space has intensified due to its significance for wildlife conservation, human well-being, and climate change adaptation. The future of transforming Delhi's green forests and spaces lies in looking beyond protected areas. There is a need for innovative concepts in ecological restoration and rehabilitation that focus on enhancing ecosystem functions, services, and biodiversity conservation within urban environments. Eco-restoration through the removal of invasive species and the reintroduction of native species throughout the Ridge is a key component of the FLR (Forest Landscape Restoration) approach. This strategy not only enhances green cover but also takes into account the social and environmental significance of tree patches and forests within the city (Kowarik 2023). Like in Copenhagen, Tokyo, and Singapore, this requires a reorientation of how we think about the space available for nature. Environmental conditions and species composition in many urban ecosystems have been significantly altered, leading to the emergence of novel and hybrid ecosystems. To support the restoration and rehabilitation of these urban ecosystems, the use of established species-rich and well-functioning urban ecosystems as references is recommended. Urban ecological restoration and rehabilitation efforts encompass interventions in both historic and novel ecosystems. Small ecosystems like road verges, hedgerows, green roofs, and traffic islands can serve as valuable green spaces that improve habitat connectivity (Pierer and Creutzig 2019). Ecosystem restoration and rehabilitation in cities effectively improve ecosystem services, such as climate change mitigation or recreation. Engaging citizens and other stakeholders in the successful redesign and reimagination of urban green spaces will contribute to making our cities more sustainable in the future (Aronson et al. 2017). Encouraging the development of innovative concepts for multifunctional and biodiverse urban green spaces requires strengthening research efforts and effectively implementing flagship projects by teams of ecologists, landscape designers, and urban planners (Ali 2023).

Delhi has considerable green cover but should not become complacent. This is an opportune moment to elevate urban forestry management strategies by exploring innovative models and design approaches, embracing technology, and identifying incentives to enhance coordination among diverse stakeholders. Under the Joint Declaration of Intent (JDI) on FLR, signed by India and Germany, GIZ India (Deutsche Gesellschaft für Internationale Zusammenarbeit India) collaborates with various stakeholders in Delhi to seek opportunities for restoring and protecting India's forests, alleviating poverty, preserving and restoring biodiversity, enhancing resilience and ecosystem services such as water availability, and mitigating the impacts of climate change. Other states have started following Delhi Biodiversity Park models for their cities. The Union Minister of Environment, Forest and Climate Change has directed all state chief ministers to create biodiversity parks in all states. The staff at Delhi's biodiversity parks are providing guidance to the officials of different states.

The international community has established ambitious restoration targets, such as achieving land degradation neutrality by 2030. India's forest policies, which emphasize restoration, have developed over time to ensure that the benefits of restoration activities are sustained for local communities and ecosystems in the long term. These policies create a supportive framework for the country to fulfil its national and international restoration commitments.

References

Abhir J (2024) The spirit of conservation: the case of Mangar Bani. Ashoka University. Retrieved from https://www.ashoka.edu.in/the-spirit-of-conservation-the-case-of-mangar-bani/

Agarwal C (2020) Mangar Bani and other forests. Sanctuary Asia, 40(2). Retrieved from https://sanctuarynaturefoundation.org/article/mangar-bani-and-other-forests

Agarwal P (2021, February 18) Eyeing variety: Delhi's Tilpath Valley park ready to welcome herbivores. The Times of India. Retrieved from https://timesofindia.indiatimes.com/city/delhi/eyeing-variety-tilpath-valley-park-gets-ready-to-welcome-herbivores/articleshow/81078914.cms

Agarwal P (2022) At Tilpath biodiversity park, a playground for lost species. The Times of India. Retrieved from https://timesofindia.indiatimes.com/city/delhi/delhi-at-tilpath-biodiversity-valley-park-a-playground-for-lost-species/articleshow/93307401.cms

Ali A (2023) Linking forest ecosystem processes, functions, and services under an integrative social–ecological research agenda: current knowledge and perspectives. Sci Total Environ 2023(892):164768

Amir S, Shannon K (2023) The taking and making of a forest—socio-ecological transformations of the Aravalli Hills, India. Landsc Archit Front 11(1):16

Aronson MFJ, Christopher AL, Evan KL, Goddard MA, Lerman SB, MacIvor JS, Nilon CH, Timothy V (2017) Biodiversity in the city: key challenges for urban green space management. Front Ecol Environ 15(4):189–196

Arora S (2018a) Saving Mangar Bani: why returning grove to original custodians may be only solution. The Times of India. Retrieved from https://timesofindia.indiatimes.com/city/gurgaon/saving-mangar-bani-why-returning-grove-to-original-custodians-may-be-only-solution/articleshow/66061056.cms

Arora S (2018b) Gurugram: work on Bandhwari check dam begins. The Times of India. Retrieved from https://timesofindia.indiatimes.com/city/gurgaon/work-on-bandhwari-check-dam-begins/articleshow/63436527.cms

Babu S, Singh A (2024) The future of Delhi's Green Spaces, Delhi beyond protected areas, Mongaybay series. Retrieved from https://india.mongabay.com/2024/01/commentary-the-future-of-delhis-green-spaces/

Bani U (2015) Down to earth. Retrieved from https://www.downtoearth.org.in/environment/unhappy-bani-34433

Bansal P, Rao AS, Yadav SS, Bhandoria MS, Dash SS (2023) Floristic diversity of native wild ornamental plants of Aravalli Hill range: a case study from district Rewari, Haryana, India. J Threat Taxa 15(1):22479–22493

Bawa KS, Nawn N, Chellam R, Krishnaswamy J, Mathur V, Olsson SB, Pandit N, Rajagopal P, Sankaran M, Shaanker RU, Shankar D (2020) Envisioning a biodiversity science for sustaining human well-being. Proc Natl Acad Sci 117(42):25951–25955

Chandra A, Sinha PG, Joshi T, Sharma P, Chawla M, Gupta S, Kohli PS, Suyesh R, Rao GPC, Rao KS (2018) Pollution in Sanjay Van Lake: a restored waterbody of Delhi. DU J Undergrad Res Innov 4(1):1–13

Chhangani AK, Robbins P, Mohnot SM, Sharma BK (2013) Threats to faunal diversity of the Aravalli Hills with special reference to Kumbhalgarh Wildlife Sanctuary. In: Faunal Heritage of Rajasthan, India: conservation and Management of Vertebrates. Springer International Publishing, Cham, pp 93–112

Correspondent HT (2018a) Delhi gets its third biodiversity park in Tilpath Valley. Hindustan Times. Retrieved from https://www.hindustantimes.com/delhi-news/delhi-gets-its-third-biodiversity-park-in-tilpat-valley/story-22c7t3YnkSk3bqW2jMMpsJ.html

Correspondent TOI (2018b) Two years on, Tilpath Valley biodiversity park breathes again. The Times of India. Retrieved from https://timesofindia.indiatimes.com/city/delhi/two-years-on-tilpath-valley-biodiversity-park-breathes-again/articleshow/63035604.cms

Dar AA, Jamal K, Alhazmi A, El-Sharnouby M, Salah M, Sayed S (2021) Moth diversity, species composition, and distributional pattern in Aravalli Hill range of Rajasthan, India. Saudi J Biol Sci 28(9):4884–4890

Delhi biodiversity Foundation, DDA- Tughlaqabad Biodiversity Park. Retrieved from https://www.delhibiodiversityparks.org/tughlaqabad-biodiversity-park.html

Delhi Biodiversity foundation DDA. Kamala Nehru Biodiversity Park. Retrieved from https://delhibiodiversityparks.org/kn-ridge.html

Delhi Biodiversity foundation DDA. Tilpath Valley Biodiversity Park. Retrieved from https://www.delhibiodiversityparks.org/tilpath-valley.html

Delhi Biodiversity Foundation, DDA-Yamuna Biodiversity Park. Retrieved from https://www.delhibiodiversityparks.org/index.html

Delhi Development Authority. Biodiversity Parks and Nature Reserves of Delhi (2021). Retrieved from: https://dda.gov.in/sites/default/files/Landescape/BiodiversityParksNatureReservesofDelhi.pdf

Dixit K (2021) Lost world: how reforestation plan has sown seeds of hope. The Times of India. Retrieved from http://timesofindia.indiatimes.com/articleshow/88378789.cms?from=mdr&utm_source=contentofinterest&utm_medium=text&utm_campaign=cppst

Dutta J, Choudhury MM, Environmentalist G, Sen MT, Banerjee A, Tiwari C, Nagar EHK. Voice of Environment Newsletter 2019

Edrisi SA, Bundela AK, Verma V, Dubey PK, Abhilash PC (2023) Assessing the impact of global initiatives on current and future land restoration scenarios in India. Environ Res 216:114413

Gandhiok J (2019) Aravali Biodiversity Park tells a tale of Years. The Times of India. Retrieved from: http://timesofindia.indiatimes.com/articleshow/68880325.cms?utm_source=contentofinterest&utm_medium=text&utm_campaign=cppst

Gandhiok J (2022) Delhi government to plant 14 varieties of grass at Asola Bhatti wildlife sanctuary. Hindustan Times. Retrieved from https://www.hindustantimes.com/cities/delhi-news/delhi-govt-to-plant-14-varieties-of-grass-at-asola-bhatti-wildlife-sanctuary-101643568745325.html

Gandhiok J (2023) Dragonfly, damselfly count dips in Delhi after extreme weather. Hindustan Times. Retrieved from https://www.hindustantimes.com/cities/delhi-news/dragonfly-damselfly-count-dips-in-delhi-after-extreme-weather-101696787962403.html

Goel V, Rath R (2022) Asola Bhatti Wildlife Sanctuary—a case study of ecological restoration: its success and present status. Vantage: J Themat Anal 3(2):104–116

Gupta S (2022) From Stockholm 1972 to 2022: holistic and multidisciplinary approach of India to attain sustainability. Vantage: J Themat Anal 3(2):75–87

Harigovind A (2022) Regeneration of 12 hectares at Asola Bhatti to be roadmap to restore other sites. The Indian Express. Retrieved from https://indianexpress.com/article/cities/delhi/regeneration-of-12-hectares-at-asola-bhatti-to-be-roadmap-torestore-other-sites-7926759/

Harigovind A (2024) Why efforts to reclaim the ridge- Delhi's lungs -is no easy task. The Indian Express September 9, 2024. Retrieved from https://indianexpress.com/article/cities/delhi/delhis-southern-ridge-forest-land-indian-forest-act-1927-ignou-maidangarhi-9557360/

Harjani A (2013) Biodiversity Park Revives Yamuna's Pristine Glory. 4:6

IUCN (2024) Strengthening capacities in Forest landscape restoration in India. Ministry of Environment, Forest and Climate Change (MoEFCC), India. Retrieved from https://iucn.org/story/202406/strengthening-capacities-forest-landscape-restoration-india

Jacob N (2019) Local governance and participative water management in urban contexts. In: Water governance and management in India: issues and perspectives, vol 1. Springer, pp 137–167

Jain AK (2009) Low Carbon City policy: planning and practice India. Discovery Publishing House. ISBN: 978-8183564687

Jain P (2022) Delhi LG, CM, Dy CM undertake joint visit to Asola-Bhatti forest area. India Today. Retrieved from https://www.indiatoday.in/india/story/delhi-arvind-kejriwal-v-k-saxena-manish-sisodia-viist-asola-bhatti-forest-1961501-2022-06-12

Klaus VH, Kiehl K (2021) A conceptual framework for urban ecological restoration and rehabilitation. Basic Appl Ecol 52:82–94

Kowarik I (2011) Novel urban ecosystems, biodiversity, and conservation. Environ Pollut 159(8–9):1974–1983

Kowarik I (2023) Urban biodiversity, ecosystems, and the city: insights from 50 years of the Berlin School of urban ecology. Landsc Urban Plan 240:104877. https://doi.org/10.1016/j.landurbplan.2023.104877

Kowarik I, Hiller A, Planchuelo G, Seitz B, von der Lippe M, Buchholz S (2019) Emerging urban forests: opportunities for promoting the wild side of the urban green infrastructure. Sustainability 11(22):6318

Kumar R, Baishya R, Gawdiya S, Verma A (2022) Phytosociology and diversity assessment of tree species of Northern Delhi Ridge Forest. Ecol Environ Conserv 28(Suppl. Issue). https://doi.org/10.53550/EEC.2022.v28i03s.029

Liu Z, He C, Zhou Y, Wu J (2014) How much of the world's land has been urbanized, really? A hierarchical framework for avoiding confusion. Landsc Ecol 29:763–771

Liu Y, Ou X, Zheng X (2022) Urban biodiversity conservation planning integrating the analysis of green space structure and functional connection. Eco Cities 3(1):14

Madan S (2021) First wildlife census in Delhi's Asola Bhatti Sanctuary in July; leopards focal point: Officials. The Tribune. Retrieved from https://www.tribuneindia.com/news/delhi/first-wildlife-census-in-delhis-asola-bhatti-sanctuary-in-july-leopards-focal-point-officials-235886

Madan S, Prakash S, Yadav G, Kohli L, Singh R. Avifaunal diversity of Asola Bhatti wildlife sanctuary. EIACP: Geodiversity and Impact on Environment. 2022;28(1). Retrieved from: https://www.researchgate.net/publication/368918129_Avifaunal_Diversity_of_Asola_Bhatti_Wildlife_Sanctuary_Delhi

Mani G (2022) Delhi govt plans to build 45-km-long 'patrol passage' alongside Asola Bhatti Wildlife Sanctuary. The Indian Express. Retrieved from https://indianexpress.com/article/cities/delhi/delhi-govt-plans-to-build-45-km-long-patrol-passage-alongside-asola-bhatti-wildlife-sanctuary-7930546/

McCleery RA, Moorman CE, Peterson MN (eds) (2014) Urban wildlife conservation: theory and practice. Springer. https://doi.org/10.1007/978-1-4899-7500-3

Muhammad M, Batool S, Hivare V, Li W, Waheed A, Sinha D (2024) Chapter 1 – Bioremediation techniques—classification, principles, advantages, limitations, and prospects. In: Microbiome research in plants and soil, pp 1–23. https://doi.org/10.1016/B978-0-443-21911-5.00003-9

NGT (2017) NGT asks CPCB to test groundwater sample near Bandhwari plant. India Today. Retrieved from https://www.indiatoday.in/pti-feed/story/ngt-asks-cpcb-to-test-groundwater-sample-near-bandhwari-plant-1008325-2017-08-01

Panwar H, Dhote M (2022) Degraded land management in urban areas: the case of biodiversity parks in Delhi. In: Sustainable urbanism in developing countries. CRC Press, pp 291–304

Patel S (2019, May 23) From barren land to rich ecosystem: story behind Yamuna Biodiversity Park. Retrieved from https://indianexpress.com/article/cities/delhi/from-barren-land-to-rich-ecosystem-story-behind-yamuna-biodiversity-park-5743228/

Pati I (2023) Land grab bid in Mangar again, 8 acres in Aravalis encroached. The Times of India. Retrieved from https://timesofindia.indiatimes.com/city/gurgaon/land-grab-bid-in-mangar-again-8-acres-in-aravalis-encroached/articleshow/98685569.cms

Pati I (2024) Over 100 trees near Mangar Bani buffer zone set afire, vigil upped. The Times of India. Retrieved from https://timesofindia.indiatimes.com/city/gurgaon/over-100-trees-near-mangar-bani-buffer-zone-set-afire-vigil-upped/articleshow/110096095.cms

Pierer C, Creutzig F (2019) Star-shaped cities alleviate trade-off between climate change mitigation and adaptation. Environ Res Lett 14:085011. https://doi.org/10.1088/1748-9326/ab2081

Raghurama M, Dutta S, Nair T, Sankaran M, Raman TRS, Mudappa D, Osuri AM, Krishnaswamy J, Ramakrishnan U, Pandit N, Krishen P, Nerlekar A, Agrawal I, Blanchflower P, Borawake N, Bosco GV, Das A, Datar MN, Datta A, Dixit AM, Ghate K, Hiremath A, Joshi P, Joshi A, Joshua J, Khan D, Kumar V, Nandikar M, Naniwadekar R, Narayan G, Negi VS, Rao S, Saberwal V, Saxena V, Varghese A, Watve A, Yadav SR, Chellam R, Bawa K (2023) Restoring India's terrestrial ecosystems: needs, challenges, and policy recommendations. Retrieved from: https://era-india.org/wp-content/uploads/2023/10/Restoring-Indias-Terrestrial-Ecosystems-Needs-Challenges-and-Policy-Recommendations.pdf

Rajput A (2016) Delhi DDA to create biodiversity park at Tughlakabad. Hindustan Times. Retrieved from https://www.hindustantimes.com/cities/delhi-dda-to-create-bio-diversity-park-at-tughlakabad/story-G6fGFyS6vnw3Xvj7hYxM6O.html

Saxena R, Rath R, Gupta S, Sood N (2021) A review on ecological degradation, its causes and sustainable development in Delhi, India. J Appl Nat Sci 13(4):1294. https://doi.org/10.31018/jans.v13i4.2978

Schwarz N, Moretti M, Bugalho MN, Davies ZG, Haase D, Hack J, Hof A, Melero Y, Pett TJ, Knapp S (2017) Understanding biodiversity-ecosystem service relationships in urban areas: a comprehensive literature review. Ecosyst Serv 27:161–171

Seth N, Vats S, Lakhanpaul S, Arafat Y, Babu CR (2024) Performance of horizontal surface free flow integrated constructed wetland developed for treatment of sewage water at Neela Hauz Biodiversity Park, Delhi, India. Int J Ecol Environ Sci 50(4):631–643

Shahabuddin G (2015) Mangar Bani forest: preserve biodiversity to ensure ecological security. Hindustan Times. Retrieved from https://www.hindustantimes.com/ht-view/mangar-bani-forest-preserve-biodiversity-to-ensure-ecological-security/story-6a4KTKSlkHkI3qtiHpzmaO.html

Sharma SN (2012) Yamuna Biodiversity Park, Delhi. EduPedoa Publications, New Delhi

Sharma S (2019) Haryana Government moots buffer zone to save Asola sanctuary. The Times of India. Retrieved from Haryana Government moots buffer zone to save Asola sanctuary Gurgaon News – Times of India (indiatimes.com)

Singh G, Sharma A, Gupta I, Baveja P (2016) Sustainability appraisal and economic valuation of Northern Delhi ridge using participatory research approach. J Innov Incl Devl 1(1):11–19

Singh K, Singh RP, Tewari SK (2021) Ecosystem restoration: challenges and opportunities for India. Restor Ecol 29(3):e13341

Singh K, Abhilash PC, Maiti SK, Venkataraman R (2022) Ecosystem restoration in India during the United Nations decade on ecosystem restoration: the way forward. Restor Ecol 30(5):13692

Singh K, Agrawala M, Dhyani S, Mishra N, Bauddh K, Byun C (2024) Land degradation and development in India: status quo, challenges and opportunities. Land Degrad Dev 35(8):2675–2683

Sinha GN (2014) An introduction to the Delhi Ridge. Department of Forests and Wildlife, Government of National Capital Territory of Delhi, New Delhi

Sinha S (2015) Haryana's Mangar Bani forest demarcated at last. Hindustan Times. Retrieved from https://www.hindustantimes.com/gurgaon/haryana-s-mangar-bani-forest-demarcated-at-last/story-OqTvqpQIMoffH85fpYNc7H.html

Sinha A, Nath A, Lahkar BP, Brahma N, Sarma HK, Swargowari A (2022) Understanding the efficacy of different techniques to manage *Chromolaena odorata* L., an invasive alien plant in the sub-Himalayan tall grasslands: toward grassland recovery. Ecol Eng 179:106618

Sivaramakrishnan K (2017) Courts, public cultures of legality, and urban ecological imagination in Delhi. In: Places of nature in ecologies of urbanism. Hong Kong University Press, pp 137–161

Sonowal K, Kumar P, Dwivedi N, Mondal R (2019) The dying wetlands of Delhi: an overview on the threats and conservation strategies of wetlands. G-J Environ Sci Technol 7(1):1–5

State of Forest Report (2011) Published by Forest Survey of India (Ministry of Environment & Forests) Kaulagarh Road, P.O -IPE, Dehradun – 248195 Website: www.fsi.nic.in

State of Forest Report (2021) Published by Forest Survey of India (Ministry of Environment & Forests) Kaulagarh Road, P.O -IPE, Dehradun – 248195 Website: www.fsi.nic.in

Tewari SK, Negi H, Kaushal R (2019) Status of bamboo in India. Int J Econ Plants 6(1):030–039. https://doi.org/10.23910/IJEP/2019.6.1.0288. https://www.researchgate.net/publication/334255710_Status_of_Bamboo_in_India

Verchot M, Biswas S (2021) On the outskirts of New Delhi, Mangar looks to its past to protect its future. Mongabay. Retrieved from https://india.mongabay.com/2021/09/mangar-on-the-outskirts-of-new-delhi-looks-to-its-past-to-protect-its-future/

Zari MP (2018) Regenerative urban design and ecosystem biomimicry. In: Routledge research in sustainable urbanism series. Routledge

Ecological Restoration of Grasslands in Saurashtra

11

Sunil Kumar Berwal

Abstract

Grasslands are essential ecosystems with less than 10% tree and shrub cover, ranging from temperate to tropical regions. They support diverse species, provide water security, and support biodiversity. Saurashtra's grasslands, once vibrant, are being degraded due to overgrazing, agricultural expansion, and climate change. These activities cause degradation, habitat loss, and increased carbon release, contributing to climate change and loss of biodiversity. Urbanization and resource extraction fragment grasslands, making it difficult for wildlife to access resources. To mitigate these impacts, efforts are needed to restore grasslands in Saurashtra through a holistic approach to ensure the long-term health and sustainability of these ecosystems. This article provides a framework for ecorestoration, focusing on ecological principles, restoration techniques, community engagement, and policy recommendations.

Keywords

Grasslands · Land degradation · Ecorestoration · Community engagement · Biodiversity

11.1 Introduction

Grasslands are vital ecosystems characterized by the dominance of grasses and other herbaceous plants with less than 10% tree and shrub cover (UNESCO). Ranging from temperate to tropical regions, these diverse ecosystems occupy a significant portion of the Earth's land surface. They are habitats to a large number of

S. K. Berwal (✉)
Forest Department, Government of Gujarat, Morbi, Gujarat, India

D. Mittal, D. K. Gupta (eds.), *Ecological Restoration in India: Perspective and Practices*, https://doi.org/10.1007/978-981-96-8943-9_11

Fig. 11.1 Grassland area in Porbandar District

species, including grassland birds, mammals, reptiles, insects, and plants. They provide water security through the recharge of groundwater resources and support biodiversity and local livelihoods. They are large carbon stores and are critical to the global carbon cycle and climate regulation, acting as sources and sinks of carbon dioxide, a major greenhouse gas responsible for climate change. The role of grasslands as carbon sinks is primarily due to the large amount of carbon stored in the soil and vegetation, which helps reduce the amount of carbon dioxide in the atmosphere. They also provide crucial ecosystem services, such as carbon sequestration, soil formation, and water regulation.

The grasslands of Saurashtra, a peninsular region in Gujarat, India, are ecologically significant, supporting a unique biodiversity (Fig. 11.1). However, these grasslands face severe threats from anthropogenic activities and climate change, necessitating urgent ecorestoration efforts.

Once a tapestry of vibrant grasslands, Saurashtra's landscape is now scarred by degradation. Vast stretches, once teeming with life, have succumbed to the relentless pressures of overgrazing, agricultural expansion, and climate change. These grasslands, critical for biodiversity and local livelihoods, are vanishing, leaving behind a stark reminder of ecological imbalance. This alarming decline necessitates immediate and comprehensive ecorestoration efforts, not merely to recover lost flora and fauna but to restore the foundational ecological functions that sustain the region.

11.2 Understanding Saurashtra Grasslands

Saurashtra's grasslands are shaped by its unique geological and climatic contexts. The peninsula, formed by Deccan Trap basalt, features an undulating terrain and a coastal belt (The Preparatory Survey on Project for Ecosystem Restoration 2020).

This geology influences soil formation, resulting in shallow soils, with low organic matter content, susceptible to erosion. The semiarid climate, characterized by a distinct dry season and erratic monsoon rainfall, dictates the grassland's composition and productivity.

Saurashtra is characterized by a semiarid climate with seasonal rainfall, influencing the grassland ecosystem's composition and dynamics. The region's diverse flora includes dominant grass species like *Dichanthium annulatum*, *Cenchrus ciliaris*, and *Sehima nervosum*, along with various herbaceous plants and shrubs. These grasslands support a rich variety of wildlife, including ungulates, carnivores, and avifauna. The soil characteristics of Saurashtra grasslands are typically shallow, with low organic matter content, influenced by the region's geology and climate. Hydrological features include seasonal streams and ephemeral water bodies, playing a crucial role in maintaining ecosystem functions.

11.2.1 Geographical and Climatic Contexts

Saurashtra's geography is characterized by an undulating terrain and a coastal belt. The climate is semiarid, with a distinct dry season and a monsoon season that brings limited rainfall. The variability in rainfall patterns significantly impacts grassland productivity and species composition.

Saurashtra is located in the peninsular region of Gujarat state in Western India, encompassing an area of 47,000 km^2, flanked by the Gulf of Kachchh and coastal plains in the North, Arabian Sea in the West and South, the Gulf of Khambhat in the South East, and the alluvial plains of Bhavnagar in the East. The Saurashtra peninsula forms a rocky tableland comprising Deccan lava. The region is dominated by an undulating surface, broken by hills and various dissecting rivers that flow out in various directions, thus rendering a rugged topography. Most of the area lies below 300 m asl.

Saurashtra's topography, ranging from coastal plains to inland plateaus, creates microclimates that influence local vegetation. The region's mean annual rainfall is about 580 mm, concentrated during the monsoon season (June–September). However, rainfall variability is high, leading to frequent droughts. Temperature extremes range from 10 °C in winter to 45 °C in summer. These climatic conditions favor the growth of hardy, drought-resistant grass species.

11.2.2 Flora and Fauna Diversity

Saurashtra does not have true grasslands of the climatic climax stage; instead, they are found as tertiary seral communities, where the successional processes are arrested due to continued biotic interference like livestock grazing and, to some extent, by fire. The entire region has sparse thorny and scrub forests. The openness of the area, coupled with low and variable rainfalls, provides ideal ecological conditions for the development of grasslands in this region. Interestingly, on the basis of climate, soil, and more importantly, the physiographic differentiation, it supports diverse grassland communities.

Grasslands of Saurashtra are critical for the conservation of biodiversity values and the extraction of fodder through open grazing, as well as by cutting, storing, and distributing during periods of scarcity. For administrative purposes, most of these grasslands are under the jurisdiction of the Gujarat Forest Department and are managed as reserved *vidis* and non-reserved *vidis* (Table 11.1).

The reserved *vidis* are those grasslands where the annual average production of grass is more than 100,000 kg, and in these *vidis*, grass is protected, collected, and stored by the Forest Department and distributed during the period of scarcity. Non-reserved *vidis* have less than one hundred thousand kilograms of annual average grass production, and in most of the cases, local people are allowed to collect the grass to fulfill their needs. These are managed by the Forest Department by leasing out the grass through auctions to various agencies. Apart from the Forest Department, there are substantial tracts of grasslands that are under private control, locally known as "private *vidis*," and are managed by their owners.

Some of the most frequently recorded grass species include *Dichanthium annulatum, Sehima sulcatum, Chrysopogon fulvus, Heteropogon contortus, Sehima nervosum, Cenchrus setigerus, Eremopogon foveolatus, Cenchrus ciliaris, Eleusine compressa, Sporobolus marginatus, Cynodon dactylon, Cymbopogon martini, Desmostachya bipinnata, Iseilema laxum*, etc.

Legumes like *Alysicarpus* spp. and *Indigofera* spp. fix nitrogen, enriching soil fertility. Shrubs like *Ziziphus* sp. provide food and shelter for wildlife. The grasslands support diverse fauna, including the endangered Indian wolf (*Canis lupus pallipes*), the chinkara (*Gazella bennettii*), the Indian fox (*Vulpes bengalensis*), and numerous bird species like the lesser florican (*Sypheotides indicus*). Trophic interactions are complex, with herbivores shaping vegetation structure and carnivores

Table 11.1 Information on *vidis* in Saurashtra

		Reserved *vidis*		Non-reserved *vidis*		
S. no.	District	No.	Area (Ha)	No.	Area (Ha)	Total area (Ha)
1	Amreli	13	5158.00	27	3373.00	**8531.00**
2	Bhavnagar	18	7089.64	28	1383.59	**8473.23**
3	Botad	5	622.36	17	455.67	**1078.03**
4	Devbhoomi Dwarka	1	90.14	8	370.60	**460.74**
5	Gir Somnath	2	337.49	21	3719.44	**4056.93**
6	Jamnagar	10	3967.81	52	3360.85	**7328.66**
7	Junagadh	32	7866.19	80	2663.64	**10529.83**
8	Morbi	5	4450.39	39	2568.74	**7019.13**
9	Porbandar	1	51.44	18	2029.81	**2081.25**
10	Rajkot	19	6829.25	123	5056.05	**11885.30**
11	Surendranagar	6	3071.90	12	486.55	**3558.45**
	Grand total	**112**	**39534.61**	**425**	**25467.94**	**65002.55**

Gujarat Forest Statistics: 2023–24 (Source: Development & Management Wing, PCCF & HoFF Office, Gujarat State)

regulating herbivore populations. The grasslands undergo dramatic seasonal changes. During the monsoon, rapid growth occurs, followed by senescence in the dry season. This seasonality influences wildlife movements and foraging patterns.

11.2.3 Soil Characteristics

The soils of Saurashtra grasslands are typically shallow, with low organic matter content and high susceptibility to erosion. The soil texture varies, with sandy loam and clay loam being common types. Saurashtra's soils are predominantly characterized by low organic matter content, high pH, and sandy loam–to–clay loam texture. Nutrient levels, particularly those of nitrogen and phosphorus, are often limiting. The soils are prone to wind and water erosion, especially in degraded areas.

11.2.4 Hydrological Features

Seasonal streams, ephemeral water bodies, and groundwater resources are crucial hydrological features that support the grassland ecosystem. The availability of water significantly influences grassland productivity and species distribution. These water bodies recharge groundwater, support aquatic biodiversity, and provide water for livestock and wildlife. Groundwater resources are vital during the dry season, but overextraction is a growing concern.

11.3 Threats

The grasslands of Saurashtra face numerous threats, primarily driven by anthropogenic activities. Overgrazing, agricultural expansion, invasive species, climate change, mining, and habitat fragmentation are significant factors contributing to grassland degradation.

11.3.1 Overgrazing and Livestock Pressure

Excessive livestock grazing is a major threat, leading to soil compaction, loss of vegetation cover, and reduced biodiversity (Fig. 11.2). The high livestock population in Saurashtra places immense pressure on the grasslands. Livestock density in Saurashtra exceeds the carrying capacity of the grasslands, leading to severe overgrazing. Data shows a livestock density of over 300 animals per square kilometer in some areas. Overgrazing results in soil compaction, reduced vegetation cover, and loss of palatable grass species.

Overgrazing tends to favor certain plant species that are more resistant to grazing, leading to a decline in overall plant diversity and the loss of habitats for other animals. In arid and semiarid regions, overgrazing can contribute to desertification,

Fig. 11.2 Grass collection from a *vidi*

where grasslands turn into deserts due to soil degradation and loss of vegetation. Degraded grasslands have less capacity to store carbon, which contributes to climate change.

11.3.2 Agricultural Expansion and Land Conversion

The conversion of grasslands to agricultural land for crops and plantations has resulted in significant habitat loss and fragmentation. They are rich in biomass and have good carbon content in the soil. Therefore, they are found suitable for cultivation and can easily be converted into farmland owing to the moderate terrain. This conversion leads to the destruction of the grassland ecosystem by replacing native plants with commercial crops. In Saurashtra, the expansion of cotton, groundnut, and other crops has led to the conversion of vast grassland areas.

11.3.3 Invasive Species

Several invasive species, such as *Prosopis juliflora*, *Lantana camara*, *Parthenium hysterophorus*, *Cassia tora*, and *Cassia uniflora*, have spread rapidly, outcompeting native vegetation and altering ecosystem functions in areas that have not been scientifically managed over the years (Fig. 11.3). Many of these species have invaded large tracts of grasslands, forming dense thickets that suppress native grasses and alter soil properties with reduced biodiversity.

Fig. 11.3 *Lantana* removal through manual/mechanical means

11.3.4 Climate Change Impacts

Climate change–induced droughts and temperature changes are exacerbating grassland degradation, reducing productivity, and altering species composition. Climate change affects grassland soil's organic carbon storage by modifying the processes of plant carbon inputs and microbial catabolism and anabolism. Improved grazing management and biodiversity restoration can provide low-cost and/or high-carbon-gain options for natural climate solutions in global grasslands (Yongfei and Cotrufo 2022). Climate models predict increased frequency and intensity of droughts in Saurashtra. Rising temperatures and altered rainfall patterns disrupt grassland productivity and species composition.

11.3.5 Mining and Industrial Activities

Mining and industrial activities contribute to habitat destruction and pollution, negatively impacting grassland ecosystems. Limestone mining and industrial development contribute to habitat destruction and pollution. Dust from mining operations contaminates soil and vegetation.

11.3.6 Fragmentation and Habitat Loss

Habitat fragmentation due to infrastructure development and land conversion disrupts ecological connectivity and reduces wildlife habitat. Infrastructure development and land conversion fragment grassland habitats, isolating wildlife populations. The changing fire regimes, with less frequent but more intense fires, due to invasive species and climate change, are also a major threat.

11.4 Ecorestoration Principles and Practices

Ecological restoration is a process that aims to assist in the recovery of degraded, damaged, or destroyed ecosystems. This can include a wide range of activities, such as removing invasive species, reintroducing native species, and altering management practices to support the restoration of natural processes. Grasslands are undervalued assets in adapting to climate change. Conserving and restoring grasslands can help mitigate the impacts of climate change by increasing their carbon sink capacity. This can be achieved through the protection and restoration of native grasslands, the promotion of sustainable land use practices, and the reduction of disturbances from human activities. Restoration can be achieved through practices such as planting native vegetation, controlling invasive species, and reducing disturbance from human activities. The restoration of degraded grasslands can increase the carbon sink capacity of these ecosystems. This can be achieved through the planting of native vegetation, the control of invasive species, and the reduction of disturbance from human activities. These efforts can help increase the carbon stored in the soil and vegetation of grasslands, reducing the release of carbon dioxide into the atmosphere and mitigating the impacts of climate change.

Ecorestoration of Saurashtra grasslands involves restoring degraded ecosystems to their natural state or a desired ecological condition.

11.4.1 Ecological Succession and Restoration Goals

Understanding ecological succession is crucial for establishing realistic restoration goals. The aim is to restore the ecosystem's structure, function, and biodiversity. Restoration efforts should aim to facilitate natural ecological succession, restoring the ecosystem to a desired state.

11.4.2 Site Assessment and Planning

Remote sensing, Geographic Information System (GIS), and field surveys are essential for assessing site conditions and identifying restoration potential. Surveys are to be conducted to assess the existing vegetation, soil, and wildlife conditions, with on-site observations to assess vegetation, soil, and water conditions. Satellite imagery or aerial photography may be utilized to map vegetation cover and identify areas of concern. Historical data on climate, land use, and grazing patterns is to be analyzed to understand trends and potential problems. Local communities and landowners have to be positively engaged to gather knowledge and perspectives to determine the feasible restoration goals based on site conditions and ecological principles.

11.4.3 Restoration Techniques

Implementing sustainable grazing management through rotational grazing, adjusting stocking rates, and monitoring grazing pressure can help prevent overgrazing. Rotational grazing, deferred grazing, and livestock exclusion are key strategies, along with improving livestock breeds and providing better nutrition to reduce the need for grazing on grasslands. Restoring degraded grasslands through techniques like replanting and soil conservation can also help improve their health and productivity, wherein native grass and herbaceous species are used for seeding and planting to restore vegetation cover. Mechanical, chemical, and biological methods have to be employed to control invasive species, apart from integrated pest management strategies. Soil and moisture conservation measures like contour bunding, trenching, and water harvesting help reduce soil erosion. Ecological connectivity can be enhanced by creating artificial wetlands, restoring habitat corridors, and creating wildlife refuges. Policies that promote sustainable grazing practices and involve local communities in grassland management can help address threats to the landscape in the long run.

11.4.4 Monitoring and Evaluation

Vegetation, soil, and wildlife populations need to be monitored to assess restoration success. It helps in tracking changes in vegetation cover, species composition, and productivity, apart from monitoring soil organic matter content, nutrient levels, and microbial activity. Assessment of the population dynamics and distribution of key wildlife species is also an important aspect of the same. These objectives may be achieved through the use of remote sensing technologies to monitor large areas and access data for difficult-to-reach locations.

11.5 Grassland Management

Grasslands are an important aspect of the Saurashtra region and cater to the needs of local pastoral communities, apart from being an excellent ecosystem that supports wildlife and avian biodiversity. The grasslands, apart from having their own ecological significance, form one of the most important sources of fodder, especially during years of fodder scarcity. Earlier, these grasslands were managed by the Revenue Department, with the objective of collecting and distributing grass during periods of scarcity, which are common in this part of Gujarat. Realizing the ecological and socioeconomic importance of the grasslands, the Government of Gujarat transferred these lands to the Forest Department to be managed on a scientific basis.

11.5.1 Management Practices

Most of the grasslands in the Saurashtra region are primarily managed by the Gujarat Forest Department, apart from some areas that are under private ownership and management. Working plans are the main basis for working direction in forest areas under the State Forest Department and, thus, the management of its resources, including the grasslands. These working plans are prepared in accordance with the national working plan code and are generally valid for 10 years. While the working plans decide the objectives of management for respective forest divisions, the treatment plans for specific areas are known as "working circles"—the basic forest management units in the working plan. The grassland working circle is one of the key working circles in most of the forest divisions in the Saurashtra region. Accordingly, out of a total working plan area of about 2,00,000 Ha in its grassland-dominated forest divisions, about 42,000 Ha is covered under grassland working circles (*as per Preparatory Survey on Project for Ecosystem Restoration in Gujarat in India Final Report*). Annually, an area of about 2000 Ha has been treated as per the working plan prescriptions for the past 10 years. The Project for Ecosystem Restoration has also been undertaken under financial assistance from the Japan International Cooperation Agency (JICA) and has been categorically designed toward the restoration of such landscapes with grasslands.

11.5.2 Community Involvement and Stakeholder Engagement

Engaging local communities and stakeholders is crucial for ensuring long-term restoration success. This approach ensures that projects align with local needs, foster sustainable practices, and build ownership for long-term success. Involving local communities ensures that restoration efforts are sustainable and aligned with local needs and priorities, leading to long-term ecological and socioeconomic benefits. Community participation fosters a sense of ownership and responsibility for the restored grasslands, leading to better long-term management and conservation efforts. Open and transparent communication and engagement can help prevent and resolve potential conflicts related to land use, resource access, and other issues.

11.5.3 Integrating Traditional Ecological Knowledge

Incorporating traditional ecological knowledge into restoration planning and management enhances the effectiveness of restoration efforts. Local communities possess valuable knowledge about the environment and its resources, which is essential for effective restoration planning and implementation. It helps communities adapt to socio-ecological changes, improving the long-term sustainability of their livelihood strategies and fostering social-ecological resilience. Landscape restoration has the ability to promote self-determination while safeguarding the livelihoods, beliefs, culture, and biodiversity of local people.

11.5.4 The Role of Technology in Grassland Restoration

Drones for seed distribution, remote sensing for monitoring, and GIS mapping are all useful tools. Technology plays a vital role in grassland restoration by enabling efficient monitoring, assessment, and implementation of restoration strategies, including remote sensing, data analysis, and innovative techniques like soil microbiome manipulation and grazing management.

Recent advances in technology can facilitate the efficient planting and establishment of native grasses and other vegetation, including the use of drones for seeding and fertilizer application. Research is exploring the use of beneficial soil microbes to improve soil health and enhance plant growth, potentially accelerating grassland restoration. It can also help optimize grazing practices, ensuring that livestock grazing does not negatively impact grassland health and promoting sustainable land use. Combining different technologies, such as remote sensing, data analysis, and grazing management, can lead to more effective and sustainable grassland restoration outcomes.

11.6 Policy and Management Recommendations

Effective policies and management strategies are essential for sustainable grassland restoration, and some of the important aspects are as follows:

11.6.1 Strengthening Legal Frameworks

There is a need for a policy framework for grassland conservation and restoration that integrates the values of grasslands and other non-forest ecosystems, including both their use for conservation and the many benefits they bring to people into the broader legal framework (Rawat and Adhikari 2015).

11.6.2 Promoting Sustainable Land Use Practices

Promoting sustainable land use practices, especially in grassland ecorestoration, involves managing land resources to maintain ecological integrity while supporting economic and social needs, prioritizing long-term resource efficiency, and minimizing degradation. This includes practices like sustainable grazing management, reforestation, and responsible agricultural practices.

11.6.3 Incentivizing Restoration Efforts

A multifaceted approach combining financial incentives, community engagement, and sustainable land management practices is crucial, focusing on both ecological and socioeconomic benefits to effectively incentivize grassland ecorestoration efforts. Programs that directly compensate landowners or communities for restoring and maintaining grasslands, recognizing the value of ecosystem services like carbon sequestration, water filtration, and biodiversity conservation, are needed. The payments must be tied to measurable outcomes, such as increased native plant cover, improved soil health, or enhanced wildlife habitat. Government grants and subsidies to support restoration efforts must be offered to the local communities, particularly in areas with high ecological value or those facing significant degradation.

11.6.4 Collaborative Management and Capacity Building

In 2015, the government issued a carrying capacity–based livestock grazing system and also a Joint Forest Management (JFM) system for non-reserved *vidis* (Govt. order no. MFP/12/2015/C.F. 28/M dt. 14/9/2015). Joint Forest Management committees can also be engaged in the collaborative management of reserved *vidis*, the order stipulated. JFM committee members have rights on 20% of total grass production of the *vidis*, which committee members can harvest. In case there are no JFM committees near reserved *vidis*, local people can engage in regular grass cutting practice and are eligible to take 10% of the cut grass at nominal rates.

There is a need to educate stakeholders about the importance of grasslands and promote restoration skills by providing theoretical and practical knowledge through field activities, conferences, courses, participatory studies, workshops, and facilitated visits to areas of importance.

11.7 Conclusion

Natural grasslands support high levels of biodiversity, including many rare and threatened species. Despite their significance, grasslands face increasing pressure from human activities, including agriculture, urbanization, and resource extraction, which have caused degradation and habitat loss. Grasslands undergo degradation by human actions that cause conversion and intensification, including changes in fire frequency and intensity and types and intensity of grazing, introduction of non-native grasses, landscape fragmentation, application of agrochemicals, spread of invasive plant and animal species, and air pollution. Impacts like these are complex and enhance fire risk in such ecosystems, releasing more carbon. The fragmentation of these ecosystems releases this carbon back into the atmosphere, contributing to climate change and leading to declines in biodiversity and ecosystem functions. Urbanization and resource extraction have fragmented grasslands into smaller, isolated patches, reducing the size and connectivity of grassland habitats and making it more difficult for wildlife to move and access resources. Climate change is affecting the carbon balance of grasslands, altering precipitation patterns, increasing temperatures, and causing shifts in vegetation communities. In light of these challenges, grasslands offer a unique opportunity for mitigating the impacts of climate change through the conservation, restoration, and management of these ecosystems that additionally support associated wildlife.

Ecorestoration of grasslands in Saurashtra is crucial for conserving biodiversity and maintaining ecosystem services. Continued efforts are needed to implement effective restoration techniques, engage stakeholders, and promote sustainable land management practices. By adopting a holistic and collaborative approach, we can ensure the long-term health and sustainability of Saurashtra's grasslands.

References

Gujarat Forest Statistics: Year 2023–24, Forest Department Government of Gujarat

Rawat GS, Adhikari BS (2015) Ecology and management of grassland habitats in India. ENVIS Bull Wildl Prot Area 17

The Preparatory Survey on Project for Ecosystem Restoration in Gujarat in India (FINAL REPORT) January 2020 Japan International Cooperation Agency (JICA)

Yongfei B, Cotrufo MF (2022) Grassland soil carbon sequestration: current understanding, challenges, and solutions. Science 377(6606):603–608

Ecosystem Restoration Through Umbrella Species: Tracking Ecological Response of Tiger Recovery in Panna Tiger Reserve, Central India

12

Ramesh Krishnamurthy, Supratim Dutta, and Sankarshan Chaudhuri

Abstract

Ecosystem restoration is a key strategy for biodiversity conservation, with umbrella species playing a crucial role in maintaining ecological balance. This chapter evaluates the ecological response to the recovery of tigers (*Panthera tigris*) in Panna Tiger Reserve (PTR), India, highlighting their influence on trophic interactions, prey dynamics, and co-predator relationships. The concept of umbrella species is explored, emphasizing the significance of large carnivores in structuring ecosystems.

Following the local extinction of tigers in PTR in 2009, a reintroduction programme was implemented, leading to a significant population recovery. Using systematic camera trapping, occupancy modelling, and spatial capture–recapture techniques, this chapter examines tiger–leopard (*Panthera pardus*) coexistence, the role of striped hyenas (*Hyaena hyaena*) as scavengers, and changes in prey availability. The findings indicate that tiger recovery has induced spatial segregation among co-predators, with leopards shifting to denser habitats and hyenas exhibiting high conditional occupancy alongside tigers, underscoring their reliance on scavenging tiger kills. Prey density trends show seasonal fluctuations, with stronger tiger–prey associations observed in summer, highlighting the

R. Krishnamurthy (✉)
Wildlife Institute of India, Dehradun, Uttarakhand, India

Faculty of Forestry, University of British Columbia, Vancouver, BC, Canada
e-mail: ramesh@wii.gov.in

S. Dutta · S. Chaudhuri
Wildlife Institute of India, Dehradun, Uttarakhand, India
e-mail: supratim@wii.gov.in

D. Mittal, D. K. Gupta (eds.), *Ecological Restoration in India: Perspective and Practices*, https://doi.org/10.1007/978-981-96-8943-9_12

necessity for prey base management. Additionally, this chapter emphasizes the role of habitat connectivity in sustaining long-term carnivore populations. Landscape-level conservation strategies, including stepping stones and satellite core areas, are essential for facilitating dispersal and maintaining genetic flow.

Despite successful tiger population recovery, challenges persist due to human-induced habitat fragmentation and prey depletion. This chapter underscores the importance of integrating ecological monitoring with adaptive management approaches to ensure sustained conservation outcomes. PTR serves as a model for reintroduction success, demonstrating how science-based interventions can restore apex predator populations while fostering ecosystem resilience. The findings provide valuable insights for future large carnivore conservation programmes across fragmented landscapes.

Keywords

Community · Large carnivore · Population recovery · Reintroduction · Sustainability · Trophic cascade

12.1 Ecological and Ecosystem Restoration

Ecological restoration refers to the process of restoring one or more valued processes or attributes within a landscape (Davis and Slobodkin 2004). Similarly, ecosystem restoration involves assisting the recovery of ecosystems that have been degraded, damaged, or destroyed, aiming to return them to a condition that ensures their structural, functional, and compositional integrity (SER 2004). This process includes active interventions designed to re-establish ecological processes, enhance biodiversity, and improve ecosystem services (IPBES 2019). Ecosystem restoration is vital for addressing global challenges such as biodiversity loss, climate change, and desertification, as it promotes resilience and sustainability within natural systems (IPCC 2019). Additionally, it contributes significantly to international conservation goals, including the Aichi Biodiversity Targets and the Sustainable Development Goals (SDGs) (CBD 2010; United Nations 2015).

Ecosystem restoration has immense potential to reverse environmental degradation, mitigate climate change impacts, and support livelihoods. Restored ecosystems strengthen biodiversity by creating habitats for threatened species and maintaining genetic diversity. Socio-economically, restoration initiatives often generate employment opportunities, particularly in rural and Indigenous communities. For instance, large-scale projects such as the African Great Green Wall have successfully mitigated desertification while promoting economic development and enhancing community resilience (UNCCD 2022).

Ecosystem restoration encompasses diverse ecological contexts/approaches:

1. *Passive Restoration*: This approach relies on natural recovery processes by eliminating stressors such as overgrazing or deforestation. For example, natural for-

est regeneration in the Amazon has led to biodiversity recovery and carbon sequestration without human intervention (Chazdon 2019).
2. *Active Restoration:* Involves deliberate actions like replanting native vegetation, reintroducing species, or controlling invasive species to accelerate recovery. An example is the active riparian reforestation projects in Mexico, where planting native trees has restored degraded riparian ecosystems (Meli et al. 2013).
3. *Rewilding:* Focuses on reintroducing keystone species and apex predators to restore natural processes and trophic cascades. A classic example is the reintroduction of grey wolves (*Canis lupus*) in Yellowstone National Park, which rebalanced ecosystems by controlling herbivore populations and benefitting plant and animal diversity (Ripple and Beschta 2012).
4. *Rehabilitation:* Aims to improve ecosystem functionality rather than restoring ecosystems to their original state. The rehabilitation of cheetahs (*Acinonyx jubatus*) in Namibian nature reserves exemplifies this approach (Walker et al. 2022).

The selection of species is a crucial aspect of ecosystem restoration, shaping and balancing ecological processes. Species with a high ecological impact, such as flagship species, keystone species, indicator species, and umbrella species, are often prioritized in restoration planning. This chapter aims to provide a comprehensive overview of species restoration in the form of rewilding, with reference to the reintroduction and recovery of tigers (*Panthera tigris*) as an umbrella species in Panna Tiger Reserve (PTR). The chapter also examines the broader ecological impacts of tiger reintroduction on ecosystem restoration within the surrounding landscape.

12.2 Concept of Umbrella Species

The term "umbrella species" was introduced by Frankel (1981) and later refined by Wilcox (1984), who suggested focusing conservation efforts on species whose habitat requirements encompass to other species' community, thereby providing a "protective umbrella" for co-occurring species. Initially used as a theoretical concept in conservation ecology (Ehrlich and Murphy 1987), the approach gained attention after the 1990s, with critical evaluations of its potential and the development of criteria for selecting umbrella species (Caro and O'Doherty 1999; Fleishman et al. 2000, 2001). Over time, multiple definitions of this concept have emerged (Zacharias and Roff 2001). Fleishman et al. (2000) defined umbrella species as those "whose conservation confers a protective umbrella to numerous co-occurring species," emphasizing their role in ecosystem-wide conservation efforts.

The concept of umbrella species has emerged as a key approach in guiding ecosystem management by addressing the ecological requirements of a focal species. For a species to qualify as an umbrella species, it must meet certain fundamental criteria. One primary criterion is the requirement for a large home range, which ensures the conservation of extensive habitats and broader spatial scales (McNab 1963). Consequently, large-bodied carnivores or herbivores are often favoured as umbrella species due to their expansive spatial needs and ecological significance

(Noss et al. 1996). Additionally, protecting the populations of a wide array of beneficiary species within a protected area (PA) network is essential to achieve effective conservation outcomes for the umbrella species (Fleishman et al. 2000, 2001).

However, a single species often fails to meet all the necessary criteria, leading to criticisms of the umbrella species concept (Noss et al. 1997). To address this limitation, scientists have proposed a multi-species strategy that considers broader conservation goals and ecological processes (Fleishman et al. 2001). For example, studies have demonstrated that grizzly bears (*Ursus arctos horribilis*) are effective umbrella species for explaining species richness, while wolves and cougars (*Puma concolor*) are better suited for capturing food web dynamics in North American landscapes (Steenweg et al. 2023). Similarly, lions (*Panthera leo*) and buffaloes (*Syncerus caffer*) serve as umbrella species in African ecosystems (Caro 2003), whereas jaguars (*Panthera onca*) play a similar role in the landscapes of Latin America (Thornton et al. 2016).

In Asia, particularly in the Indian subcontinent, tigers and Asian elephants (*Elephas maximus)* have been identified as highly effective umbrella species (Penjor et al. 2024; Vasudeva et al. 2022). Tigers, in particular, have traditionally been prioritized due to their large home ranges, specific habitat requirements, and their status as a flagship species. A focused study on the Greater Panna Landscape (GPL) emphasized the significance of tiger conservation in promoting biodiversity. While tigers play a critical role in conserving biodiversity, their habitat does not always need to align with those of other taxa, potentially leading to under-representation of certain species and ecological niches (Seddon and Leech 2008; Wang et al. 2021). It has been estimated that 62% of the GPL requires restoration interventions to ensure adequate representation of both the tiger habitat and the broader biodiversity of the region (Vasudeva et al. 2022). Consequently, relying exclusively on tigers as an umbrella species may not comprehensively address the diverse ecological requirements of the GPL. A landscape-level conservation approach incorporating additional focal species is essential for achieving more inclusive and effective biodiversity conservation in this region.

12.3 Global Status of Tigers

In the Anthropocene era, habitat loss, fragmentation, poaching, and prey depletion pose significant challenges to the long-term survival of tigers (Seidensticker et al. 1999). These threats have caused a drastic reduction in tiger populations, with their habitat shrinking by approximately 93% of their historical range (Goodrich et al. 2015). Similar trends have been observed in other large carnivores, such as lions (94%), cheetahs (92%), leopards (79%), and Asiatic black bears (64%; Wolf and Ripple 2017). Consequently, tiger populations have become confined to small, isolated forest patches, which are insufficient for sustaining viable populations of wide-ranging carnivores in the long run (Wikramanayake et al. 1998; Ranganathan et al. 2008). Such fragmented populations face an increased risk of extinction (Caughley 1994; Purvis et al. 2000). Prey depletion further exacerbates this problem, with approximately 50% of tiger prey species being classified as threatened and 81% showing declining trends (Wolf and Ripple 2016). These factors have led to alarming declines in global tiger populations, with fewer than 3200 wild tigers reported in 2010. By 2015, the number increased slightly to 3890 individuals (Third Asia Ministerial Conference on Tiger Conservation 2016), and as of 2022, the global tiger population was estimated at 4694 (Global Tiger Forum 2022).

The tiger is globally recognized as one of the most charismatic and ecologically significant species. Serving as an indicator species, it plays a critical role in maintaining ecosystem health. Tigers inhabit 13 countries across the Indian subcontinent, South-East Asia, and parts of China and Russia (Mazák 1981; Sunquist and Sunquist 2002; Luo et al. 2004). However, habitat destruction, illegal wildlife trade, poaching, and declining prey populations have significantly impacted tiger populations (Nowell and Jackson 1996; Seidensticker et al. 1999). Despite sporadic conservation initiatives, the lack of science-based management has contributed to the local extinction of large carnivores in various regions, including Asian ecosystems (Michalski and Peres 2005; Brugière et al. 2015; Louys 2014; Mahmood et al. 2021). In some regions, collective conservation efforts have been insufficient, leading to the local extinction of tigers in Laos (Rasphone et al. 2019), Vietnam, and Cambodia (Goodrich et al. 2015).

12.4 Overview of Reintroduction Programmes

Local extinctions are increasingly being addressed through active recovery efforts, including conservation translocations. Reintroduction, a critical conservation tool, aims to re-establish viable populations of species within their native ranges (IUCN/SSC 2013). The International Union for Conservation of Nature (IUCN) defines translocation as "a deliberate and mediated movement of wild individuals or populations from one part of their range to another," and reintroduction as "an attempt to establish a species in an area that was once part of its historical range" (IUCN 1998). Despite the promise of these approaches, the success rate of large carnivore reintroductions has historically been low (Wemmer and Sunquist 1988; Breitenmoser 2001). Translocation success rates for carnivores are at around 48% (Griffith et al. 1989), and tiger translocations have mostly been limited to sporadic events and rescue cases (Seidensticker 1976). Consequently, the fate of released Siberian tigers from the Otis Rehabilitation Centre into the wild remains unknown (Goodrich and Miquelle, 2005). The lack of post-release monitoring has further hampered the success and understanding of reintroduction outcomes (Hunter 1998).

Reintroduction programmes face numerous challenges, including demographic and genetic issues associated with small effective population sizes (O'Brien and Evermann 1988), habitat isolation (Slough 1994), and reduced genetic variation (Stockwell et al. 1996). Ecological factors, such as poaching, habitat degradation, and fragmentation, significantly influence the success of tiger populations (Matyushkin 1998; Karanth and Stith 1999).

12.5 Large Carnivore Recovery Programmes

In response to the global tiger population decline, 13 tiger range countries, supported by conservation agencies, committed to a 12-year goal of doubling wild tiger numbers by 2022 (Global Tiger Initiative 2011). The Global Tiger Recovery Program (GTRP) was established to implement an action plan aimed at strengthening policies and institutional frameworks. A total of 76 Tiger Conservation Landscapes (TCLs) were identified, representing clusters of "potential effective habitat." Studies indicate that 18 tiger population recovery sites had tiger densities far below the habitat's potential carrying capacity (Harihar et al. 2018). Notably, in September 2024, two Amur tigers were reintroduced in Ile-Balkhash, Kazakhstan, marking the return of tigers to the region after 70 years. Simultaneously, Cambodia plans to reintroduce tigers from India under a bilateral MoU with the technical support of the National Tiger Conservation Authority (NTCA), Wildlife Institute of India, Global Tiger Forum, and Wildlife Alliance (Kesaro et al. 2024).

Reintroduction efforts are not limited to tigers. In Africa, five apex predators have been reintroduced across 30 projects, while two species in Asia and six species across 28 projects in North America have undergone reintroductions (Breitenmoser 2001). Under the Endangered Species Act (ESA) of 1973, grey wolves were reintroduced from Canada to the United States in 1995–1996 (Fritts et al. 1997). Similarly, eight female pumas were translocated from Texas to Florida in 1995 to enhance genetic diversity. Other notable efforts include the reintroduction of lions in Akagera National Park, Rwanda (Briers-Louw et al. 2019), and ongoing plans to rewild the South China Tiger through the Save China's Tigers initiative. The jaguar, facing habitat loss and fragmentation, has become extinct in El Salvador and Uruguay, yet experts plan to release 20 jaguars in Iberá, Argentina. The evaluation of short- and long-term success indicators for ongoing reintroduction programmes is a critical aspect of conservation practice and knowledge, serving as a foundation for guiding and improving future reintroduction initiatives.

12.6 Tiger Reintroduction Programmes in India

India, historically home to approximately 40,000 tigers during the nineteenth century (Gee 1964), witnessed drastic declines in tiger populations due to post–World War II industrialization and human population expansion (Kummer and Sham 2023). Over 57% of historic tiger habitats have been converted into urban areas, agricultural lands, and settlements (Sanderson et al. 2010). To combat these declines, the Government of India launched the "Project Tiger" initiative in 1973 (Panwar 1987). Despite these efforts, tiger extinctions were declared in Sariska Tiger Reserve (STR), Rajasthan, in 2004 (Sankar et al. 2010) and in Panna Tiger Reserve, Madhya Pradesh, in 2009 (Wildlife Institute of India 2009), mostly due to poaching.

The reintroduction and translocation of large carnivores is a well-established conservation practice in Africa (Hofmeyr and Van Dyk 1998; Hofmeyr et al. 2003). However, such efforts have been less frequent in India. Following a feasibility study conducted in accordance with IUCN/NTCA guidelines, five adult tigers (three females and two males) were translocated from Ranthambhore to Sariska during mid-2008 (Sankar et al. 2010). Subsequently, three additional tigers were introduced to Sariska as founder individuals. Regarding the reintroduction in Panna Tiger Reserve (PTR), two adult female tigers were reintroduced in March 2009. The first female (T1) was translocated from Bandhavgarh Tiger Reserve, while the second female (T2) was moved from Kanha Tiger Reserve. Later, a male tiger (T3) was captured in Pench Tiger Reserve and translocated to Panna. In addition, two orphaned female tigers (T4 and T5), raised in a semi-wild environment in Kanha Tiger Reserve, were introduced into Panna in March and September 2011, respectively. The final female tiger (T6) was reintroduced from Pench Tiger Reserve to Panna. In 2015, a rescued male tiger (T7) near Bhopal was captured and translocated to PTR (Ramesh et al. 2013).

12.7 Status of Tigers in Sariska and Panna

Sariska Tiger Reserve (STR), located in the Aravalli hill range of Rajasthan, holds significant importance for tiger conservation. A recent camera-trapping survey conducted during 2021–22 documented 19 adult tigers residing in Sariska (Qureshi et al. 2023), with a poor population growth rate of 16% (Jhala et al. 2021). Population supplementation efforts in the dry deciduous forests of Sariska have faced significant setbacks, primarily due to substantial human-induced disturbances that have resulted in habitat degradation within key tiger habitats. As a result, the tiger recovery programme in STR continues to encounter considerable challenges in achieving its conservation objectives.

Conversely, PTR is situated in the Vindhya hill range of Central India and encompasses two administrative units, with a core area of 542 km^2 and a buffer zone of 1032 km^2. PTR also includes the adjacent Gangau Wildlife Sanctuary (33.47 km^2) and spans the Panna, Chhatarpur, and Damoh districts of Madhya Pradesh. This tropical dry deciduous forest ecosystem (Champion and Seth 1968) has been under scientific observation for over four decades. In 1982, PTR's tiger population was estimated at 22 individuals using the conventional pugmark method (Dwivedi 2003). However, a robust camera-trapping study in 2002 estimated the population at 25–30 individuals (Fig. 12.1, Karanth et al. 2004). By 2006, the population had declined to nine tigers within an intensive study area (Jhala and Qureshi 2006). A rapid camera trap assessment in 2009 revealed the presence of only one tiger, which subsequently disappeared, leading to the official declaration of functional extinction in February 2009 (Wildlife Institute of India 2009).

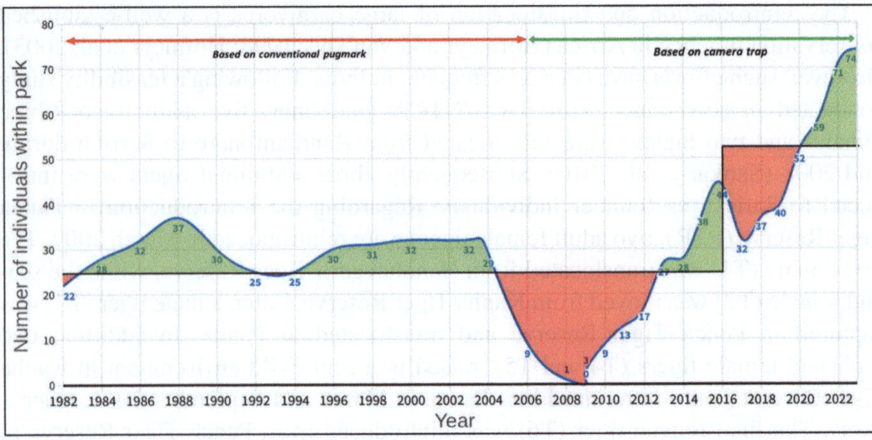

Fig. 12.1 Graph exhibiting the population trend of tigers in Panna Tiger Reserve between 1982 and 2023. The green and red shaded areas show the population numbers above and below the carrying capacity, respectively

12.8 Population Recovery in Panna

From the initial group of seven reintroduced individuals, the tiger population in PTR has demonstrated remarkable growth, exceeding eight times its initial size and reaching the reserve's carrying capacity of 58 individuals. The population growth rate (r) of 26% signifies a highly positive trajectory, far surpassing the national tiger population growth rate of approximately 6% (Jhala et al. 2020a). The annual multiplicative growth rate (λ), representing the annual change in population size, was calculated at 1.29, indicating a 29% yearly population increase (Dutta and Krishnamurthy 2024a). Between 2010 and 2021, 120 cubs were recorded from 45 litters produced by 18 females. The mean litter size of 2.66 ± 0.1 (SE) in PTR's reintroduced population was higher than the mean litter size of the native Panna tiger population (2.3) reported by Chundawat et al. (2002). This increase is likely attributable to genetic admixture from diverse source populations, enhancing genetic potency (Reddy et al. 2016). Similarly, the mean inter-birth interval (IBI) in PTR was 19.16 ± 0.5 (SE) months, ranging from 14 to 24 months (Dutta and Krishnamurthy 2024a). Breeding females abandoned 12 litters, delivering subsequent litters within 5–8 months of abandonment. Reintroduced tigers in PTR exhibited shorter IBIs compared to stable Indian tiger populations in similar biogeographical ranges (Singh et al. 2013; Sadhu et al. 2017) and the historic Panna tiger population (Chundawat et al. 2002). This shorter interval facilitated rapid population recovery. Survival rates, a key metric for assessing reintroduction success, were high in PTR. The overall survival rate of reintroduced tigers was 82% (± 0.02) due to different life history traits (Smith 1993), with males exhibiting lower survival rates (79% ± 0.03) compared to females (84% ± 0.02; Dutta and Krishnamurthy 2024a).

Positive population growth rates, shorter IBIs, and high cub survival rates collectively contributed to the rapid recovery of the PTR tiger population. Simultaneously, strong management and conservation interventions have played a critical role in stabilizing the reintroduced tiger population. As a result, the tiger population in PTR now demonstrates a high probability of long-term viability, reflecting a successful recovery effort. Reintroduction represents a pivotal ecological strategy for reviving near-extinct populations, making it a critical component of species-specific conservation efforts. The success of such programmes depends heavily on meticulous post-release monitoring and thorough documentation of the reintroduction process, which are essential for effective conservation and the advancement of ecological knowledge (Bubac et al. 2019). The success story of tiger reintroduction in PTR has served as a beacon of inspiration across the Indian subcontinent. In recent years, two additional tiger reserves within the Central Indian landscape have initiated tiger reintroduction programmes. In April 2008, two tigers (one male and one female) were translocated from Kanha and Bandhavgarh Tiger Reserves to the newly designated Veerangana Durgawati Tiger Reserve (Qureshi et al. 2023). This reserve now supports a population of 15 tigers, including young adults. Similarly, Madhav National Park witnessed the reintroduction of tigers after a gap of 27 years. In March 2023, tigers were translocated to this park from Bandhavgarh and Satpura Tiger Reserves. The reintroduced populations in these landscapes are showing promising signs of establishment, presenting park managers with the significant challenge of adopting a landscape-level conservation approach within a metapopulation framework.

12.9 Role of Large Carnivores in the Ecosystem: Trophic Cascade

The order "Carnivora" comprises 245 terrestrial species distributed across all major habitats globally (Hunter 2018). Carnivores of varying body sizes are integral to ecosystem regulation (Beschta and Ripple 2009; Prugh et al. 2009; Estes et al. 2011). Among them, large carnivores, as apex predators, play a pivotal role by exerting top-down regulatory effects on lower trophic levels (Elmhagen et al. 2010; Estes et al. 2011; Ripple et al. 2014; Brose et al. 2019). Through predation, they control herbivore populations, while intra-guild competition enables them to regulate mesocarnivores, thereby influencing ecosystem dynamics along multiple food web pathways (Ripple et al. 2014). Apex predators also drive prey species to avoid areas with high predation risk, altering behaviours and interactions across the food chain (Le Roux et al. 2018; Pringle 2020; Gaynor et al. 2019; Sheriff et al. 2020). By modulating herbivore density and behaviour, large carnivores contribute significantly to the maintenance of global plant biomass, supporting the "green world hypothesis" (Bond 2005; Wilkinson and Sherratt 2016). These predator–prey dynamics can initiate trophic cascades that shape the ecosystem structure (Werner and Peacor 2003; Peckarsky et al. 2008; Preisser and Bolnick 2008), although such interactions are often overlooked in conservation and wildlife management practices (Creel and Christianson 2008).

The interplay between trophic dynamics and local climatic factors establishes distinct equilibria between bottom-up and top-down forces within food webs, thereby influencing ecosystem diversity, structure, and functionality (Hopcraft et al. 2010; Galetti et al. 2018). The loss of apex predators disrupts this balance, leading to profound shifts in ecosystem structure and function, with cascading effects on biodiversity and species composition (Prugh et al. 2009). A direct consequence of apex predator decline is the unchecked proliferation of mesopredators, as described by the "mesopredator release hypothesis," which often results in severe consequences for biodiversity and the integrity of plant and animal communities (Ripple et al. 2013; Brashares et al. 2010; Terborgh 2015).

In ecosystems regulated by top-down processes, the removal of apex predators can drive herbivore populations beyond sustainable levels, leading to a transition towards resource-limited, bottom-up control (Owen-Smith 1987; Ripple and Van Valkenburgh 2010; Letnic and Ripple 2017). According to the exploitation ecosystem hypothesis, herbivores respond most strongly to the absence of apex predators in highly productive habitats. These changes trigger cascading effects that disrupt the dynamics of lower trophic levels, causing significant biodiversity loss and altering species composition (Hebblewhite et al. 2005; Beschta and Ripple 2009).

12.10 Regulation of Herbivore Population and Vegetation

The reintroduction of grey wolves in Yellowstone National Park is a trademark example of top-down trophic rewilding and its profound ecological impacts (Fortin et al. 2005). The extirpation of wolves from the Greater Yellowstone ecosystem initiated a trophic cascade, disrupting the balance between herbivores and vegetation (Boyce 2018). During the 70-year absence of wolves, herbivory increased significantly, hindering the regeneration of key plant species such as aspen and cottonwood (Beschta 2005; Halofsky and Ripple 2008). The reintroduction of wolves mitigated these adverse effects by restoring critical herbivore behavioural patterns, including elk population regulation and the establishment of a "landscape of fear," thus reinstating a top-down trophic structure (Fortin et al. 2005; Boyce 2018; Halofsky and Ripple 2008; Ripple and Beschta 2012). Beyond the Yellowstone wolves, predator reintroductions have demonstrated their broader potential for ecosystem restoration. In the tropical savanna ecosystems of Australia, dingoes (*Canis dingo*) have been shown to regulate herbivore and mesopredator populations, significantly influencing the ecosystem structure (Leo et al. 2019). Similarly, in European landscapes, the reintroduction of the Iberian lynx (*Lynx pardinus*) has had a cascading effect, notably controlling the population of mesocarnivores such as red foxes (Jimenez et al. 2019). In African savannas, studies reveal that buffalo populations actively avoid areas with high lion activity (Davies et al. 2016). Additionally, the presence of cheetahs has been linked to reduced visitation rates of medium-sized prey species, highlighting their role in regulating prey populations (Ruble et al. 2022).

12.11 Ecological Responses in PTR

Prey density is a fundamental determinant of tiger density, as the availability of adequate prey biomass directly affects tiger population size, distribution, and survival. Numerous studies have established a strong positive correlation between prey abundance and tiger density across various landscapes, highlighting the necessity of a minimum threshold of large-bodied prey, such as deer and wild boars, to sustain viable tiger populations (Karanth et al. 2004; Lovari et al. 2015). Karanth et al. (2004) provided compelling evidence supporting the functional relationship between prey density and tiger density across diverse landscapes. Their findings indicate that in prey-rich habitats, tigers predominantly target medium- to large-sized ungulates, whereas in prey-deficient areas, tiger populations are restricted due to heightened inter-specific competition (Miller et al. 2014). In India, reserves with high prey densities, such as Corbett Tiger Reserve, have demonstrated the capacity to support substantial tiger populations (Jhala et al. 2020a).

Previous data have revealed significant fluctuations in prey populations over the years, with species such as the chital, sambar, and nilgai exhibiting an initial surge in density during 2008–2009, followed by a subsequent decline in 2020–21 (Table 12.1). The highest collective prey density was recorded in 2008–2009 when the tiger population was functionally extinct (Wildlife Institute of India 2009). The native tiger population was estimated at 6.94 ± 3.23/100km^2; following reintroduction efforts, the tiger population showed a consistent increase from an estimated 1.54 ± 0.44 individuals per 100 km^2 in 2012–2013 to 5.61 ± 0.95 in 2020–2021 (Table 12.1), indicating a successful recovery. Subsequently, the leopard population, which was not recorded in the initial years, increased steadily from 12.2 ± 0.02 individuals per 100 km^2 in 2012–2013 to 25.12 ± 2.09 in 2020–2021, suggesting successful coexistence of the two apex predators (Table 12.1).

The comparative analysis of the relationship between the relative abundance index (RAI) of all prey species and the RAI of tigers across two distinct seasons—winter (depicted in blue) and summer (depicted in brown; Fig. 12.2)—revealed a positive correlation between prey availability and tiger presence. The shaded regions around the regression lines represent confidence intervals, indicating a relatively consistent trend across seasons with some variability. The steeper slope of the regression line in summer suggests a stronger dependence of tiger presence on prey availability during this period, which could be attributed to seasonal variations in prey behaviour and water availability (Gupta and Krishnamurthy 2023).

These findings underscore the importance of effective prey management in sustaining tiger populations. It further emphasizes that habitat quality, protection measures, and human disturbances play secondary yet significant roles in influencing prey availability and, subsequently, tiger densities. The observed decline in prey density post-reintroduction is concerning, especially given the concurrent twofold increase in both tiger and leopard populations. Thus, the tiger recovery programme must prioritize prey population management to ensure long-term success. Additionally, a dedicated recovery plan for grassland indicator species, such as the chinkara, is essential to maintain habitat heterogeneity. The stronger prey–tiger

Table 12.1 Year-wise density estimation of prey, tigers, and leopards in Panna Tiger Reserve

	Year-wise prey (per km^2) and carnivore density (per 100 km^2) in Panna Tiger Reserve (core)					
Species	2001	2003–2004	2008–2009	2012–2013	2016–2017	2020–2021
Chital	10.8	16.6	13.9	5.0	13.8	10.67
Sambar	9.2	10.3	12	8.7	11.6	6.98
Nilgai	6.0	13.1	17.6	9.6	12.4	7.46
Wild boar	1.3	1.51	1.6	7.5	5.98	3.99
Chinkara	0.9	1.97	6.2	4.2	–	0.47
Tiger	–	***6.94 ± 3.23***	***0***	***1.54 ± 0.44***	***3.18 ± 0.7***	***5.61 ± 0.95***
Leopard	–	–	–	***12.2 ± 0.02***	***17.91 ± 0.2***	***25.12 ± 2.09***

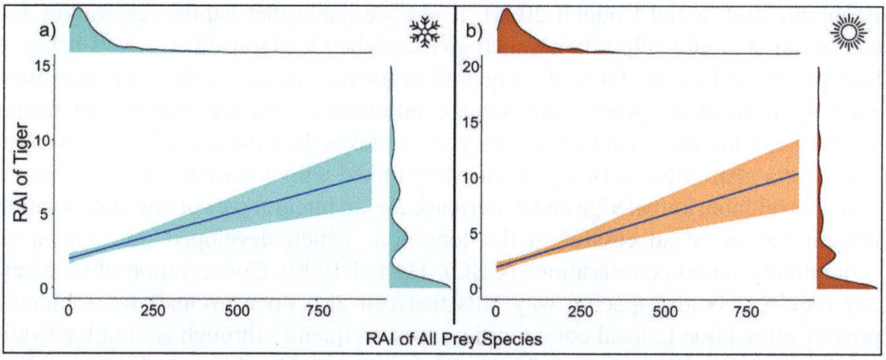

Fig. 12.2 Relation between the relative abundance index (RAI) of tiger and prey species (the chital, sambar, nilgai, and wild boar) obtained from camera trap data from Panna Tiger Reserve across (**a**) winter and (**b**) summer

association observed in summer suggests the need for seasonal prey management strategies to sustain prey resources during periods of environmental stress. Maintaining a stable prey base is crucial for conserving predator–prey dynamics, achieving ecological balance, and ensuring the long-term success of large carnivore conservation efforts in the reserve.

12.12 Human Dimensions of Conservation: Community-Based Conservation

Humanity has impacted biodiversity in a myriad of ways. The ever-increasing human expansion has imperilled biodiversity to a large extent, both directly and indirectly, which is substantiated by rapid land use changes, climate change, exploitation of natural resources, pollution, and invasive alien species (IPBES 2019; Pereira et al. 2012; Tilman et al. 2017). A few decades ago, the Convention on Biological Diversity (CBD) was formed to reduce the escalation of biodiversity loss by 2010, but unfortunately, biodiversity extirpation continues unabated. South-East Asian countries, including the Indian subcontinent and China, are some of the last remaining regions with high biodiversity, yet approximately two-thirds of mammals found here have been threatened with extinction (IUCN 2024) in the past 50 years due to human population explosion, land use changes, and per capita consumption (Van Vuuren and Bouwman 2005; IUCN 2024). Therefore, the fate of current biodiversity depends on the ever-rising global demands for food, energy, and other natural resources.

In 2015, the United Nations' member states unanimously agreed to achieve "Sustainable Development Goals" (SDGs) by 2030 to restore and maintain peace and prosperity for people and nature. However, the progress report for achieving these SDGs shows a rather marginal development in enhancing people's quality of life as well as integrating the health of the natural ecosystems (General Assembly

Economic and Social Council 2024). It is undeniable that habitat restoration and conservation would help achieve such goals, as they lead towards the well-being of both people and nature. Globally, a quarter of the land mass is either owned or managed by Indigenous people and local communities, and the intertwined nature between communities and natural ecosystems makes their participation in conservation efforts instrumental. Being inextricably linked with natural resources, communities' invaluable knowledge and experience are an integral part of any conservation programme to be successful in the long run, which developed the context of "community-based conservation" (CBC) (Hackel 1999). Conservation of biodiversity requires context-specific way outs that can also open avenues for achieving poverty alleviation in local communities, most frequently through alternative livelihood generation (Gadgil et al. 2000). With such a wider focus and applicability, especially in achieving conservation and ensuring the development of local people alongside, the CBC concept has gained significant attention in the past few decades (Berkes 2004; Kothari et al. 2000).

12.13 Genesis of Conservation Displacement

While involving communities in conservation has shown promise, the challenge grows high, especially in regions with high biodiversity that are also exposed to the pressure of the upsurging human population in the past decades (Williams 2011). Research says that to achieve conservation targets for species (e.g. terrestrial mammals), strategic expansion of the PA network, reduction of per capita consumption, and progressive legislation are necessary (Visconti et al. 2016). Now the question arises of how to achieve such hard-to-come-by and conflicting dual conservation goals in these situations. Conventional practices for the conservation of biodiversity and habitat involve managing PAs with human-free zones to ensure the least anthropogenic impact on the environment (Terborgh et al. 2002). As also recognized by the IUCN, one of the categories of legal protection is the removal or "displacement" of human interference from PAs, enabling the conservation of large mammals, such as tigers and elephants, which require large, undisturbed habitats (Watson et al. 2014). The primary objectives of maintaining such a human-free, "pristine habitat" are to ensure the survival of species of interest (often endangered ones) and reduce the human–wildlife conflict (Karanth and Karanth 2007; Rodrigues et al. 2004; Watson et al. 2014). However, such displacement events often do not fully comply with the interests of the local communities, and rigid adherence to eliminate human interference from PAs, otherwise known as "fortress conservation" (Schmidt-Soltau 2003), frequently caused further impoverishment of the relocated people (Shahabuddin et al. 2006). It is also worth noticing that cases where people's interests were not incorporated into the decision-making or policymaking system failed to address the overarching goal of conservation (Cernea and Schmidt-Soltau 2003; McLean and Straede 2003; Pinedo-Vasquez and Padoch 1993). In India, the situation is even more challenging as the conservation of flagship species like tigers requires large, protected habitat patches in PA networks, but often, those PAs possess human settlements, which can create "obstacles" for the conservation. Moreover, <5% of the landmass in India comes under the PA network, and those relatively small-sized and fragmented PAs (average 210 km^2) certainly need reinforcements for long-term conservation of such flagship species, whose former range has already shrunk up to 40–60% in the past century (Karanth et al. 2010). Providentially, in the past decade, due to the growing evidence for the untenability of the rigid "fortress conservation" (Schmidt-Soltau 2003) and the rise of the concept of "Free, Prior and Informed Consent" (FPIC) (Hanna and Vanclay 2013), the long-due human rights were recognized in conservation displacement (Corson et al. 2020). These reforms were reflected in the revised guidelines issued by the NTCA for the village relocations from the core zone of tiger reserves, which brought a significant change in the conservation approach for biodiversity in India (Sarma and Barpujari 2023). In the Indian scenario of conservation, switching from the earlier need-based approach to the "rights-based" approach (RBA) holds significant promise to achieve dual goals of conservation during the conservation displacement of communities (Sarma and Barpujari 2020). As mentioned, instead of evicting people from the PAs, such voluntary relocations of human settlements have resulted in effective outcomes with community participation and improved livelihood opportunities (Dash and Behera 2015; Harihar et al. 2014; Karanth et al. 2018; Karanth and Karanth 2007).

12.14 Conservation Displacement in India

In India, the conservation displacement or relocation of villages received mixed responses; some ended up with failed community participation, whereas some gained conservation benefits from both human and biodiversity perspectives. In Bhadra Tiger Reserve, voluntary village relocation sets an exceptionally promising example, where the displacement of people resulted in their upliftment of socio-economic status, thus benefitting both people and nature (Karanth 2007; Karanth and Karanth 2007). In Melghat Tiger Reserve, it was found that most of the relocated people consider being mainstreamed as either better than or at least as good as being in the forest (Sekar 2016). Similarly, in Simlipal Tiger Reserve, Odisha, social scientists found that education level, working class, and gender played important roles in voluntarily agreeing to be relocated, which was primarily associated with income opportunities (Dash and Behera 2018). However, food security concerns, arising from limited biophysical and administrative opportunities tend to cause unease among the people on the larger benefits of relocation (Dash and Behera 2018), and in some cases, the evacuation was involuntary, without obtaining consent from the oustees (Shahabuddin and Bhamidipati 2014). Satpura Tiger Reserve, situated in the Central Indian landscape, was among the first tiger reserves that followed the NTCA's guidelines for "voluntary" village relocation in notified critical tiger habitats of tiger reserves, where a total of 42 villages were relocated (National Tiger Conservation Authority 2017; Sarma and Barpujari 2023). Following the rights-based approach, Satpura Tiger Reserve sets another successful example of conservation displacement, by ensuring community participation in every stage and post-relocation hand-holding (Sarma and Barpujari 2023). Additionally, situational adoption of the "hybrid model," where families can avail of both monetary compensation and cultivable land, may have explained Satpura's successful village relocations (Sarma and Barpujari 2020, 2023). However, the socio-economic status of the relocated communities could largely determine the post-relocation success, as the more marginalized the community is, the less likely they are to receive a better livelihood after relocation, for which post-relocation hand-holding may be necessary for the people to sustain (Kabra 2009). Thus, broadly, most of the case studies of conservation displacement in India reported multifaceted responses that call for site-specific actions to make the process successful.

When it comes to crises for conservation, such as the extinction of iconic species due to human interference, they warrant desperate measures to reduce the anthropogenic impact. In India, tigers became extinct from two of the prime tiger reserves, namely Sariska Tiger Reserve and Panna Tiger Reserve. The decimation of tiger populations naturally created a lot of hue and cry, which possibly led to the creation and retention of inviolate spaces in the core zones of tiger reserves through village relocation (Narain et al. 2005; Shahabuddin and Bhamidipati 2014). In Sariska Tiger Reserve, tigers became locally extinct, primarily due to poaching, and were subsequently reintroduced (Sankar et al. 2010). The village relocation process received criticism due to a lack of people engagement and poor institutional framework (Torri 2011) and inadequate addressing of gender-specific (females) negative responses towards tiger reintroduction (Doubleday 2018). Also, for several households, the resettlement was claimed to be "induced" by the forest department, rather than being voluntary (Shahabuddin and Bhamidipati 2014).

12.15 Case Study of Panna Tiger Reserve: Reintroduction of Tigers and Village Relocation

However, unlike Sariska, PTR gained global attention for its successful revival of tigers in the past decade (Dutta and Krishnamurthy 2024a). This sharp rise in tiger recovery in PTR was partly attributed to the habitat quality in terms of less human pressure, at least in the core zone (Malviya et al. 2018). In PTR, 17 villages were situated in the core zone, out of which 14 were relocated during 1987–2015. The rate of village relocation was expedited along with the gradual disappearance of tigers from Panna and the launch of the revised guidelines for village relocation, which slowed down once the tigers again started flourishing. During the early 2000s, villagers living inside the core zone faced more and more hardships due to the stringent rules and regulations for daily activities, namely, grazing, farming, and collection of timber and non-timber forest produce (NTFP), which eventually got intensified at the time (2008–2009) of demise and subsequent reintroduction of tigers. This decade-long effort of conservation displacement led to reduced livelihood opportunities for those communities, and more people started to explore the feasibility and financial benefits of relocation. During this period, several households could not reap the benefits of the revised scheme due to their displacement before its enactment (Runacres 2023). Villages relocated after 2008 (e.g. Talgaon, Bundhror, and Pipartola) received the better "package" with their proper engagement during the process of displacement. However, despite the revised scheme aimed at addressing the rights of the communities, such norms are often "effectively extinguished" in order to conduct a hassle-free relocation, which was also the case for Panna, like other tiger reserves (Indian Social Institute 2013; Shahabuddin and Bhamidipati 2014). Nevertheless, such loopholes could undermine the rights and welfare of the communities, jeopardize achieving dual goals of conservation, and directly or indirectly affect the SDGs related to socio-economic status, health, and education for communities (1–5). Researchers found that people living in the Panna landscape are mostly tribes and other communities that are mostly dependent on farming and manual labour, with significant reliance on natural resources (WII 2022). The communities are also beset with several challenges, including water scarcity, health, education, non-functional community centres (or lack of institutional mechanism), and human–wildlife conflict (WII 2022). Such practical issues must be addressed before displacement by effectively incorporating communities' concerns and priorities during the relocation process. Additionally, and perhaps more relevantly, the communities living in the buffer zone (or in the forested landscape) of Panna are going to encounter more interactions with wild animals, as the active conservation efforts have resulted in an abundance of wildlife. The recovery of tigers and sizable populations of other species (e.g. leopards and sloth bears) has already caused conflicts with people living in and around PTR (Jhala et al. 2020b; Malviya and Krishnamurthy 2022; WII 2022). Interestingly, despite the successful recovery of tigers in Panna, people perceived tigers as a threat to human life and livestock (Malviya et al. 2022), which indicated conflictual situations. As a matter of fact, it is the marginalized people residing inside or in close vicinity to the PAs like PTR who face the recurring damage caused by the wildlife, which is detrimental to their well-being. In this context, managing conflict has become necessary, ensuring the coexistence between humans and wildlife on shared landscapes (IUCN SSC HWCTF 2022). Hence, in Panna and other PAs, where the conservation of flagship species and other forms of biodiversity goes hand in hand with community participation, adopting the previously mentioned rights-based approach can find effective solutions for human–wildlife conflict. While such a mechanism is followed for a "justified" conservation displacement by putting forward the rights and concerns of the people during relocation, it also has the potential to manage the conflict by promoting awareness about the empowerment of communities for their rights and bearing accountability. For example, recently in Madhya Pradesh, the compensation provided for livestock loss by the forest department can now be claimed through the Public Service Centres (*Lok Sewa Kendras*), which not only guarantees to quicken the process but also develops a sense of right to get timely compensation for the loss (Sarma and Barpujari 2024). Such steps would surely bring positive changes in the attitudes of those marginalized communities by efficaciously incorporating their concerns into the decision-making systems.

Assuming the inclusion of human dimensions or perspectives would only improve the decision-making system for biodiversity conservation. Better decisions help us get closer to obtaining the desired benefits of conservation. The genesis of conservation displacement in the Indian context starts with the need-based approach, which recognized little or no rights of communities on lands/resources in PAs, but with time, it further evolved to be a rights-based approach where people's participation in the decision-making system was ensured. However, it took some time to understand that in places where people and wildlife are inextricably linked, conservation cannot successfully happen without engaging communities. Unfortunately, such poor and marginalized people bear the heavy burden of conservation through their displacement from their native lands, and if their concerns remain unaddressed, they become further impoverished after relocation. Also, people living in PAs like PTR, where the wildlife population is increasing due to conservation, are always exposed to human–wildlife conflict, in addition to other issues. Thus, altogether, these situations are acting as limiting factors to poverty alleviation, food security, economic well-being, and good health, thus creating obstacles in achieving the SDGs in India. Nevertheless, it is believed that the careful observations and lessons learnt from past experiences of conservation-oriented actions, i.e. village relocations, would be invaluable for devising or modifying site-specific strategies. Currently, given the significant number of human habitations that are still inside PAs, the ongoing and future relocation programmes need to follow the norms of revised policies thoroughly to recognize the rights of people at every step. Such successful engagement of communities in each conservation action through transparency, good governance, accountability, and equity will ensure biodiversity conservation on the one hand and prevent the communities from being further impoverished while ensuring their well-being on the other. Lastly, we conclude the section with a quote from the philosopher *Baruch (Benedict de) Spinoza*, which perhaps depicts the cardinal theme of the human dimensions of biodiversity conservation:

> I have striven not to laugh at human actions, not to weep at them, nor to hate them, but to understand them.

12.16 Landscape-Level Conservation Approach

Movement is a fundamental ecological trait essential for the survival of large carnivores, enabling them to access critical resources such as food, shelter, and mates. The optimal balance of these resources is crucial for maximizing individual fitness (Morales et al. 2010; Whoriskey et al. 2017). Due to their ecological requirements, large carnivores often require extensive areas to meet their life requisites (Tucker et al. 2014), which can drive them to disperse from their natal ranges into multi-use landscapes, particularly those outside PAs (Gantchoff et al. 2020). Research suggests that even areas exceeding 1000 km^2 may sometimes be insufficient to sustain viable populations of these species (Shaffer and Samson 1985; Schonewald-Cox

1988; Woodroffe and Ginsberg 1998). In contrast, smaller reserves (<500 km^2) often support small populations of large carnivores and are characterized by high perimeter-to-area ratios, leading to pronounced "edge effects" (Woodroffe and Ginsberg 1998). In such scenarios, the survival of large carnivores increasingly depends on their ability to adapt to human-modified landscapes (Krishnamurthy et al. 2016; Dutta and Krishnamurthy 2024b). Small habitable patches or "stepping stones" are critical for maintaining connectivity across fragmented landscapes (Baum et al. 2004). These intermediate habitats facilitate wildlife movement between larger habitat areas, ensuring genetic exchange and reducing the risks of population isolation (Sharma et al. 2013).

Recognizing the challenges of biodiversity conservation at a landscape level, the Strategic Plan for Biodiversity was adopted in 2010 by the Parties to the Convention on Biological Diversity (CBD 2010). Aichi Target 11 of this plan aimed to conserve at least 17% of terrestrial areas by 2020 through "ecologically representative and well-connected systems of PAs and other effective area-based conservation measures" (OECMs) (CBD 2018). OECMs are defined as areas that achieve long-term, effective, in situ conservation of biodiversity outside PAs (Marnewick et al. 2021; MoEFCC, NBA, and UNDP 2022; Jonas et al. 2023).

Since the inception of the tiger reintroduction initiative in PTR, a total of 37 tigers had dispersed into the GPL by 2021. This number increased to over 50 individuals by 2024, with tigers expanding their ranges into adjacent forest divisions, including North and South Panna, Chhatarpur, Damoh, Sagar, and Satna, as well as PAs such as Ranipur Tiger Reserve and Veerangana Durgawati Tiger Reserve (Dutta and Krishnamurthy 2024b). PTR remains the sole source population within the GPL. The continued spillover of this growing tiger population underscores the urgent need for enhanced protection and management strategies across the landscape. A least-cost pathway analysis has revealed that PTR maintains strong connectivity with Ranipur Tiger Reserve, Madhav National Park, and Nauradehi Wildlife Sanctuary (Dutta et al. 2016). A recent study identified 11 satellite core areas (2405.9 km^2), ten stepping stones (385.8 km^2), and 67 potential corridors (1277.3 km^2) within this landscape (Makwana et al. 2023). Notably, 21.5% of the dispersal routes were delineated within the boundaries of stepping stones. Further degradation of these stepping stones can significantly disrupt animal movement, potentially leading to failed colonization and rendering corridors non-functional (Kramer-Schadt et al. 2011). Satellite core areas also play a critical role by temporarily supporting tiger populations and facilitating dispersal, thus promoting a metapopulation framework for species conservation (Dolrenry et al. 2014).

The Central Indian landscape, including the GPL, represents one of the most fragmented habitats for tigers. Human-dominated areas outside protected zones, including agricultural fields and village matrices, impose significant constraints on tiger dispersal (Krishnamurthy et al. 2016). Fragmented forest patches and nocturnal movements enable tigers to navigate these challenges during dispersal (Dutta and Krishnamurthy 2024b). Consequently, maintaining landscape-level connectivity remains a critical concern in the GPL. This involves integrating multiple PAs and identifying functional or potential corridors to ensure safe passage for dispersing tigers (WII 2022). The establishment of the Greater Panna Landscape Council (GPLC) marks a pivotal step towards implementing the Integrated Landscape Management Plan (ILMP) for the GPL. ILMP successfully identified several satellite core areas and other PAs, recommending their designation as conservation reserves or the expansion of existing PAs. Notably, the Kaldah–Saleha range of the South Panna Forest Division exhibits significant potential to support tiger populations. This forest patch also serves as a critical corridor, facilitating tiger movement towards Bandhavgarh Tiger Reserve. Furthermore, Ranipur and Nauradehi wildlife sanctuaries have been upgraded to Ranipur Tiger Reserve and Veerangana Durgawati Tiger Reserve, respectively. Such conservation actions are vital for promoting and ensuring the long-term survival of apex predators in the region, particularly to support a metapopulation framework.

12.17 Conclusion

The successful recovery of the tiger population in PTR stands as a testament to the effectiveness of science-based conservation efforts. The integration of rigorous monitoring techniques, habitat management interventions, and active stakeholder participation has enabled the reserve to overcome past challenges and achieve remarkable population growth. The reintroduced tigers have demonstrated adaptive capabilities in the dry deciduous ecosystem, utilizing available resources efficiently and establishing stable populations. In PTR, wild-caught tigers contributed more effectively compared to the hand-reared tigers (orphaned) introduced into the wild. However, sustaining this success requires continued vigilance in mitigating emerging threats such as habitat fragmentation and human–wildlife conflict. As the populations of tigers and other wildlife grow, the interface between human and wildlife also becomes intense, which needs to be addressed to achieve conservation goals, namely, the betterment of both people and biodiversity. Inclusion of human dimensions into conservation policymaking by recognizing and prioritizing their rights with transparency would only lead to better judgement to secure the future of man and biodiversity. Hence, in a nutshell, future conservation efforts should prioritize landscape connectivity, adaptive management strategies, and community engagement to ensure the long-term persistence of tigers and the overall ecological integrity of the region. The insights gained from Panna's recovery serve as a valuable model for similar conservation initiatives across the tiger's range.

12.18 Future Aspects

Post-release monitoring is a vital component of reintroduction programmes for large carnivores, providing essential insights into their survival, behaviour, adaptation, and ecological impacts in new environments (Briers-Louw et al. 2024). This process ensures that conservation objectives are achieved, challenges are identified and addressed promptly, and valuable data are generated to inform future reintroduction efforts (Thomas et al. 2023). Key aspects of monitoring include evaluating habitat use, movement patterns, prey availability, health, and reproduction and mitigating human–wildlife conflicts. In PTR, all reintroduced tigers were intensively monitored using radio collars to systematically gather data on their behaviour, interactions, habitat use, and home range (Sarkar et al. 2016). Collaring also played a critical role in mitigating human–wildlife conflicts by enhancing security measures and preventing casualties. As the tiger population grew, dominant individuals were fitted with advanced (Very High Frequency) VHF and GPS/ Ultra High Frequency (UHF) collars. Subsequently, a systematic continuous camera trap monitoring system (CCMS) was later implemented across core and buffer zones using a 2-km^2 grid design to expand surveillance.

Technological advancements have further enhanced monitoring efforts. Drones equipped with thermal imaging cameras are now used for non-invasive monitoring in dense vegetation or at night, proving especially effective in conflict situations for rapid action and response. Plans are underway to deploy wireless camera traps for real-time data transfer, which promises to revolutionize monitoring capabilities. Additionally, Radio Frequency Identification (RFID) tagging (Rafiq et al. 2021) and passive acoustic monitoring (Browning et al. 2017) are emerging as valuable tools for species monitoring. The integration of these advanced technologies ensures comprehensive data collection, strengthens adaptive management practices, and enhances conservation outcomes for large carnivores in reintroduction programmes.

References

3rd Asia ministerial conference on Tiger conservation April 12–14, (2016) New Delhi

Baum KA, Haynes KJ, Dillemuth FP, Cronin JT (2004) The matrix enhances the effectiveness of corridors and stepping stones. Ecology 85(10):2671–2676

Berkes F (2004) Rethinking community-based conservation. Conserv Biol 18(3):621–630

Beschta RL (2005) Reduced cottonwood recruitment following extirpation of wolves in Yellowstone's northern range. Ecology 86(2):391–403

Beschta RL, Ripple WJ (2009) Large predators and trophic cascades in terrestrial ecosystems of the western United States. Biol Conserv 142(11):2401–2414

Bond WJ (2005) Large parts of the world are brown or black: a different view on the 'green world' hypothesis. J Veg Sci 16(3):261–266

Boyce MS (2018) Wolves for Yellowstone: dynamics in time and space. J Mammal 99(5):1021–1031

Brashares JS, Prugh LR, Stoner CJ, Epps CW (2010) Ecological and conservation implications of mesopredator release. In: Trophic cascades: predators, prey, and the changing dynamics of nature, pp 221–240

Breitenmoser U (2001) Assessment of carnivore reintroductions. In: Gittleman JL, Funk SM, Macdonald DW, Wayne RK (eds) Carnivore conservation, pp 241–281

Briers-Louw WD, Verschueren S, Leslie AJ (2019) Big cats return to Majete wildlife reserve, Malawi: evaluating reintroduction success. Afr J Wildl Res 49(1):34–50

Briers-Louw WD, Lindsey P, Gaylard A, Cristescu B, Verschueren S, du Plessis C et al (2024) Guidelines for evaluating the success of large carnivore reintroductions. bioRxiv:2024–2006

Brose U, Archambault P, Barnes AD, Bersier LF, Boy T, Canning-Clode J et al (2019) Predator traits determine food-web architecture across ecosystems. Nat Ecol Evol 3(6):919–927

Browning E, Gibb R, Glover-Kapfer P, Jones KE (2017) Passive acoustic monitoring in ecology and conservation, WWF conservation technology series 1(2). WWF-UK, Woking, p 76. https://doi.org/10.25607/OBP-876

Brugière D, Chardonnet B, Scholte P (2015) Large-scale extinction of large carnivores (lion Panthera leo, cheetah Acinonyx jubatus and wild dog Lycaon pictus) in protected areas of west and Central Africa. Trop Conserv Sci 8(2):513–527

Bubac CM, Johnson AC, Fox JA, Cullingham CI (2019) Conservation translocations and post-release monitoring: identifying trends in failures, biases, and challenges from around the world. Biol Conserv 238:108239

Caro TM (2003) Umbrella species: critique and lessons from East Africa. In Animal conservation forum (vol. 6, no. 2, pp. 171–181). Cambridge University Press

Caro TM, O'Doherty G (1999) On the use of surrogate species in conservation biology. Conserv Biol 13(4):805–814

Caughley G (1994) Directions in conservation biology. J Anim Ecol 63:215–244

CBD (2010) Decision adopted by the conference of the parties to the convention on biological diversity at its tenth meeting, Nagoya, pp 1–13

CBD (2018) Decision adopted by the conference of the parties to the convention on biological diversity: protected areas and other effective area-based measures. Sharm El-Sheikh, Egypt, pp 1–19

Cernea MM, Schmidt-Soltau K (2003) Biodiversity conservation versus population resettlement: risks to nature and risks to people. In: International conference on rural livelihoods, forests and biodiversity, pp 19–23

Champion HG, Seth SK (1968) A revised survey of the forest types of India. Manager Publications

Chazdon RL (2019) Second growth: the promise of tropical forest regeneration in an age of deforestation. University of Chicago Press

Chundawat RS, Gogate N, Malik PK (2002) Understanding tiger ecology in the tropical dry deciduous forests of Panna Tiger Reserve. Final report. Wildlife Institute of India, Dehradun

Corson C, Worcester J, Rogers S, Flores-Ganley I (2020) From paper to practice? Assembling a rights-based conservation approach. J Polit Ecol 27(1):1128–1147

Council, General Assembly Economic and Social (2024) Progress towards the Sustainable Development Goals

Creel S, Christianson D (2008) Relationships between direct predation and risk effects. Trends Ecol Evol 23(4):194–201

Dash M, Behera B (2015) Local institutions, collective action and forest conservation: the case of Similipal Tiger Reserve in India. J For Econ 21(4):167–184

Dash M, Behera B (2018) Biodiversity conservation, relocation and socio-economic consequences: a case study of Similipal Tiger Reserve, India. Land Use Pol 78:327–337

Davies AB, Tambling CJ, Kerley GI, Asner GP (2016) Limited spatial response to direct predation risk by African herbivores following predator reintroduction. Ecol Evol 6(16):5728–5748

Davis MA, Slobodkin LB (2004) The science and values of restoration ecology. Restor Ecol 12(1):1

Dolrenry S, Stenglein J, Hazzah L, Lutz RS, Frank L (2014) A metapopulation approach to African lion (Panthera leo) conservation. PLoS One 9(2):e88081

Doubleday KF (2018) Rewilding expectations: human-environmental relations in context of apex predator reintroduction in Rajasthan, India. The University of Texas at Austin

Dutta S, Krishnamurthy R (2024a) Successful conservation translocation: population dynamics of tiger recovery in Panna Tiger Reserve, Central India. Ecol Solut Evid 5(2):e12337

Dutta S, Krishnamurthy R (2024b) Multiphasic movement and step-selection patterns of dispersed tigers in the central Indian landscape. PLoS One 19(10):e0309517

Dutta T, Sharma S, McRae BH, Roy PS, DeFries R (2016) Connecting the dots: mapping habitat connectivity for tigers in Central India. Reg Environ Chang 16:53–67

Dwivedi AP (2003) Protected areas of Madhya Pradesh: government printing press, Bhopal, p 254

Ehrlich PR, Murphy DD (1987) Monitoring populations on remnants of native vegetation. In: Nature conservation: the role of remnants of native vegetation. Surrey Beatty and Sons, Chipping Norton, pp 210–210

Elmhagen B, Ludwig G, Rushton SP, Helle P, Lindén H (2010) Top predators, mesopredators and their prey: interference ecosystems along bioclimatic productivity gradients. J Anim Ecol 79(4):785–794

Estes JA, Terborgh J, Brashares JS, Power ME, Berger J, Bond WJ et al (2011) Trophic downgrading of planet Earth. Science 333(6040):301–306

Fleishman E, Murphy DD, Brussard PF (2000) A new method for selection of umbrella species for conservation planning. Ecol Appl 10(2):569–579

Fleishman E, Blair RB, Murphy DD (2001) Empirical validation of a method for umbrella species selection. Ecol Appl 11(5):1489–1501

Fortin D, Beyer HL, Boyce MS, Smith DW, Duchesne T, Mao JS (2005) Wolves influence elk movements: behavior shapes a trophic cascade in Yellowstone National Park. Ecology 86(5):1320–1330

Frankel OH (1981) Conservation and evolution. Cambridge University Press

Fritts SH, Bangs EE, Fontaine JA, Johnson MR, Phillips MK, Koch ED, Gunson JR (1997) Planning and implementing a reintroduction of wolves to Yellowstone National Park and Central Idaho. Restor Ecol 5(1):7–27

Gadgil M, Seshagiri Rao PR, Utkarsh G, Pramod P, Chhatre A (2000) New meanings for old knowledge: the people's biodiversity registers program. Ecol Appl 10(5):1307–1317

Galetti M, Moleón M, Jordano P, Pires MM, Guimaraes PR Jr, Pape T et al (2018) Ecological and evolutionary legacy of megafauna extinctions. Biol Rev 93(2):845–862

Gantchoff MG, Hill JE, Kellner KF, Fowler NL, Petroelje TR, Conlee L et al (2020) Mortality of a large wide-ranging mammal largely caused by anthropogenic activities. Sci Rep 10(1):8498

Gaynor KM, Brown JS, Middleton AD, Power ME, Brashares JS (2019) Landscapes of fear: spatial patterns of risk perception and response. Trends Ecol Evol 34(4):355–368

Gee EP (1964) The wildlife of India. UK. Collins, London

Global Tiger Forum (2022) Wild Tiger conservation across Tiger range countries: an appraisal of financial shortfalls, 4th Asia ministerial conference (19–21 January, 2022), Kuala Lumpur

Goodrich JM, Miquelle DG (2005) Translocation of problem Amur tigers Panthera tigris altaica to alleviate tiger-human conflicts. Oryx 39(4):454–457

Global Tiger Initiative Secretariat (2011) Global tiger recovery program 2010–2022, Washington (DC)

Goodrich J, Lynam A, Miquelle D, Wibisono H, Kawanishi K, Pattanavibool A, … Karanth U (2015) Panthera tigris. The IUCN red list of threatened species 2015: e. T15955A50659951

Griffith B, Scott JM, Carpenter JW, Reed C (1989) Translocation as a species conservation tool: status and strategy. Science 245(4917):477–480

Gupta D, Krishnamurthy R (2023) Spatial density patterns of herbivore response to seasonal dynamics in the tropical deciduous forest of Central India. Biotropica 55(2):430–443

Hackel JD (1999) Community conservation and the future of Africa's wildlife. Conserv Biol 13(4):726–734

Halofsky J, Ripple W (2008) Linkages between wolf presence and aspen recruitment in the Gallatin elk winter range of southwestern Montana, USA. Forestry 81(2):195–207

Hanna P, Vanclay F (2013) Human rights, indigenous peoples and the concept of free, prior and informed consent. Impact Assess Proj Apprais 31(2):146–157

Harihar A, Ghosh-Harihar M, MacMillan DC (2014) Human resettlement and tiger conservation–socio-economic assessment of pastoralists reveals a rare conservation opportunity in a human-dominated landscape. Biol Conserv 169:167–175

Harihar A, Chanchani P, Borah J, Crouthers RJ, Darman Y, Gray TN et al (2018) Recovery planning towards doubling wild tiger Panthera tigris numbers: detailing 18 recovery sites from across the range. PLoS One 13(11):e0207114

Hebblewhite M, White CA, Nietvelt CG, McKenzie JA, Hurd TE, Fryxell JM et al (2005) Human activity mediates a trophic cascade caused by wolves. Ecology 86(8):2135–2144

Hofmeyr M, Van Dyk G (1998) Cheetah introductions to two north-west parks: case studies from Pilanesburg National Park and Madikwe Game Reserve. In: Cheetahs as game ranch animals, Onderstepoort, pp 60–71

Hofmeyr M, Davies R, Nel P, Dell S (2003) Operation phoenix—the introduction of larger mammals to Madikwe game reserve. In: Madikwe Game Reserve: a decade of progress, pp 8–20
Hopcraft JGC, Olff H, Sinclair ARE (2010) Herbivores, resources and risks: alternating regulation along primary environmental gradients in savannas. Trends Ecol Evol 25(2):119–128
Hunter L (2018) Carnivores of the world, vol 117. Princeton University Press
Hunter LTB (1998) Early post-release movements and behaviour of reintroduced cheetahs and lions, and technical considerations in large carnivore restoration. In Proceedings of a Symposium on Cheetahs as game ranch animals (Vol. 23).
Indian Social Institute (2013) Summary report of national consultation on forest rights act and protected areas. 8
IPBES (Intergovernmental Science-Policy Platform on Biodiversity and Ecosystem Services) (2019) Summary for policymakers of the IPBES global assessment report on biodiversity and ecosystem services. IPBES, Bonn
IPCC (The Intergovernmental Panel on Climate Change) (2019) Climate change and land: an IPCC special report on climate change, desertification, land degradation, sustainable land management, food security, and greenhouse gas fluxes in terrestrial ecosystems. IPCC, Geneva
IUCN (1998) IUCN guidelines for re-introductions. IUCN, Gland/Cambridge
IUCN (2024) The IUCN red list of threatened species. Version 2024-2
IUCN SSC HWCTF (2022) Perspectives on human-wildlife coexistence
IUCN/SSC (2013) Guidelines for reintroductions and other conservation translocation. Version 1.0. IUCN Species Survival Commission, Gland, p viiii + 57
Jhala YV, Qureshi Q (2006) Habitat occupancy and abundance of tigers in Panna Tiger Reserve. Report submitted to Supreme Court of India. Wildlife Institute of India, Dehradun, p 10
Jhala YV, Qureshi Q, Nayak AK (eds) (2020a) Status of tigers, copredators and prey in India, 2018. National Tiger Conservation Authority, Government of India, New Delhi, and Wildlife Institute of India, Dehradun. ISBN No. 81-85496–50-1
Jhala YV, Qureshi Q, Yadav SP (2020b) Status of leopards in India, 2018. National Tiger Conservation Authority, government of India, New Delhi, and wildlife Institute of India. Dehradun. Technical report TR/2020/16
Jhala Y, Gopal R, Mathur V, Ghosh P, Negi HS, Narain S et al (2021) Recovery of tigers in India: critical introspection and potential lessons. People Nat 3(2):281–293
Jimenez MF, Laverty TM, Bombaci SP, Wilkins K, Bennett DE, Pejchar L (2019) Underrepresented faculty play a disproportionate role in advancing diversity and inclusion. Nat Ecol Evol 3(7):1030–1033
Jonas HD, MacKinnon K, Marnewick D, Wood P (2023) Site-level tool for identifying other effective area-based conservation measures (OECMs). IUCN (International Union for Conservation of Nature), Gland
Kabra A (2009) Conservation-induced displacement: a comparative study of two Indian protected areas. Conserv Soc 7(4):249–267
Karanth KK (2007) Making resettlement work: the case of India's Bhadra wildlife sanctuary. Biol Conserv 139(3–4):315–324
Karanth UK, Karanth KK (2007) Protected areas and human displacement: a conservation perspective (vol. 29). In: Redford KH, Fearn E (eds) Conservation and voluntary resettlement in the Western Ghats of Karnataka, India. Wildlife Conservation Society, New York
Karanth KU, Stith BM (1999) Prey depletion as a critical determinant of tiger population viability. In: Riding the tiger: tiger conservation in human-dominated landscapes, pp 100–113
Karanth KU, Chundawat RS, Nichols JD, Kumar NS (2004) Estimation of tiger densities in the tropical dry forests of Panna, Central India, using photographic capture–recapture sampling. Anim Conserv 7(3):285–290
Karanth KK, Nichols JD, Karanth KU, Hines JE, Christensen NL Jr (2010) The shrinking ark: patterns of large mammal extinctions in India. Proc R Soc B Biol Sci 277(1690):1971–1979
Karanth KK, Kudalkar S, Jain S (2018) Re-building communities: voluntary resettlement from protected areas in India. Front Ecol Evol 6:183

Kesaro L, Sam Ang C, Bhardwaj GS, Ramesh K, Singh H, Khobragade D, Gauntlett S, Yadav SP, Gopal R, Gray T, Borah J (2024) Action plan for Tiger reintroduction in cardamom Forest, Cambodia. Ministry of Environment, Royal Government of Cambodia, Ministry of Environment, Forest and Climate Change, Government of India, National Tiger Conservation Authority, Global Tiger Forum, Wildlife Institute of India and Wildlife Alliance

Kothari A, Pathak N, Vania F (2000) In: Roe D (ed) Where communities care: community-based wildlife and ecosystem Management in South Asia. Kalpavriksh and International Institute of Environment and Development, Pune

Kramer-Schadt S, Kaiser TS, Frank K, Wiegand T (2011) Analyzing the effect of stepping stones on target patch colonisation in structured landscapes for Eurasian lynx. Landsc Ecol 26:501–513

Krishnamurthy R, Cushman SA, Sarkar MS, Malviya M, Naveen M, Johnson JA, Sen S (2016) Multi-scale prediction of landscape resistance for tiger dispersal in Central India. Landsc Ecol 31:1355–1368

Kummer D, Sham CH (2023) The causes of tropical deforestation: a quantitative analysis and case study from The Philippines. In: The causes of tropical deforestation, pp 146–158

Le Roux E, Kerley GI, Cromsigt JP (2018) Megaherbivores modify trophic cascades triggered by fear of predation in an African savanna ecosystem. Curr Biol 28(15):2493–2499

Leo V, Reading RP, Gordon C, Letnic M (2019) Apex predator suppression is linked to restructuring of ecosystems via multiple ecological pathways. Oikos 128(5):630–639

Letnic M, Ripple WJ (2017) Large-scale responses of herbivore prey to canid predators and primary productivity. Glob Ecol Biogeogr 26(8):860–866

Louys J (2014) The large terrestrial carnivore guild in quaternary Southeast Asia. Quat Sci Rev 96:86–97

Lovari S, Pokheral CP, Jnawali SR, Fusani L, Ferretti F (2015) Coexistence of the tiger and the common leopard in a prey-rich area: the role of prey partitioning. J Zool 295(2):122–131

Luo SJ, Kim JH, Johnson WE, Walt JVD, Martenson J, Yuhki N, Miquelle DG, Uphyrkina O, Goodrich JM, Quigley HB, Tilson R (2004) Phylogeography and genetic ancestry of tigers (Panthera tigris). PLoS Biol 2(12):e442

Mahmood T, Vu TT, Campos-Arceiz A, Akrim F, Andleeb S, Farooq M, Hamid A, Munawar N, Waseem M, Hussain A, Fatima H (2021) Historical and current distribution ranges and loss of mega-herbivores and carnivores of Asia. PeerJ 9:e10738

Makwana M, Vasudeva V, Cushman SA, Krishnamurthy R (2023) Modelling landscape permeability for dispersal and colonization of tigers (Panthera tigris) in the greater Panna landscape, Central India. Landsc Ecol 38(3):797–819

Malviya M, Krishnamurthy R (2022) Multiscale spatially explicit modelling of livestock depredation by reintroduced tiger (Panthera tigris) to predict conflict risk probability. Global Ecol Conserv 40:e02313

Malviya M, Kumar V, Mandal D, Sarkar MS, Nigam P, Gopal R et al (2018) Correlates of physiological stress and habitat factors in reintroduction-based recovery of tiger (Panthera tigris) populations. Hystrix 29(2):195

Malviya M, Kalyanasundaram S, Krishnamurthy R (2022) Paradox of success-mediated conflicts: analysing attitudes of local communities towards successfully reintroduced tigers in India. Front Conserv Sci 2:783467

Marnewick D, Stevens CM, Jonas H, Antrobus-Wuth R, Wilson N, Theron N (2021) Assessing the extent and contribution of OECMs in South Africa. Parks 27:57–70

Matyushkin EN (1998) Amurskii tigr v Rossii 1925–1997 WWF. EN Matyushkin

Mazák V (1981) Panthera tigris, Mammalian species, no. 152. American Society of Mammologist, pp 1–8

McLean J, Straede S (2003) Conservation, relocation, and the paradigms of park and people management–a case study of Padampur villages and the Royal Chitwan National Park, Nepal. Soc Nat Resour 16(6):509–526

McNab BK (1963) Bioenergetics and the determination of home range size. Am Nat 97(894):133–140

Meli P, Martínez-Ramos M, Rey-Benayas JM (2013) Selecting species for passive and active riparian restoration in southern Mexico. Restor Ecol 21(2):163–165

Michalski F, Peres CA (2005) Anthropogenic determinants of primate and carnivore local extinctions in a fragmented forest landscape of southern Amazonia. Biol Conserv 124(3):383–396

Miller CS, Hebblewhite M, Petrunenko YK, Seryodkin IV, Goodrich JM, Miquelle DG (2014) Amur tiger (Panthera tigris altaica) energetic requirements: implications for conserving wild tigers. Biol Conserv 170:120–129

MoEFCC, NBA, UNDP (2022) Criteria and guidelines for identifying other effective area based conservation measures (OECMs) in India

Morales JM, Moorcroft PR, Matthiopoulos J, Frair JL, Kie JG, Powell RA et al (2010) Building the bridge between animal movement and population dynamics. Philos Trans R Soc B: Biol Sci 365(1550):2289–2301

Narain S, Panwar HS, Gadgil M, Thapar V, Singh S (2005) Joining the dots: the report of the Tiger task force. Project Tiger Directorate, Union Ministry of Environment, Government of India, New Delhi

National Tiger Conservation Authority. (2017). Tiger conservation in India and Voluntary Village relocation from critical Tiger habitat of Tiger reserves

Noss RF, Quigley HB, Hornocker MG, Merrill T, Paquet PC (1996) Conservation biology and carnivore conservation in the Rocky Mountains. Conserv Biol 10(4):949–963

Noss RF, O'Connell MA, Murphy DD (1997) The science of conservation planning: habitat conservation under the endangered species act. Island Press

Nowell K, Jackson P (eds) (1996) Wild cats: status survey and conservation action plan (vol. 382). IUCN, Gland

O'Brien SJ, Evermann JF (1988) Interactive influence of infectious disease and genetic diversity in natural populations. Trends Ecol Evol 3(10):254–259

Owen-Smith N (1987) Pleistocene extinctions: the pivotal role of megaherbivores. Paleobiology 13(3):351–362

Panwar HS (1987) Project Tiger: the reserves, the tigers and their future. Tigers of the world: the biology, biopolitics, management and conservation of an endangered species. Noyes Publications, Park Ridge, pp 110–117

Peckarsky BL, Kerans BL, Taylor BW, McIntosh AR (2008) Predator effects on prey population dynamics in open systems. Oecologia 156:431–440

Penjor U, Kaszta ZM, Macdonald DW, Cushman SA (2024) Identifying umbrella and indicator species to support multispecies population connectivity in a Himalayan biodiversity hotspot. Front Conserv Sci 5:1306051

Pereira HM, Navarro LM, Martins IS (2012) Global biodiversity change: the bad, the good, and the unknown. Annu Rev Environ Resour 37(1):25–50

Pinedo-Vasquez M, Padoch C (1993) Community and governmental experiences protecting biodiversity in the lowland Peruvian Amazon. In: Perspectives on biodiversity: case studies in genetic conservation and development. American Association for the Advancement of Science Press, Washington, DC

Preisser EL, Bolnick DI (2008) The many faces of fear: comparing the pathways and impacts of nonconsumptive predator effects on prey populations. PLoS One 3(6):e2465

Pringle RM (2020) Untangling food webs. In: Unsolved problems in Ecology, pp 225–238

Prugh LR, Stoner CJ, Epps CW, Bean WT, Ripple WJ, Laliberte AS, Brashares JS (2009) The rise of the mesopredator. Bioscience 59(9):779–791

Purvis A, Gittleman JL, Cowlishaw G, Mace GM (2000) Predicting extinction risk in declining species. Proc R Soc London B: Biol Sci 267(1456):1947–1952

Qureshi Q, Jhala YV, Yadav SP, Mallick A (eds) (2023) Status of tigers, co-predators and prey in India, 2022. National Tiger Conservation Authority, Government of India, New Delhi, and Wildlife Institute of India, Dehradun. ISBN No: 81-85496-92-7

Rafiq K, Appleby RG, Edgar JP, Radford C, Smith BP, Jordan NR et al (2021) Wild WID: an open-source active RFID system for wildlife research. Methods Ecol Evol 12(9):1580–1587

Ramesh K, Johnson JA, Sen S, Murthy RS, Sarkar MS, Malviya M et al (2013) Status of tiger and prey species in Panna Tiger Reserve, Madhya Pradesh: capture-recapture and distance sampling estimates. Technical Report. Wildlife Institute of India, Dehradun and Panna Tiger Reserve, Madhya Pradesh, p 33

Ranganathan J, Chan KM, Karanth KU, Smith JLD (2008) Where can tigers persist in the future? A landscape-scale, density-based population model for the Indian subcontinent. Biol Conserv 141(1):67–77

Rasphone A, Kéry M, Kamler JF, Macdonald DW (2019) Documenting the demise of tiger and leopard, and the status of other carnivores and prey, in Lao PDR'S most prized protected area: Nam et-Phou Louey. Global Ecol Conserv 20:e00766

Reddy AP, Ramesh K, Shekhar Sarkar M, Srivastava A, Bhavanishankar M, Shivaji S (2016) Significance of mate selection and adult sex ratio in tiger reintroduction/reinforcement programs. J Zool 299(2):132–141

Ripple WJ, Beschta RL (2012) Trophic cascades in Yellowstone: the first 15 years after wolf reintroduction. Biol Conserv 145(1):205–213

Ripple WJ, Van Valkenburgh B (2010) Linking top-down forces to the Pleistocene megafaunal extinctions. Bioscience 60(7):516–526

Ripple WJ, Wirsing AJ, Wilmers CC, Letnic M (2013) Widespread mesopredator effects after wolf extirpation. Biol Conserv 160:70–79

Ripple WJ, Estes JA, Beschta RL, Wilmers CC, Ritchie EG, Hebblewhite M et al (2014) Status and ecological effects of the world's largest carnivores. Science 343(6167):1241484

Rodrigues AS, Andelman SJ, Bakarr MI, Boitani L, Brooks TM, Cowling RM et al (2004) Effectiveness of the global protected area network in representing species diversity. Nature 428(6983):640–643

Ruble DB, Verschueren S, Cristescu B, Marker LL (2022) Rewilding apex predators has effects on lower trophic levels: cheetahs and ungulates in a woodland savanna. Animals 12(24):3532

Runacres A (2023) Relocated tigers and relocated villagers: Ferality and human–animal entanglement in Indian conservation. Mod Asian Stud 57(1):222–251

Sadhu A, Jayam PPC, Qureshi Q, Shekhawat RS, Sharma S, Jhala YV (2017) Demography of a small, isolated tiger (Panthera tigris tigris) population in a semi-arid region of western India. BMC Zool 2:1–13

Sanderson EW, Forrest J, Loucks C, Ginsberg J, Dinerstein E, Seidensticker J et al (2010) Setting priorities for tiger conservation: 2005–2015. In: Tigers of the world. William Andrew Publishing, pp 143–161

Sankar K, Qureshi Q, Nigam P, Malik PK, Sinha PR, Mehrotra RN, Gopal R, Bhattacharjee S, Mondal K, Gupta S (2010) Monitoring of reintroduced tigers in Sariska Tiger Reserve, Western India: preliminary findings on home range, prey selection and food habits. Trop Conser Sci 3:301–318

Sarkar MS, Ramesh K, Johnson JA, Sen S, Nigam P, Gupta SK et al (2016) Movement and home range characteristics of reintroduced tiger (Panthera tigris) population in Panna Tiger Reserve, Central India. Eur J Wildl Res 62:537–547

Sarma UK, Barpujari I (2020) Relocation from tiger reserves under the NTCA guidelines. Perspectives 55(30):33

Sarma UK, Barpujari I (2023) Realizing a rights-based approach to resettlement from protected areas: lessons from Satpura Tiger Reserve, Madhya Pradesh (India). Land Use Pol 125:106494

Sarma UK, Barpujari I (2024) Coexistence and the SDGs: an argument for a rights-based approach to human-wildlife conflict in India. Hum Dimens Wildl 30:248–256

Schmidt-Soltau K (2003) Conservation–related resettlement in Central Africa: environmental and social risks. Dev Chang 34(3):525–551

Schonewald-Cox CM (1988) Boundaries in the protection of nature reserves: translating multidisciplinary knowledge into practical conservation. Bioscience 38(7):480–486

Seddon PJ, Leech T (2008) Conservation short cut, or long and winding road? A critique of umbrella species criteria. Oryx 42(2):240–245

Seidensticker J (1976) On the ecological separation between tigers and leopards. Biotropica 8:225–234
Seidensticker J, Jackson P, Christie S (eds) (1999) Riding the tiger: tiger conservation in human-dominated landscapes. Cambridge University Press
Sekar N (2016) Tigers, tribes, and bureaucrats: the voluntariness and socioeconomic consequences of village relocations from Melghat Tiger Reserve, India. Reg Environ Chang 16:111–123
Shaffer ML, Samson FB (1985) Population size and extinction: a note on determining critical population sizes. Am Nat 125(1):144–152
Shahabuddin G, Bhamidipati PL (2014) Conservation-induced displacement: recent perspectives from India. Environ Justice 7(5):122–129
Shahabuddin G, Kumar R, Shrivastava M (2006) Policy and process of relocation from conservation areas: the case of the Sariska Tiger Reserve, Rajasthan. Soc Change 36(1):130–140
Sharma R, Manda R, Gupta S, Kumar S, Kumar V (2013) Isolation and characterization of osmotolerant bacteria from Thar Desert of Western Rajasthan (India). Rev de Biol Trop 61(4):1551–1562
Sheriff MJ, Peacor SD, Hawlena D, Thaker M (2020) Non-consumptive predator effects on prey population size: a dearth of evidence. J Anim Ecol 89(6):1302–1316
Singh R, Kumar A, Singh PS (2013) Spatial analysis of tiger habitat suitability in Panna Tiger Reserve of Madhya Pradesh. India Life Sci Bull 10:95–98
Slough BG (1994) Translocations of American martens: an evaluation of factors in success. In: Martens, sables, and fishers: biology and conservation. Cornell University Press, Ithaca, pp 165–178
Smith JLD (1993) The role of dispersal in structuring the Chitwan tiger population. Behaviour 124(3–4):165–195
Society for Ecological Restoration International (2004) SER international primer on ecological restoration. Society for Ecological Restoration International
Steenweg R, Hebblewhite M, Burton C, Whittington J, Heim N, Fisher JT et al (2023) Testing umbrella species and food-web properties of large carnivores in the Rocky Mountains. Biol Conserv 278:109888
Stockwell CA, Mulvey M, Vinyard GL (1996) Translocations and the preservation of allelic diversity. Conserv Biol 10(4):1133–1141
Sunquist M, Sunquist F (2002) Wild cats of the world. University of Chicago press
Terborgh JW (2015) Toward a trophic theory of species diversity. Proc Natl Acad Sci 112(37):11415–11422
Terborgh J, Schaik CV, Davenport L, Madhu R (2002) Making parks work: strategies for preserving tropical nature, p xix+−511
Thomas S, van der Merwe V, Carvalho WD, Adania CH, Gomerčić T, Krofel M et al (2023) Evaluating the performance of conservation translocations in large carnivores across the world. Biol Conserv 279:109909
Thornton D, Zeller K, Rondinini C, Boitani L, Crooks K, Burdett C et al (2016) Assessing the umbrella value of a range-wide conservation network for jaguars (Panthera onca). Ecol Appl 26(4):1112–1124
Tilman D, Clark M, Williams DR, Kimmel K, Polasky S, Packer C (2017) Future threats to biodiversity and pathways to their prevention. Nature 546(7656):73–81
Torri MC (2011) Conservation, relocation and the social consequences of conservation policies in protected areas: case study of the Sariska Tiger Reserve, India. Conserv Soc 9(1):54–64
Tucker MA, Ord TJ, Rogers TL (2014) Evolutionary predictors of mammalian home range size: body mass, diet and the environment. Glob Ecol Biogeogr 23(10):1105–1114
UNCCD (2022) The great green wall initiative. In: United Nations convention to combat desertification (UNCCD). The Great Green Wall Initiative
United Nations (2015) Sustainable development goals. United Nations
Van Vuuren DP, Bouwman LF (2005) Exploring past and future changes in the ecological footprint for world regions. Ecol Econ 52(1):43–62

Vasudeva V, Upgupta S, Singh A, Sherwani N, Dutta S, Rajaraman R et al (2022) Conservation prioritization in a tiger landscape: is umbrella species enough? Land 11(3):371
Visconti P, Bakkenes M, Baisero D, Brooks T, Butchart SH, Joppa L et al (2016) Projecting global biodiversity indicators under future development scenarios. Conserv Lett 9(1):5–13
Walker EH, Verschueren S, Schmidt-Küntzel A, Marker L (2022) Recommendations for the rehabilitation and release of wild-born, captive-raised cheetahs: the importance of pre-and post-release management for optimizing survival. Oryx 56(4):495–504
Wang F, Winkler J, Vina A, McShea WJ, Li S, Connor T et al (2021) The hidden risk of using umbrella species as conservation surrogates: a spatio-temporal approach. Biol Conserv 253:108913
Watson JE, Dudley N, Segan DB, Hockings M (2014) The performance and potential of protected areas. Nature 515(7525):67–73
Wemmer C, Sunquist ME (1988) Felid reintroductions: economic and energetic considerations. In: Freeman H (ed) Proceedings of the 5th international snow leopard symposium, pp 193–205
Werner EE, Peacor SD (2003) A review of trait-mediated indirect interactions in ecological communities. Ecology 84(5):1083–1100
Whoriskey K, Auger-Méthé M, Albertsen CM, Whoriskey FG, Binder TR, Krueger CC, Mills Flemming J (2017) A hidden Markov movement model for rapidly identifying behavioral states from animal tracks. Ecol Evol 7(7):2112–2121
WII (2022) Integrated landscape management plan for greater Panna landscape, Madhya Pradesh and Uttar Pradesh. Wildlife Institute of India, Dehradun, p 577
Wikramanayake ED, Dinerstein E, Robinson JG, Karanth U, Rabinowitz A, Olson D et al (1998) An ecology-based method for defining priorities for large mammal conservation: the tiger as case study. Conserv Biol 12(4):865–878
Wilcox BA (1984) In situ conservation of genetic resources: determinants of minimum area requirements
Wildlife Institute of India (2009) Status of tiger and its prey species in Panna Tiger Reserve, Madhya Pradesh. Final report submitted to Madhya Pradesh Forest Department, Bhopal, vol 36. Wildlife Institute of India, Dehra Dun
Wilkinson DM, Sherratt TN (2016) Why is the world green? The interactions of top–down and bottom–up processes in terrestrial vegetation ecology. Plant Ecol Divers 9(2):127–140
Williams JN (2011) Human population and the hotspots revisited: a 2010 assessment. In: Biodiversity hotspots: distribution and protection of conservation priority areas. Springer Berlin Heidelberg, Berlin, pp 61–81
Wolf C, Ripple WJ (2016) Prey depletion as a threat to the world's large carnivores. R Soc Open Sci 3(8):160252
Wolf C, Ripple WJ (2017) Range contractions of the world's large carnivores. R Soc Open Sci 4(7):170052
Woodroffe R, Ginsberg JR (1998) Edge effects and the extinction of populations inside protected areas. Science 280(5372):2126–2128
Zacharias MA, Roff JC (2001) Use of focal species in marine conservation and management: a review and critique. Aquat Conserv Mar Freshwat Ecosyst 11(1):59–76

An Experiment in Quasi-Eco-Restoration of a Creek Slope: A Collaborative Approach on a University Campus in India

13

Padamnabhi S. Nagar, Shishir R. Raval, Haribhai R. Kataria, Ankur Rajwadi, Snehal Chavda, Jaydeep J. Sharma, Gulzar Malek, Roshan D. Virola, and Tanmay Rohit

Abstract

This chapter presents a quasi-eco-restoration experiment conducted along the southwest bank of Bhūkhi Nālā on a university campus in India, where years of waste disposal, including construction debris and a mixture of inorganic and organic materials, have severely degraded the ecosystem. A multilayered restoration approach was employed to stabilize the streambank slope and enhance soil retention. Restoration efforts involved the removal of debris, the installation of abiotic eco-engineering measures, and the strategic planting of debris-tolerant species selected for their ecological significance and effectiveness in reducing soil erosion. A comprehensive vegetation survey revealed a diversity of species, with notable findings of *Antigonon leptopus*, *Alternanthera sessilis*, and *Ricinus communis*, among others, primarily concentrated at the toe of the slope. This chapter emphasizes the importance of a collaborative, interdisciplinary approach to ecological restoration techniques with stakeholders' involvement, fostering local ownership to ensure the viability and durability of restoration efforts. The insights gained from this project offer valuable guidance for similar restoration initiatives in degraded environments.

The original version of the chapter has been revised. A correction to this chapter can be found at https://doi.org/10.1007/978-981-96-8943-9_14

P. S. Nagar (✉) · A. Rajwadi · S. Chavda · J. J. Sharma · G. Malek · R. Virola · T. Rohit
Department of Botany, Faculty of Science, The M.S. University of Baroda, Vadodara, India
e-mail: padamnabhi.nagar-botany@msubaroda.ac.in; snehal.c-botanyphd@msubaroda.ac.in

S. R. Raval
The M.S. University of Baroda, Vadodara, India

H. R. Kataria
Department of Mathematics, Faculty of Science, The M.S. University of Baroda, Vadodara, India
e-mail: hr.kataria-maths@msubaroda.ac.in

Corrected Publication 2025
D. Mittal, D. K. Gupta (eds.), *Ecological Restoration in India: Perspective and Practices*, https://doi.org/10.1007/978-981-96-8943-9_13

Keywords

Quasi-eco-restoration · Bhūkhi Nālā · Eco-engineering · Slope stabilization · Streambank · Collaborative approach · Biodiversity · University campus · Plant species

13.1 Introduction

As we grapple with the pressing global challenges of climate change, habitat loss, pollution, and population growth, we face the imperative to make informed and often difficult decisions. Eco-restoration is crucial for addressing the widespread negative impacts of degradation of ecosystems and landscapes on global and national levels (Aradóttir and Hagen 2013). In the early twenty-first century, we have an opportunity to envision and foster healthy ecosystems and landscapes that benefit all species, including humans. Our path forward includes two potential approaches: a proactive strategy of holistic ecological planning and a reactive, yet essential, approach focused on ecological restoration (ER).

Wendell Berry, a renowned novelist, poet, essayist, environmental activist, cultural critic, and farmer, stated, "To cherish what remains of the Earth and to foster its renewal is our only legitimate hope of survival." This thought's power to ignite awareness and effect action inspired our experiment along a small, last stretch of Bhūkhi Nālā (rivulet/stream), a small tributary of the Vishwāmitri River that flows through the Mahārājā Sayājirao University (MSU) campus in Vadodarā, India (Fig. 13.1). In this chapter, we delve into the critical aspects of ER within the context of our experiment along the Bhūkhi Nālā, emphasizing the need to adapt and test restoration strategies based on the best available knowledge and practices.

The project described herein is a collaborative effort between the MSU Botany Department and a team from a professional landscape architecture firm. The goal was to demonstrate an experiment in ecological restoration (ER) for a highly degraded stretch of the bank of Bhūkhi Nālā. Repairing its bank by adapting ER knowledge and know-how helped reestablish vegetation and control soil erosion. As explained in the following section, we are introducing the term "Quasi-Ecological Restoration (QER)" for clarity of terms and purposes of what we did. This initiative is not a quick fix but rather a thoughtful experiment, grounded in both pragmatic and academic values. Over the course of the monsoon seasons in 2021–2024, we have systematically documented observations and gathered data regarding vegetation to refine our practices, ensuring that the processes are appraisable and outcomes are revealing; that is, they teach us significant lessons.

13.2 Eco-Restoration (ER) and Quasi-Eco-Restoration (QER)

ER is fundamentally the process *of assisting the recovery of an ecosystem that has been degraded, damaged, or destroyed (SER)* to a predetermined earlier state or to an appropriate local native reference ecosystem. ER strives to assist an ecosystem on a trajectory to reach full recovery.

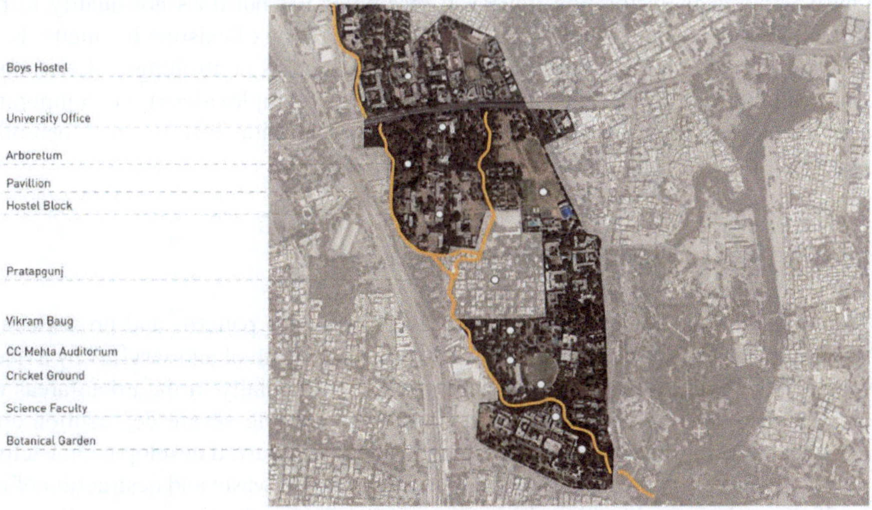

Fig. 13.1 Bhūkhi Nālā meandering through the M. S. University campus, Vadodara. (Google Earth Image Overlay, 2018) (Source: The Bhūkhi Nālā Eco-Restoration Project Preliminary Presentation by the Students and Faculty Team of the MURP Program, led by Prof. Shishir R. Raval, at Department of Architecture, M. S. University, Vadodara on Thursday, October 25, 2018)

Still, as multiple ER initiatives and projects are increasing globally and nationally, we need to go beyond such broad definitions. A deeper understanding of and agreement on what is recognized as good ER by the international scientific community is essential. Key ER principles developed by this community consist of replications or trajectories toward replications of ecological structure and function for a long period of time (Higgs 1997; Higgs et al. 2018; Higgs 2003; Weinstein et al. 2014). In addition, the Food and Agriculture Organization of the United Nations (FAO), leading the UN Decade on Ecosystem Restoration's Best Practices Task Force, the International Union for Conservation of Nature's Commission on Ecosystem Management (IUCN CEM), and the Society for Ecological Restoration (SER) have also jointly presented ten principles to form the basis for ecosystem restoration in the UN Decade on Ecosystem Restoration 2021–2030 (UNEP 2021; SER 2021).

Despite such serious global endeavors, much goes on in the name of ER with good intentions. It is critical to understand the differences among and implications of the terms such as regeneration, rehabilitation, reinhabitation, remediation, repair, reclamation, revegetation, and reparation. Even within restoration, understanding and need of reference sites, structural and functional fidelity, time scale, relationships to local communities, and such become vital (Raval 2023).

The practices and projects that use some of these other terms do exhibit good processes and outcomes close to what ER strives to achieve. They are like ER but do not consist of all or major principles and concepts of good ER. We believe that calling them ER is misleading and dilutes the intentions and measures taken at the international and national levels for ER. Therefore, we are proposing the term

Quasi-Ecological Restoration (QER). It resembles ER but does not qualify to be called ER. The concept of QER is a deliberate adaptation of existing ER methods to suit specific site conditions, specific objectives (such as controlling soil erosion), and available resources. This approach reflects the complexities of contemporary ecological challenges and acknowledges that restoration efforts are inherently experimental and need to be pragmatic in many cases.

13.3 ER and QER Issues and Approaches

The integrity (intactness, linkages, and diversity) of the patterns and processes of the riparian zone is critical for the overall ecological health of and services by a river system. However, the riparian zones everywhere, especially in the urban areas of our country, have endured anthropogenic forces, causing severe degradation and disappearance of this zone due to unplanned and ill-conceived developmental activities, especially dumping of all kinds of solid and liquid waste and destruction of its landform and vegetation cover. This, in turn, negatively affects patterns of flow and habitats upstream and downstream around the riparian zones. Continuing neglect and damage cause irreversible damage to the riparian ecosystems of many rivers and rivulets, as seen in BhūNā.

Negative effects of degradation of streams and rivers include vegetation cover removal, soil erosion, bank destabilization, channel widening or narrowing, sedimentation, reduction in water-holding capacity, and eutrophication. Soil erosion is one of the primary negative effects of riparian zone degradation. This gradual process of topsoil removal occurs due to water and wind erosion, animal movements, and various inappropriate human activities. These, in turn, lead to deteriorating conditions of water, soil, and vegetation, affecting the ecosystem services, like flood and climate mitigation, that wetlands, creeks, and rivers provide.

A prominent approach for restoration in tropical dry regions is the tree-grass system, which integrates trees with grasses and livestock management. This system can effectively harness available natural resources while enhancing soil productivity and microbial diversity. The global focus on carbon capture, as articulated in the Paris Climate Change Agreement, to which India is committed, underscores the importance of developing carbon sinks of 2.5–3 billion tons of CO_2 equivalent through expanded and apt vegetation cover (Government of India 2022). This highlights the necessity of testing tree-grass systems for eco-restoration to fulfill India's climate action commitments.

Adapting eco-engineering techniques for restoring abiotic conditions and using native plant species for revegetation and habitat enhancement are critical. They help improve ecosystem structure (like bank stabilization) and functions (like pollination). Restoration can also help reduce susceptibility to inbreeding, link fragmented habitat patches, and lower chances of local extinction, especially in small and isolated patches.

Eco-engineering measures provide a range of advantages over conventional engineering methods, particularly in ecological restoration contexts. These strategies often require minimal need for heavy equipment and personnel, making them more cost-effective and environmentally friendly. When executed properly, eco-engineering techniques cause minimal disturbance to the site and its surroundings during installation and demand less ongoing maintenance. In situations where the preservation of scenic landscapes or wildlife habitats is critical, eco-engineering offers solutions that are more compatible with the natural environment.

Utilizing natural materials for restoration is especially beneficial, as these materials enhance the establishment of vegetation compared to manufactured alternatives. Techniques such as live crib walls, vegetated rock retention structures, bio-fencing, and joint plantings in porous structures with openings designed for the insertion of live vegetative cuttings or plugs. These inert structural elements provide immediate resistance to soil sliding, erosion, and slope washout. As vegetation becomes established, roots permeate the slope and soft eco-engineering materials begin to decompose, while the canopy cover helps mitigate the impacts of rainfall and runoff. This process promotes a cohesive mass that effectively stabilizes the banks of streams and rivers and controls soil erosion.

13.4 Relevant Eco-Restoration Practices

As we planned the small research projects based on our QER experiment, we found the following examples relevant to our BhūNā experiment and follow-up studies.

From a significant 8-year study of an eco-restoration project in Eastern Quebec, Canada, we found the method of systematic line-point intercept (LPI) (Bonham 1989) to be useful for assessing differences in diversity and richness between restored and unrestored sites and determining whether the establishing plant communities conformed to the natural variation observed in reference ecosystems. The nature-based bioengineering measures in Northeast China to combat soil erosion highlighted the effectiveness of these interventions by comparing soil erodibility with natural riverbank conditions after a 4-year period (Zhang et al. 2020). This gave us an example of monitoring and learning from our project over 4 years. In addition, study by Poulin et al. (2013) was useful for analysis of species proportion and richness and to monitor community composition.

A comprehensive review of both field and experimental studies conducted in eastern Australia helped us understand the role of native vegetation in preventing mass failure of riverbanks (Hubble et al. 2010). In general, restoration efforts to protect riparian zones by reestablishing native vegetation enhanced bank substrate reinforcement, providing significant protection against erosion.

As part of the "Bonn Challenge: A Global Partnership for Restoration," various Indian governments, agencies, and private entities have undertaken initiatives to restore degraded ecological sites across the country. Restoration of tropical rainforest fragments in the landscape of the Anamalai hill range in South India, degraded forests in Nagaland, and Banni grasslands in Gujarat are notable examples of this countrywide effort (Borah et al. 2018; Bhattacharjee 2020).

A study in the Damodar River area of eastern Jharkhand near Dhanbad, India, (Srivastava et al. 2009) helped us in understanding the effectiveness of the native riparian herbs *Leonotis nepetaefolia*, *Cassia tora*, *Ageratum conyzoides*, *Parthenium hysterophorus*, and *Sida acuta* for their roles in reducing water runoff and soil conservation. A study conducted in the in the Kansai Basin of Purulia District of West Bengal (Dey et al. 2020) highlighted the importance of the naturally growing riverine plants in preventing soil erosion due to their roots' ability to slow water flow. Another study along the Rihand River near Renukoot in Eastern Uttar Pradesh (Kumar et al. 1996) found that these five herbaceous plants from riverbanks, strategically planted on sloping experimental plots, helped in understanding their potential to conserve organic carbon, sodium, potassium, and calcium. The native plant selection for our experiment benefited from such studies.

13.5 QER Experiment for a Stretch of Bhūkhi Nālā (BhūNā)

13.5.1 About Bhūkhi Nālā (BhūNā)

The BhūNā is a degraded creek that has suffered significant anthropogenic impacts over the last few decades. Its final stretch on the MSU campus presents a unique opportunity for restoration. The BhūNā story is like many such nālās in Vadodara (Munjal and Behal 2011) and other urban areas of urban India (Ahmad M. et al. 2024). It is abused with encroachment, channelization, encasement, and dumping of untreated sewage and solid waste, including construction debris. While soil erosion is a natural phenomenon, it is accelerated by human impacts, leading to slope instability. In the case of BhūNā, it is exacerbated by the removal of vegetation growth using a big machine scraper along its banks as a part of the authorities' ill-advised pre-monsoon cleanup drive to let stormwater move out during monsoonal rains.

Restoration of such degraded ecosystems offers complex challenges and unique opportunities to adapt and innovate. Healthy vegetation along riverbanks plays a crucial role in stabilizing soil and addressing many other negative impacts. Yet, relying solely on vegetation is insufficient; the specific conditions, causes, and types of erosion must be addressed through site- and context-specific solutions. Integrating low-cost eco-engineering techniques with appropriate vegetation can enhance efforts to stabilize the banks while ensuring other ecosystem benefits, as we aim to do in our experiment along the BhūNā.

13.5.2 Study Area

The study area is located along the northeast banks of BhūNā, near the Botanical Garden of the Faculty of Science at the Maharaja Sayajirao University of Baroda, India (Latitude: 22.3099, Longitude: 73.1860). Geologically, the region is characterized as barren, having been subjected to waste dumping over the last 5–10 years. The southern bank of Bhūkhi Nālā features a natural landform that has been extensively covered with layers of construction debris, as well as both inorganic and organic waste. This accumulation not only expanded the northeast edge of the Botanical Garden and thus encroached upon the southwest bank of the Nālā, resulting in a neglected and visually unappealing landscape.

For many years, the site remained barren and overlooked until an emergency restoration project was initiated in early March 2021. The formal presentation of the project to university authorities took place on March 11, 2021, coinciding with the birthday of Sri Sayajirao III Gaekwad of Baroda. The area spans approximately 160 m in length and averages 14 m in width, featuring an inclined terrain typical of a riverbank, which has suffered from erosion and flooding.

To evaluate the ecological changes resulting from the restoration efforts, the study area was divided into three distinct regions (Fig. 13.2a):

1. Control Site: This 5-m-long strip, located on the northern flank of the treated area, implemented a passive restoration approach. While cleansing of accumulated debris and waste was conducted, no native plant species were introduced.
2. Treated Site: Extending 20 m, this area underwent active restoration. In addition to the removal of waste and debris, eco-engineering techniques were employed, and native plant species were deliberately introduced to promote ecological recovery.

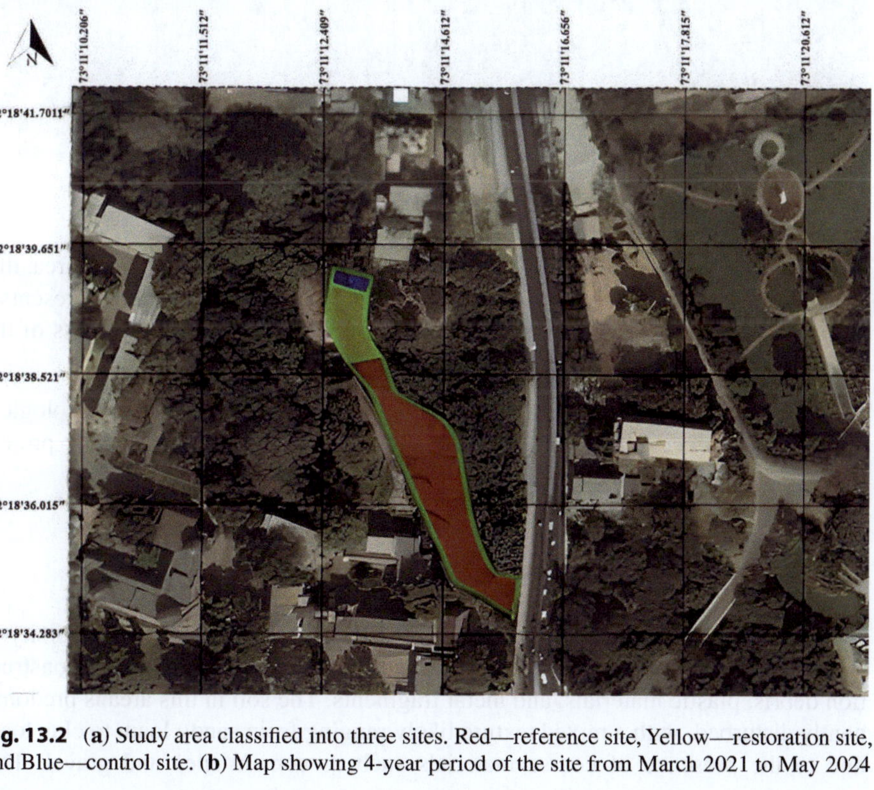

Fig. 13.2 (**a**) Study area classified into three sites. Red—reference site, Yellow—restoration site, and Blue—control site. (**b**) Map showing 4-year period of the site from March 2021 to May 2024

Fig. 13.2 (continued)

3. Reference Site: Spanning 135 m along the southern flank of the treated area, this site remained relatively untouched by previous degradation events. It presents a native condition, serving as a benchmark for assessing the effectiveness of the restoration efforts, excluding factors unrelated to over-dumping.

This structured approach allows for a comprehensive analysis of the ecological impacts and effectiveness of the restoration techniques employed in the project (Fig. 13.2b).

13.5.3 Site Characteristics

The BhūNā site near the Botanical Garden of the Maharaja Sayajirao University of Baroda has been heavily impacted by years of waste dumping, including construction debris, plastic materials, and metal fragments. The soil in this area is predominantly disturbed, with a mixed texture likely ranging from sandy loam to clay loam due to the presence of coarse materials such as cement tiles and gravel. This disturbed soil typically has a shallow depth of about 1–2 feet and exhibits poor drainage characteristics, which hinder water infiltration and retention.

The soil pH varies widely, likely ranging from 6.0 to 8.0, influenced by the alkaline nature of the cement debris, and organic matter content is expected to be low, likely below 1%. Nutrient levels in the soil are also likely deficient, with nitrogen (N) levels estimated to be below 50 kg ha^{-1}, phosphorus (P) under 10 kg ha^{-1}, and potassium (K) potentially below 50 kg ha^{-1}. The site's water-holding capacity is compromised, further complicating conditions for plant growth. Vegetation in the area is characterized by the dominance of invasive species, particularly the ice cream creeper (*Antigonon leptopus*), which has established itself in the nutrient-depleted and disturbed soil. This species can exacerbate the ecological imbalance by overshadowing native flora and further contributing to soil degradation.

The semi-arid climate, with an average annual rainfall of around 869 mm, coupled with extreme seasonal variations, adds to the environmental stress on both soil and plant life. To address these challenges, we implemented soil remediation strategies, such as the introduction of organic amendments and native vegetation, which can enhance soil quality and promote biodiversity. Additionally, managing debris and controlling invasive species will be essential to facilitate effective ecological restoration in this degraded landscape. Regular monitoring of soil and plants and soil testing provides valuable insights into the progress of restoration efforts and the evolving conditions of this site.

Experiential qualities, especially the sensory qualities, of such sites are often neglected, but they provide clues for tempering restoration measures to protect and enhance these qualities. The view of and sound from the flowing waters of BhūNā add to the value of this part of BhūNā. Being there, one almost forgets that one is so close to the hubbub of the city. The characteristics of the plant species and their locations, in addition to ecological reasons, become critical factors in the design and maintenance of such restoration work.

13.6 Eco-Restoration Goals and Strategies

Eco-engineering harmoniously integrates both ecological components (abiotic and biotic) and soft engineering methods to benefit both nature and human society (Fig. 13.3). It represents a confluence of basic and applied sciences aimed at restoring and constructing terrestrial and aquatic ecosystems. The five broad techniques utilized in eco-restoration include:

1. *Remove and Stop Bank Destabilizing and Degrading Factors*: This involves identifying and eliminating sources of degradation, such as invasive species or human activity that compromise bank stability.
2. *Live Planting*: Introducing native plant species directly into the ecosystem to enhance biodiversity and promote soil health.

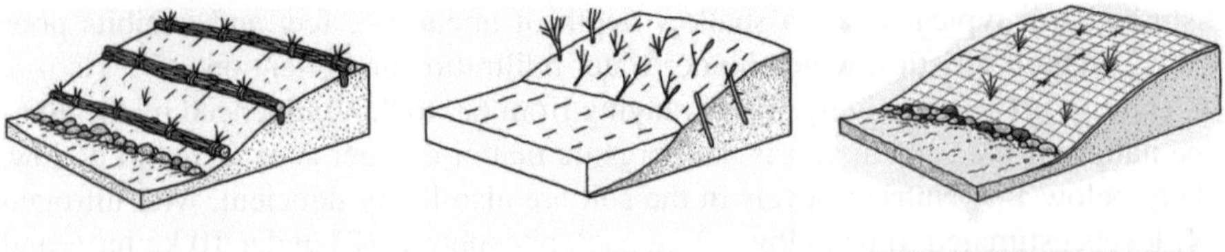

Fig. 13.3 Some biotechnical streambank stabilization techniques (Li and Eddleman 2002; USDA–NRCS 2007)

3. *Eco-Engineering*: Employing a combination of abiotic (e.g., rocks, soil) and biotic (e.g., plants, microorganisms) materials and methods to create apt and durable structures that support ecological functions.
4. *Hard Engineering*: Implementing traditional structures like retaining walls, gabions, and armoring to provide additional support to the bank while integrating with eco-engineering techniques.
5. *Management and Maintenance*: Ongoing management of banks and riparian zones after stabilization measures are implemented, ensuring the long-term success of the restoration efforts.

By adopting these biotechnical streambank stabilization techniques, we can enhance the resilience of degraded ecosystems and promote the recovery of natural habitats, ultimately benefiting both local communities and biodiversity.

Some Biotechnical Streambank Stabilization's Techniques

13.7 Materials and Method

The study site, situated along the southern bank of Bhūkhi Nālā, has been severely impacted by years of waste disposal, including layers of construction debris and a mix of inorganic and organic materials. This encroachment not only expanded the northern side of the Botanical Garden but also transformed the southern bank into a neglected and barren landscape, creating an eyesore for many years.

For the experimental phase, approximately 160 m of the riverbank were designated for observation and restoration efforts. A pilot survey was conducted to assess the existing vegetation prior to implementing the restoration models (Table 13.1). Notably, *Antigonon leptopus* was found to be highly abundant throughout the study area. At the toe of the slope, *Alternanthera sessilis* thrived, while *Ricinus communis* established itself successfully along both banks. The first patch of dense vegetation was home to *Tinospora cordifolia*, and herbaceous species such as *Abutilon indicum* and *Achyranthes aspera* were present but limited in number. Overall, most observed species were concentrated at the toe of the bank, reflecting the long-standing neglect and degradation of the BhūNā and its surroundings.

Table 13.1 The list of plants observed on the study site during the pilot survey of present vegetation before implanting the model

No.	Botanical name	Family
Herb		
1	*Abutilon indicum* (L.) Sweet.	Malvaceae
2	*Achyranthes aspera* L.	Amaranthaceae
3	*Alternanthera sessilis* (L.) R.Br. ex DC.	Amaranthaceae
4	*Barleria prionitis* L.	Asteraceae
5	*Chenopodium album* L.	Chenopodiaceae
6	*Heliotropium indicum* L.	Boraginaceae
7	*Phyllanthus amarus* Schum & Th.	Phyllanthaceae
Shrub		
8	*Capparis sepiaria* L.	Capparidiaceae
9	*Datura Metel* L	Solanaceae
10	*Jatropha gossypiifolia* L	Euphorbiaceae
11	*Ricinus communis* L.	Euphorbiaceae
12	*Solanum lycopersicum* L.	Solanaceae
Climber		
13	*Antigonon leptopus* Hook. & Arn.	Polygonaceae
14	*Bougainvillea spectabilis* Willd.	Nyctaginaceae
15	*Teramnus labialis* (L.F.) Spreng.	Papilionaceae
16	*Tinospora cordifolia* (Willd.) Miers.	Menispermaceae
17	*Acacia concinna* Linn.	Mimosaceae
Tree		
18	*Cordia dichotoma* G. Forst.	Boraginaceae
19	*Ficus hispida* L.F.	Moraceae
20	*Ficus racemosa* L.	Moraceae
21	*Holoptelea integrifolia* Planch.	Ulmaceae

Years of dumping construction debris and both solid and liquid waste—including raw sewage and hazardous chemicals—have severely degraded the ecosystem. Addressing these challenges necessitates a collaborative and interdisciplinary approach that integrates ecological restoration techniques with community involvement and expertise from various fields, including environmental science, engineering, and public health. This holistic strategy is crucial not only for the restoration of the BhūNā but also for fostering a sense of ownership among local stakeholders, ensuring the long-term durability of the restoration efforts.

Over four monsoon seasons, from March 2021 to mid-2024, we were engaged in rigorous documentation and analysis to refine our restoration techniques, enabling us to learn from both successes and setbacks. As we embark on this critical journey of ecological restoration at BhūNā, we employ a systematic framework that integrates ecological and educational aspects for all sections of society. Our approach seeks not only to restore ecological health but also to foster environmental studentship and stewardship through community engagement.

13.8 Model Establishment for Streambank Erosion Control

In this experimental study on streambank erosion control along the Bhūkhi Nālā, a multilayered eco-engineering approach was implemented to stabilize the slope and improve soil retention (Fig. 13.4). The initial phase involved removing the top layer of debris from the slope to expose the underlying alluvial soil for planting. However, due to the steepness of the slope, the sheer quantity of the dumped materials, and the limitations of available means, it was not possible to completely remove these layers. This reality necessitated making changes in our approach. Staggered linear bamboo fences were constructed at various levels along the sloping bank, which helped stabilize the soil layers both above and below, effectively preventing erosion.

The first patch was left undisturbed to serve as a control. In the second patch, a 4–6-inch layer of soil was added on top of the debris, followed by a mixture of dry mulch (comprising leaves and small twigs) and organic manure. This mulch

Fig. 13.4 Stages of eco-engineering intervention. (**a**) The top layer of debris has been removed as much as possible and leveled to maintain a uniform slope. (**b**) Pegs were inserted on the slopes at specific distances to hold the bamboo fences. (**c**) Bamboo fences were made at different levels. (**d**) The first patch serves as the control patch in the experiment. (**e**) A layer of soil is spread over the debris. (**f**) A layer of dry mulch is spread over the soil. (**g**) Locally available jute matting is spread over the first and third layers. (**h, i**) Pictures of the completion of the treated site

supports earthworms and beneficial microbes, enhancing soil aeration and moisture retention. Locally sourced jute matting (locally known as *kantān*) was then laid on top and secured with bamboo pegs. A horizontal bamboo pole anchored at the top of the slope provided additional support, while large rocks were used for back support at the base of the fence. The third patch contained only dry leaves and mulch, while the steeper slope featured a more robust bamboo fence to prevent slippage caused by the weight of the soil. *Kantān* was secured at both the top and bottom of the slope to enhance stability.

The selected plants were primarily debris-tolerant species with tap and fibrous roots, effective in reducing soil erosion. This structured approach combines innovative materials with natural vegetation to address streambank erosion effectively (Fig. 13.5). The findings from this model will offer valuable insights into durable and viable erosion control practices and their applicability in similar environments (Fig. 13.6).

To establish the model, we modified the vegetation in various patches using a selection of plant species chosen for their ecological importance and effectiveness in reducing soil erosion. The uppermost layer of planting consists of *Terminalia arjuna* (Arjun tree), *Terminalia bellirica* (Baheda tree), and *Putranjiva roxburghii* (Putranjiva tree), selected for their robust tap roots that anchor the soil, particularly along the edges, thereby helping to prevent erosion. Historically, these trees were utilized in construction prior to the widespread use of chain-link fences.

Key species selected for soil stabilization include *Pandanus* sp./*Dracaena* sp., known for their stilt and fibrous roots that enhance soil strength against heavy rainfall; *Dolichandrone falcata* and *Firmiana colorata*, which are adapted to slopes and possess bulbous and lateral fibrous roots thriving along streams; and *Vetiveria zizanioides*, *Cymbopogon citratus*, and *Cynodon dactylon*, which not only control invasive weeds but also stabilize soil through their extensive fibrous root systems.

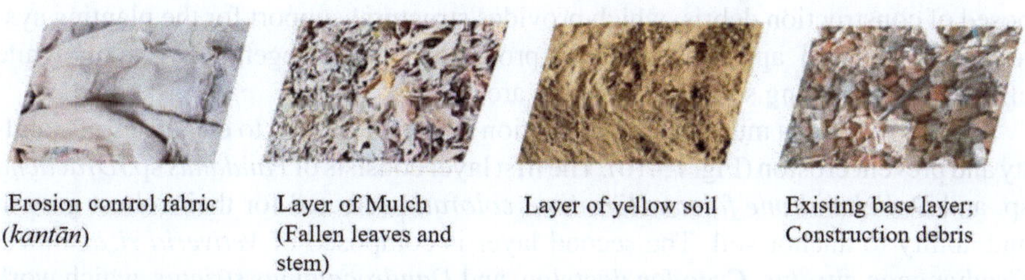

Fig. 13.5 Different materials used as different layers of soil

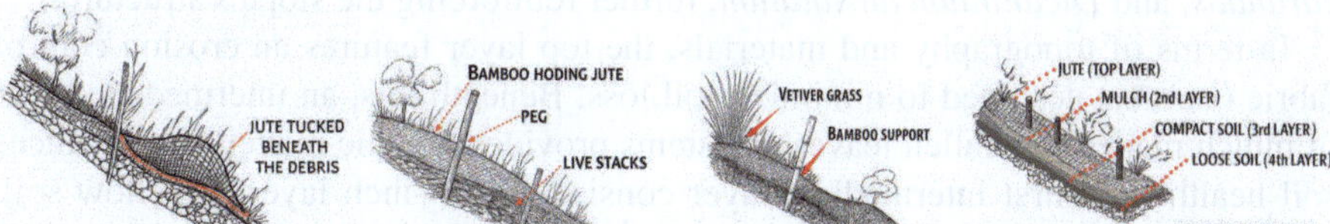

Fig. 13.6 Section of site showing different layers of soil, bamboo pegs, and supports

Additionally, *Dendrocalamus strictus* offers significant ecological value with roots that spread over large areas to reinforce soil structure, while *Carissa carandas* features thorns that deter invasive animals and thrives in rocky environments. *Pongamia pinnata* effectively holds soil on intermediate slopes, and *Dicanthium annulatum* stabilizes soil in debris-laden areas and along streambanks. *Syzygium salicifolium* grows well in debris, providing further protection against soil erosion, while grasses such as *Brachiaria ramosa*, *Sorghum halepense*, and *Eleusine indica* contribute to soil stabilization through their fibrous root systems. Through the careful selection of these plants, we aim to create a resilient ecosystem that effectively mitigates soil erosion in the targeted areas (Fig. 13.7).

In terms of topography and materials, the topmost layer features an erosion control fabric made of jute (*kantān*), designed to prevent soil loss. Beneath this, an intermediate layer of mulch made from fallen leaves and stems provides organic matter that supports soil health. The first intermediate layer consists of a 4-inch layer of yellow soil, which helps retain moisture and nutrients (Fig. 13.8). Finally, the existing base layer comprises construction debris, providing structural support for the entire planting system. This multilayered approach aims to enhance soil stability and promote vegetation growth effectively.

Patch 2 features a structured three-layer plantation system designed for effective soil stabilization and erosion control (Fig. 13.9). The first layer comprises *Pandanus* sp./*Dracaena* sp. and *Dolichandrone falcata*/*Firmiana colorata*, both of which are well-suited for establishing a strong root network. The second layer consists of *Vetiveria zizanioides*, *Cymbopogon citratus*, *Cynodon dactylon*, and *Dandrocalamus strictus*, chosen for their ability to stabilize soil and combat invasive species. The third layer includes *Vetiveria zizanioides*, *Cymbopogon citratus*, *Cynodon dactylon*, *Carissa carandas*, and *Syzygium salicifolium*, further enhancing the overall integrity and resilience of the slope.

In terms of topography and materials, the existing base layer is primarily composed of construction debris, which provides structural support for the planting system. This layered approach aims to promote healthy vegetation growth while effectively mitigating soil erosion in the area.

Patch 3 features a multilayered plantation system designed to enhance soil stability and prevent erosion (Fig. 13.10). The first layer consists of *Pandanus* sp./*Dracaena* sp. and *Dolichandrone falcata*/*Firmiana colorata*, selected for their robust growth and ability to anchor soil. The second layer is composed of *Vetiveria zizanioides*, *Cymbopogon citratus*, *Cynodon dactylon*, and *Dandrocalamus strictus*, which work together to stabilize the soil and mitigate invasive plant growth. The third layer includes *Vetiveria zizanioides*, *Cymbopogon citratus*, *Cynodon dactylon*, *Carissa carandas*, and *Dicanthium anNālātum*, further reinforcing the slope's structure.

In terms of topography and materials, the top layer features an erosion control fabric (*kantān*) designed to minimize soil loss. Beneath this, an intermediate layer of mulch made from fallen leaves and stems provides organic matter that enhances soil health. The first intermediate layer consists of a 4-inch layer of yellow soil, which aids in moisture retention and nutrient availability. The existing base layer is comprised of construction debris, offering essential structural support for the entire

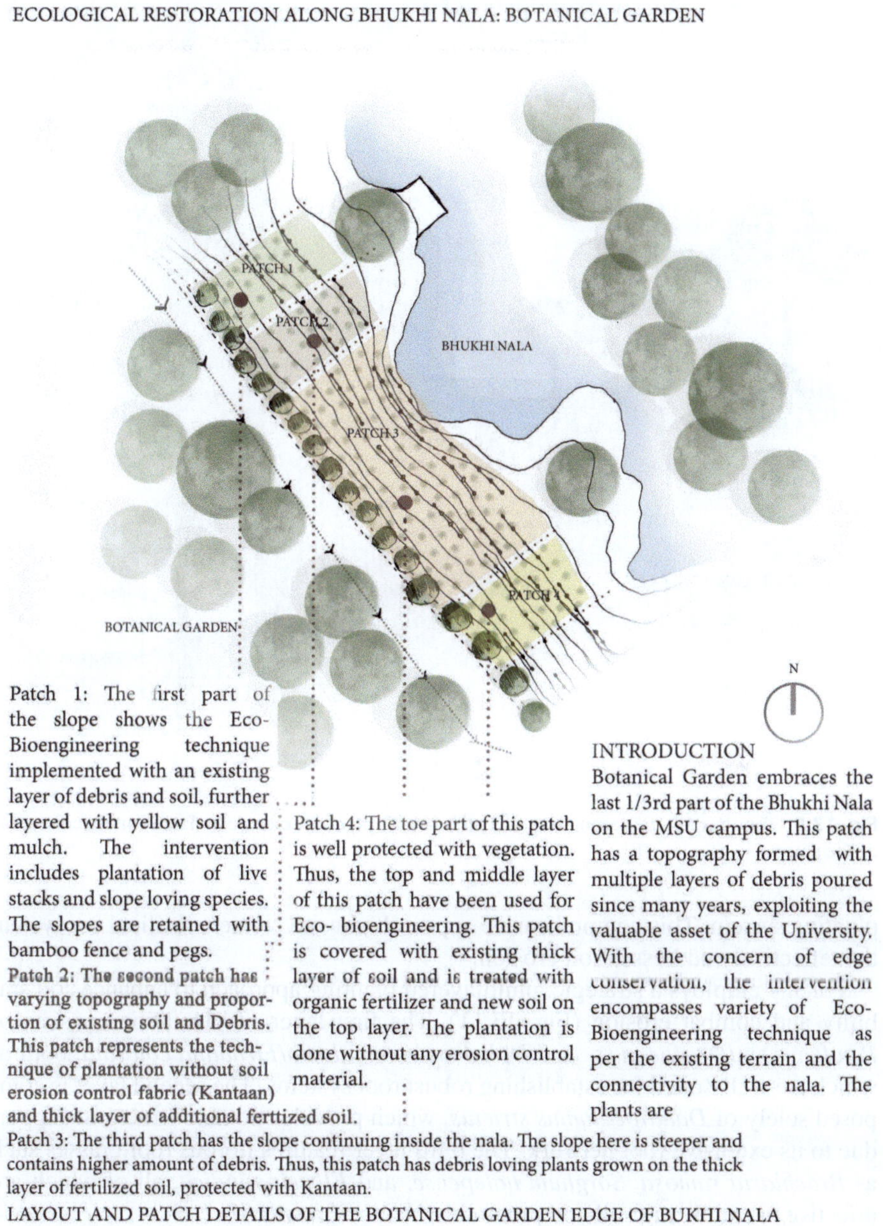

Fig. 13.7 Restoration site details

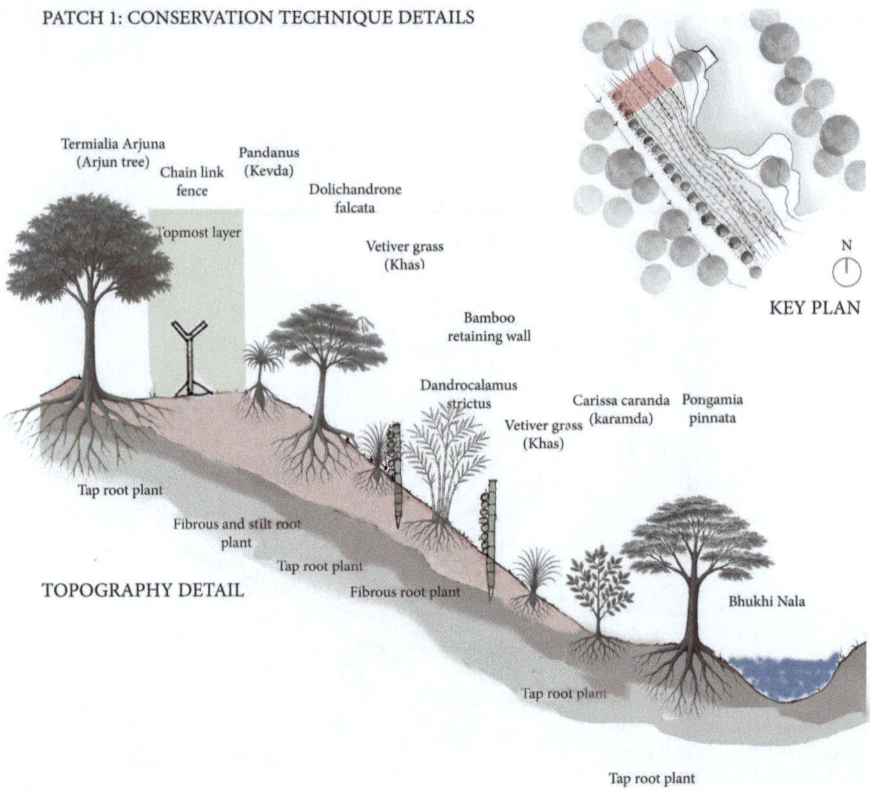

Fig. 13.8 Topographic representation of Patch 1 with plant species designated for specific areas of the slope

planting system. This comprehensive approach aims to create a resilient ecosystem that effectively addresses soil erosion.

Patch 4 employs a strategic, multilayered planting approach to enhance soil stability and combat erosion (Fig. 13.11). The first layer of plantation consists of *Pandanus* sp./*Dracaena* sp. and *Dolichandrone falcata*/*Firmiana colorata*, both of which are well-suited for establishing robust root systems. The second layer is composed solely of *Dandrocalamus strictus*, which provides notable structural support due to its extensive root network. The third layer features fibrous root grasses such as *Brachiaria ramosa*, *Sorghum halepense*, and *Eleusine indica*, all of which are effective in stabilizing soil and promoting healthy growth.

In terms of topography and materials, the intermediate layer consists of a 12-inch layer of yellow soil, which aids in moisture retention and nutrient availability. The existing base layer is primarily composed of construction debris mixed with a large proportion of soil, offering essential support for the entire planting system. This carefully designed approach aims to create a resilient ecosystem that effectively addresses soil erosion while promoting the growth of diverse vegetation. Species introduced in the restoration site are provided in Table 13.2.

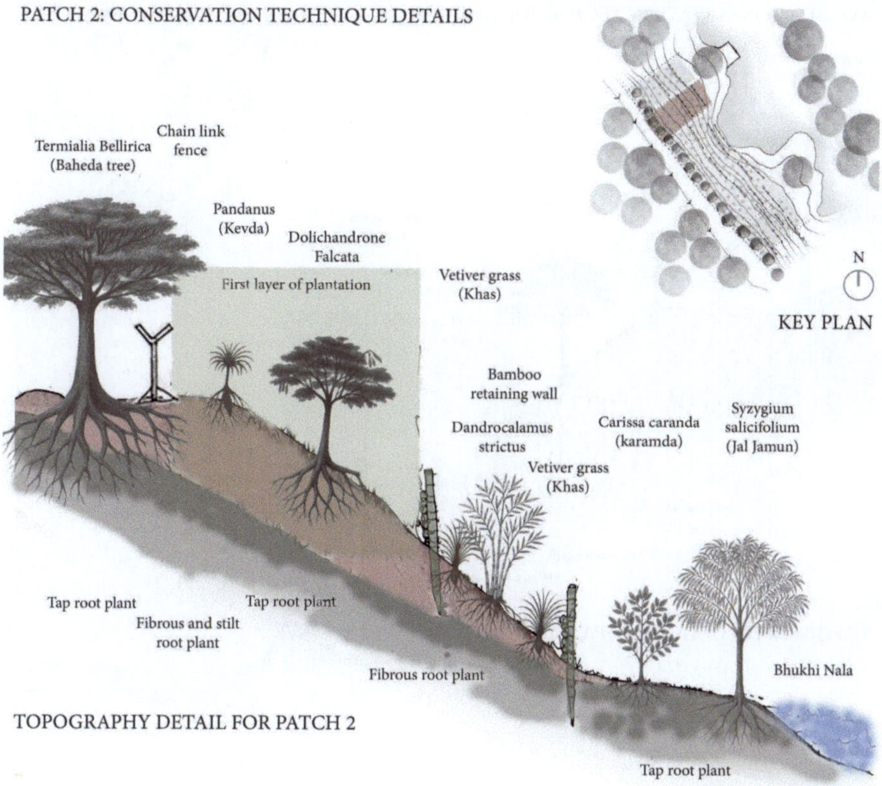

Fig. 13.9 Topographic representation of Patch 2 with plant species designated for specific areas of the slope

13.9 Result and Discussion

The eco-engineering techniques implemented along the streambank of BhūNā yielded encouraging outcomes within the first year. Excluding Patch 4, which displayed limited growth, Patches 1, 2, and 3 exhibited strong vegetative development, indicating a successful establishment of riparian vegetation. The effectiveness of this vegetation in mitigating soil erosion will be assessed more comprehensively during and after the monsoon season. Figure 13.12 illustrates plantation and drip irrigation installation (a), images (b) and (c) show plant growth progress by May 2021, reflecting successful establishment.

In February and March of 2022, a thorough survey was conducted across three areas: treated, control, and reference. The reference area was divided into 14 plots, each measuring 25 feet along the streambank, in line with previous inventory methods. Each plant species, including seedlings and newly regenerated plants, was systematically recorded.

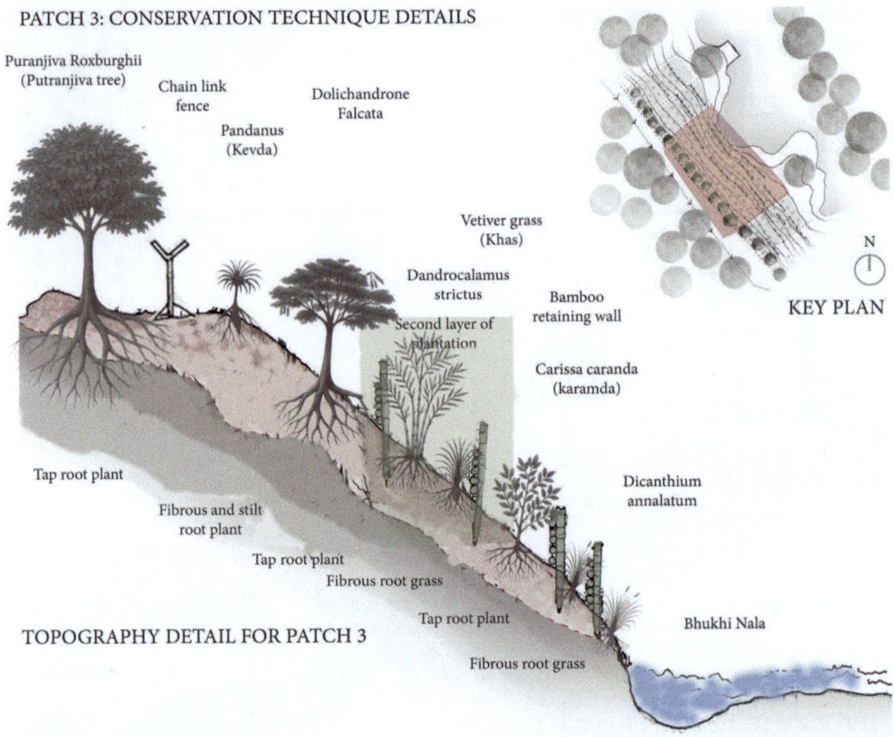

Fig. 13.10 Topographic representation of Patch 3 with plant species designated for specific areas of the slope

13.10 Species Richness

A total of 58 plant species were identified across the surveyed areas. The reference area displayed a species richness of 31, with 15 species categorized as uncommon compared to the treated and control areas. The treated area demonstrated higher species richness at 39, which included 27 species not found in the other two areas. Conversely, the control area had a species richness of 16, with four species considered uncommon relative to the treated and reference sites.

Overall, plant species richness varied remarkably among the three habitats, with the treated area showing the highest diversity. Across all study sites, 31 plant families were recorded. The Poaceae family (grasses) was predominant in both the reference and treated areas. Specifically, the reference site contained 16 families, while the treated area included 24 families, and the control area featured 14 families. Notably, 16 unique families were identified exclusively in either the treated or control sites, highlighting the ecological differentiation achieved through the eco-restoration efforts (Tables 13.3, 13.4, and 13.5).

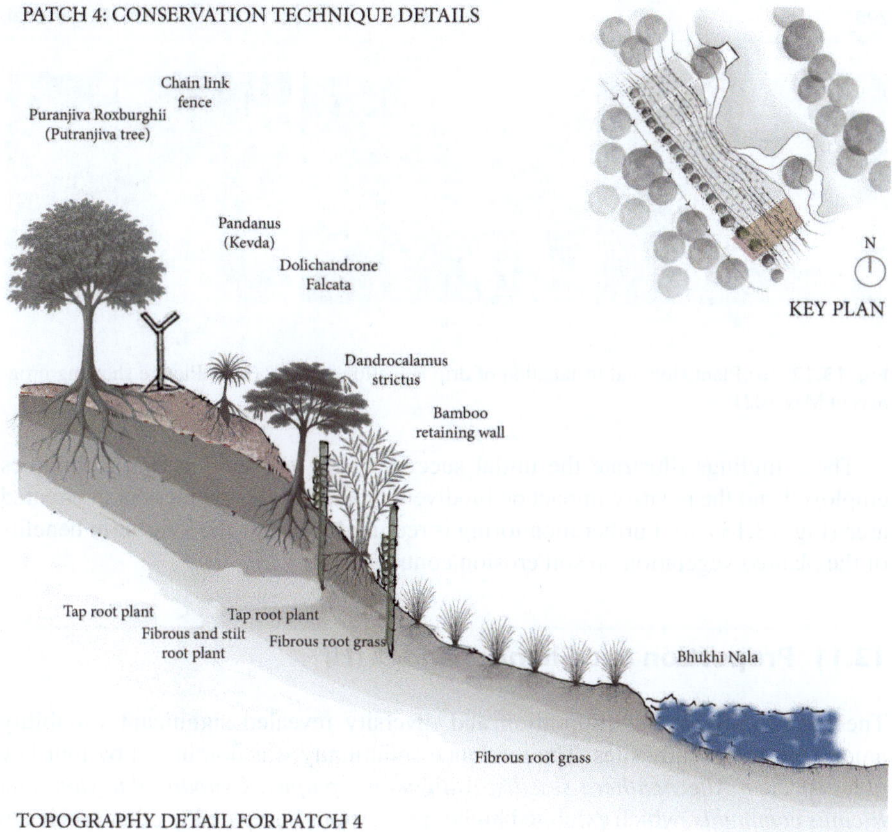

Fig. 13.11 Topographic representation of Patch 4 with plant species designated for specific areas of the slope

Table 13.2 List of species introduced in the restoration site

Layers	Plant species	No. of individuals planted
First	*Pandanus* sp.	8
	Dracaena sp.	8
	Dolichandrone falcata	24
	Firmiana colorata	24
Second	*Dendrocalamus strictus*	24
Second and third	*Vetiveria zizanioides*	24
	Cymbopogon citratus	8
	Cynodon dactylon	10
Third	*Carissa carandas*	24
	Pongamia Pinnata	8
	Syzygium salicifolium	8
	Dicanthium anNālātum	10
	Brachiaria ramosa	24
	Sorghum halepense	
	Eleusine indica	

Fig. 13.12 (**a**) Plantation and installation of drip irrigation system. (**b**, **c**) Picture showing progress in May 2021

These findings illustrate the initial success of the eco-engineering techniques employed and the positive impact on biodiversity and species richness in the treated area (Fig. 13.13a–c). Further monitoring is required to assess the long-term benefits of the planted vegetation on soil erosion control.

13.11 Proportion and Shannon Index (H)

The analysis of species proportion and diversity revealed significant variability among the three study sites. The reference community was dominated by four key plant species: *Alternanthera sessilis*, *Antigonon leptopus*, *Cynodon dactylon*, and *Ricinus communis*, which exhibited higher proportions compared to other species in this area.

In contrast, the restoration site displayed a different proportional structure, with *Dendrocalamus strictus* emerging as the most dominant species, followed by *Alternanthera sessilis* in the second position. The control area showed a higher proportion of *Cynodon dactylon*, indicating a distinct community structure.

Diversity, as measured by the Shannon index (H), was highest in the restoration site, suggesting a richer and more complex ecological community (Table 13.6). Conversely, the control site exhibited the lowest diversity among the three studied communities, highlighting the impact of the eco-restoration practices on enhancing species richness and ecological balance.

The restoration site revealed the highest diversity within it in comparison to other communities. The control site showed the lowest diversity among the three communities studied (Fig. 13.14). These findings underscore the varying dynamics of plant community structures across the study sites, with the restoration site demonstrating a positive response to the implemented eco-engineering techniques. Further analysis is needed to assess the long-term maintainability of these patterns.

Table 13.3 Recorded data of the reference community

Plant species	Family	No. of individuals in plot														
		1	2	3	4	5	6	7	8	9	10	11	12	13	14	Total
Abutilon indicum (L.) Sweet.	Malvaceae	0	2	4	0	1	0	1	3	1	0	0	3	0	1	16
Acacia intsia (L.) Willd.	Mimosaceae	5	1	0	0	0	0	0	0	0	0	0	0	0	0	6
Acacia concinna Linn.	Mimosaceae	1	0	0	0	0	0	0	0	0	0	0	0	0	0	1
Acalypha indica L.	Euphorbiaceae	1	2	3	0	2	2	3	2	1	0	4	1	2	2	25
Achyranthes aspera L.	Amaranthaceae	0	0	0	0	0	0	1	2	1	0	0	0	0	0	4
Alternanthera sessilis (L.) R.Br.	Amaranthaceae	5	31	40	19	12	13	16	10	8	16	12	13	13	18	226
Antigonon leptopus Hk. & Arn.	Polygonaceae	4	10	8	8	11	12	19	18	21	8	10	9	8	7	153
Argemone mexicana L.	Papaveraceae	0	0	0	0	0	0	0	2	1	1	0	0	0	0	4
Arundo donax L.	Poaceae	0	0	0	0	0	0	0	0	1	3	0	0	0	0	4
Barleria prionitis L.	Acanthaceae	2	0	0	0	0	0	1	0	0	0	0	1	2	0	6
Phyllanthus reticulatus Poir.	Phyllanthaceae	4	12	7	5	2	3	3	1	1	0	0	1	0	0	39
Chenopodium album L.	Chenopodiaceae	2	0	0	0	0	1	0	1	0	1	1	0	0	0	6

(continued)

Table 13.3 (continued)

Plant species	Family	No. of individuals in plot 1	2	3	4	5	6	7	8	9	10	11	12	13	14	Total
Cordia dichotoma G. Forst.	Boraginaceae	1	1	0	0	0	0	0	0	1	0	0	0	0	0	3
Cordia domestica Roth.	Boraginaceae	1	0	1	2	1	0	0	1	0	0	0	0	0	1	7
Cynodon dactylon (L.) Pers.	Poaceae	19	32	15	8	10	14	7	10	11	7	10	20	15	30	208
Datura Metel L.	Solanaceae	0	0	0	0	1	0	1	0	0	0	0	0	0	0	2
Eleusine indica (L.) Geaertn.	Poaceae	0	4	3	3	2	0	0	0	0	1	3	2	3	2	23
Ficus hispida L.f.	Moraceae	2	1	3	4	3	2	4	4	5	2	1	2	3	3	39
Ficus racemosa L.	Moraceae	1	0	0	1	0	1	0	0	0	0	0	0	1	0	4
Ficus religiosa L.	Moraceae	0	0	0	0	0	0	1	0	0	0	0	0	0	0	1
Guazuma ulmifolia Lam.	Sterculiaceae	0	0	1	0	0	1	1	0	0	0	1	0	0	0	4
Holoptelea integrifolia (Roxb.) Planch.	Ulmaceae	1	0	0	2	1	0	1	0	0	0	0	1	0	0	6
Heliotropium indicum L.	Boraginaceae	0	1	1	0	0	0	0	0	0	0	0	0	0	0	2
Isachne globosa (Thunb.) O. Kize.	Poaceae	6	5	0	3	8	5	4	5	2	1	0	0	5	4	48

Jatropha gossypifolia L.	Euphorbiaceae	0	0	0	1	0	0	0	0	2	0	0	0	0	0	3
Sida acuta Burm.f. Garcke.	Malvaceae	2	1	3	2	4	2	1	0	1	5	3	2	1	1	28
Phyllanthus amarus Schum & Th.	Phyllanthaceae	5	3	8	0	0	5	3	2	0	3	0	8	5	6	48
Polyalthia longifolia (Sonner.) Thw.	Annonaceae	0	0	0	1	0	0	0	0	0	0	0	0	0	0	1
Ricinus communis L.	Euphorbiaceae	2	1	0	8	17	10	4	6	8	3	7	2	2	4	74
Vernonia cinerea (L.) Less.	Asteraceae	0	0	1	1	0	0	0	2	0	0	0	0	1	0	5
Sorghum halepense (L.) Pers.	Poaceae	0	0	1	0	2	3	0	1	0	1	0	0	2	3	13
Total		64	107	99	68	77	74	71	70	65	52	52	65	63	82	1009

Table 13.4 Recorded data of the treated community

Plant species	Family	Number of individuals
Acalypha indica L.	Euphorbiaceae	5
Alternanthera sessilis (L.) R.Br.	Amaranthaceae	32
Amaranthus spinosus L.	Amaranthaceae	1
Anogeissus sericea Brandis.	Combretaceae	2
Antigonon leptopus Hk. & Arn.	Polygonaceae	18
Dendrocalamus strictus (Roxb.) Nees.	Poaceae	72
Barleria prionitis L.	Acanthaceae	1
Phyllanthus reticulatus Poir.	Phyllanthaceae	3
Carissa carandas Linn.	Apocynaceae	8
Corchorus aestuans L.	Tiliaceae	2
Cymbopogon citratus (DC.) Stapf.	Poaceae	5
Cynodon dactylon (L.) Pers.	Poaceae	63
Cyperus rotundus L. Rottb.	Cyperaceae	1
Diospyros montana Roxb.	Ebenaceae	1
Dolichandrone falcata Seem.	Bignoniaceae	7
Dracaena sp.	Asparagaceae	8
Eleusine indica (L.) Geaertn.	Poaceae	4
Ficus hispida L.f.	Moraceae	3
Firmiana colorata (Roxb.) R.Br.	Sterculiaceae	6
Guazuma ulmifolia Lam.	Sterculiaceae	4
Haplophragma adenophyllum (Wall) P.Dop	Bignoniaceae	6
Holoptelea integrifolia (Roxb.) Planch.	Ulmaceae	4
Ipomoea triloba L.	Convolvulaceae	2
Dichanthium annulatum (Forssk.) Stapf.	Poaceae	15
Jatropha gossypifolia L.	Euphorbiaceae	3
Sida acuta Burm.f. Garcke.	Malvaceae	4
Pongamia pinnata (L.)	Papilionaceae	1
Pandanus sp.	Pandanaceae	8
Passiflora foetida L.	Passifloraceae	1
Phyllanthus amarus Schum & Th.	Phyllanthaceae	19
Ricinus communis L.	Euphorbiaceae	4
Scoparia dulcis L.	Scrophulariaceae	5
Syzygium salicifolium (Wight). Graham.	Myrtaceae	2
Teramnus labialis (L. f.) Spreng.	Papilionaceae	1
Vernonia cinerea (L.) Less.	Asteraceae	6
Vetiveria zizanioides (L.) Nash.	Poaceae	7
Albizia saman (Jacq.) F. Muell.	Mimosaceae	1
Sorghum halepense (L.) Pers.	Poaceae	9
Brachiaria ramosa (L.) Stapf.	Poaceae	12
Total		356

Table 13.5 Recorded data of the control community

Plant species	Family	Number of individuals
Acalypha indica L.	Euphorbiaceae	3
Alternanthera sessilis (L.) R.Br.	Amaranthaceae	10
Anogeissus sericea Brandis.	Combretaceae	3
Antigonon leptopus Hk. & Arn.	Polygonaceae	4
Azadirachta indica A. Juss.	Meliaceae	1
Barleria prionitis L.	Acanthaceae	2
Cardiospermum halicacabum L.	Sapindaceae	1
Cynodon dactylon (L.) Pers.	Poaceae	26
Diospyros montana Roxb.	Ebenaceae	1
Guazuma ulmifolia Lam.	Sterculiaceae	2
Holoptelea integrifolia (Roxb.) Planch.	Ulmaceae	2
Corchorus aestuans L.	Malvaceae	4
Phyllanthus amarus Schum & Th.	Phyllanthaceae	6
Ricinus communis L.	Euphorbiaceae	4
Coccinia grandis (L.) Voigt.	Cucurbitaceae	2
Trianthema portulacastrum L.	Aizoaceae	2
Total		73

Fig. 13.13 (**a**) Image of site when we started eco-restoration in March 2021. (**b, c**) Images showing progress in February and March 2022

Table 13.6 Shannon's Index (H) of different studied communities

Community	Shannon's Index (H)
Reference	2.617
Treated	2.912
Control	2.254

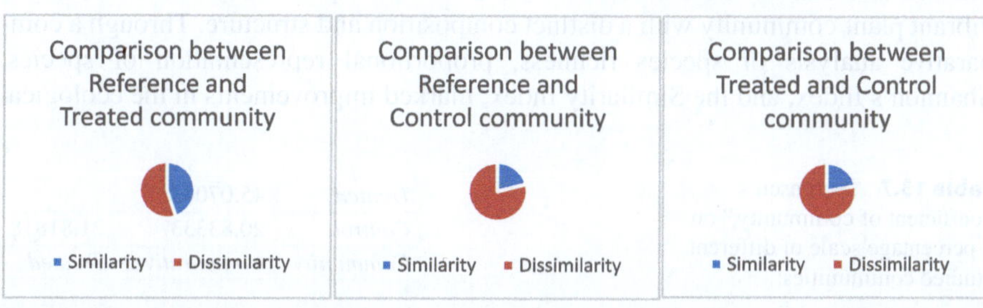

Fig. 13.14 Comparison between different communities

13.12 Sorenson's Coefficient of Community

The comparison between the reference community and the restoration site revealed dissimilarity for more than half of the evaluated criteria on a 100-point scale. The control site exhibited significant dissimilarity when compared to both the reference and restoration sites. Originally, barren land was transformed into a novel community with a targeted species composition. Metrics such as species richness, species proportion, Shannon's Diversity Index, and Similarity Index indicated that the eco-restoration program led to a unique community composition, enhanced biodiversity, and improved structure (Table 13.7).

The restored area demonstrated significantly greater species richness and diversity compared to the neighboring passively restored and native vegetation. This increase in biodiversity is critical, as it enhances ecosystem resilience and functionality, making the area more capable of withstanding environmental stressors.

Both the passively restored "control site" and the actively restored "treated site" effectively mitigated the invasiveness of species such as *Alternanthera sessilis*, *Antigonon leptopus*, and *Ricinus communis*. This success can be attributed to the physical structure and species composition of the newly established community, which provided competitive advantages that limited the establishment of invasive species. Notably, *Dendrocalamus strictus* emerged as a key species in bolstering the restored site's resistance to invasion. Its robust growth and structural complexity likely create a competitive environment that deters invasive species. Similarly, the control site benefited from a physical barrier that reinforced its defense against invasives.

ER is a gradual process that requires the identification of critical indicators to measure effectiveness and guide strategies. These findings underscore the importance of an active restoration approach, which, when combined with thoughtful species selection, can lead to positive ecological outcomes. By fostering diverse plant communities that occupy various ecological niches, we not only enhance biodiversity but also strengthen the ecosystem's ability to resist invasive threats. This has important implications for future restoration projects, suggesting that strategic species selection, planting mix and plans, and active management can yield more resilient and durable ecosystems.

13.13 Discussion

The eco-restoration initiative effectively transformed a previously barren area into a vibrant plant community with a distinct composition and structure. Through a comparative analysis of species richness, proportional representation of species, Shannon's Index, and the Similarity Index, marked improvements in the ecological

Table 13.7 "Sorensen's coefficient of community," on a percentage scale of different studied communities

Treated	45.07042	
Control	20.83333	21.81818
Community	*Comparative*	*Treated*

status of the actively restored site were evident when contrasted with the passively restored control site.

The treated site displayed a unique assemblage of species, resulting from targeted interventions like planting and habitat modification. Higher species richness and diversity indices indicate that active restoration strategies are more effective at rapidly enhancing ecosystem functions and biodiversity than passive, succession-driven approaches. In contrast, the control site, which experienced natural succession without intervention, remains in the early stages of ecological development, with considerable portions still barren. While this area holds potential for recovery, its future trajectory and community composition remain uncertain without further human involvement.

Invasive species such as *Alternanthera sessilis*, *Antigonon leptopus*, and *Ricinus communis* were found in higher proportions and larger sizes in both sites, posing threats to native species establishment. Nevertheless, both areas exhibited resilience against these invasives, indicating that the structural and compositional attributes of the newly formed communities played a vital role in limiting their spread.

The treated site demonstrated a more resilient ecological structure, likely due to the intentional selection of competitive species like *Bambusa arundinacea*. This species may have created physical barriers and competitive exclusion, thereby hindering the establishment of invasive species. Similarly, the passive restoration site appears to be in the initial phases of succession, where its future community composition will depend on natural processes and the gradual colonization by native species. Despite their different methodologies, both sites showed evidence of restricting the spread of invasive species, highlighting the importance of ecological resilience in restored habitats.

A layered planting strategy facilitated a multitiered vegetation structure, creating diverse ecological niches and promoting overall diversity. Additionally, the spatial arrangement fostered microhabitats, which further enhanced habitat heterogeneity and species richness. While the inclusion of species such as *Bambusa arundinacea* and other fast-growing, structurally dominant plants served ecological and structural purposes, these hindered the experience of the sight and sounds of the flowing nālā below and the surroundings.

The differences between the active and passive restoration sites provide valuable insights into QER measures, demonstrating how thoughtful ecological and spatial interventions can lead to more rapid, durable, and desirable ecological outcomes. This approach could serve as a model in urban or semi-urban contexts, balancing ecological restoration with human engagement and aesthetic values, such as creating green spaces for recreation, carbon sequestration, and supporting urban biodiversity.

13.14 Conclusion

The QER experiment successfully rehabilitated an over-dumped, barren area into a biodiverse plant community, illustrating the effectiveness of active restoration methods in improving species richness, diversity, and ecological resilience. The treated site exhibited higher species richness and community stability compared to both the passively restored control site and the surrounding native vegetation, suggesting that active intervention accelerates ecological recovery.

Both actively and passively restored areas demonstrated resistance to invasive species like *Alternanthera sessilis*, *Antigonon leptopus*, and *Ricinus communis*. The resilience observed in the treated site may be linked to its species composition, particularly the presence of *Bambusa arundinacea*, which likely contributed to forming a stable, invasion-resistant community. The architectural design of the vegetation, including spatial zoning and species layering, further reinforced this stability, underscoring the interconnectedness of structural diversity and ecological function. In contrast, the passive restoration site remains in a successional phase, with future community development reliant on natural processes. Although invasive species are present, the ecological trajectory of this site remains uncertain. Continued monitoring is essential to evaluate the long-term effectiveness of passive restoration in the long-term maintenance of biodiversity and preventing invasive species colonization.

From a landscape design viewpoint, this study emphasizes the need to merge ecological principles with spatial design considerations to achieve durable restoration outcomes. The design of the treated site not only bolstered biodiversity but also offered aesthetic and functional advantages that can serve as a model for future restoration projects in degraded streambanks. However, we acknowledge that we have revegetated the slopes still with debris and solid wastes. It sends the wrong message to the university community and the larger society. Though it is not an ideal practice, it was a pragmatic decision made for educational purposes, given the reality of many constraints.

The collaborative and interdisciplinary QER approach presented in this study serves as a blueprint for converting degraded environments into resilient, biodiverse ecosystems. Such local-level action must be conceived at and complemented with regional-level policies and plans. Restoration should be conceived at a watershed level to make cumulative global impacts. Since water travels downhill, ecological planning and restoration ought to be coordinated and implemented systematically from the upper parts of sub-watersheds and go downstream to ensure long-term success. Also, restoration decisions and design details affect the spatial characteristics, which, in turn, affect the experiential (mainly sensory) quality of the resulting ecosystem. Aesthetic appeal ("beauty") is elemental for ER and for humanity to exist and thrive in the larger natural order (Oelschlaeger 2007).

By adapting eco-engineering and thoughtfully selecting species to create complex communities, it is feasible to restore biodiversity, mitigate the effects of invasive species, and establish landscapes that support long-term ecological stability, aesthetic experience, and the well-being of all species. Over the last few decades, such efforts have been well recorded for us to learn from, adapt, test, and share new lessons. Ongoing research and long-term monitoring will be vital for understanding the full impact of these restoration efforts on ecosystem function and resilience, as well as their applicability in various ecological and urban contexts.

By managing our relationship with the ecosystems, we can ensure that they continue to provide essential benefits for all. This way we can cherish this wonderful planet Earth and champion its renewal and ensure the survival of all species.

Acknowledgments In this collaborative restoration work, individuals and large groups both played critical roles. We are grateful to Nurul, Kamartaha Saiyed, Agardeep, Atul, the Botanical Garden staff, the Green Army (made up of the Faculty of Science, mainly Botany Department, students), University officials, the Botany Head, the Dean of the Faculty of Science, and the collaborators from a professional landscape architecture firm, *saa..synergies:* Annagrace Babu, Sharan Vyas, Bhoomi Shah, Purva Mistry, Tatsat Shah, and Shivangi Patel.

References

Ahmad M, Ahmad WS, Ahmad SN et al (2024) Tracing the roots of wetland degradation in India: a systematic review of anthropogenic drivers, ecological consequences and conservation strategies. GeoJournal 89:24. https://doi.org/10.1007/s10708-024-10997-9

Aradóttir ÁL, Hagen D (2013) Ecological restoration: approaches and impacts on vegetation, soils and society. Adv Agron 120:173–222. https://doi.org/10.1016/B978-0-12-407686-0.00003-8

Bhattacharjee A (2020) Forest landscape restoration as a NbS strategy for achieving Bonn Challenge pledge: lessons from India's restoration efforts. In: Dhyani S, Gupta AK, Karki M (eds) Nature-based solutions for resilient ecosystems and societies. Disaster resilience and green growth. Springer, Singapore, pp 133–147. https://doi.org/10.1007/978-981-15-4712-6_8

Bonham C (1989) Measurements of terrestrial vegetation. Wiley, New York

Borah B, Bhattacharjee A, Ishwar NM (2018) Bonn Challenge and India: progress on restoration efforts across states and landscapes. IUCN and MoEFCC, Government of India, viii + 32 pp, New Delhi. https://doi.org/10.2305/IUCN.CH.2018.12.en

Dey D, Bauri U, Tikait D (2020) Riverine plants of Kansai Basin at Purulia District West Bengal & how they prevent soil erosion and corrosion. IOSR J Environ Sci Toxicol Food Technol 14(5):1–13

Government of India (2022) India's updated first Nationally Determined Contribution under Paris Agreement (2021–2030). GOI

Higgs E (1997) What is good ecological restoration? Conserv Biol 11(2):338–348

Higgs E (2003) Nature by design: people, natural processes, and ecological restoration. MIT Press, Boston

Higgs ES, Harris J, Murphy S, Bowers K, Hobbs R, Jenkins W, Kidwell J, Lopoukhine N, Sollereder B, Suding K, Thompson A, Whisenant S (2018) On principles and standards in ecological restoration. Restor Ecol 26(3):399–401

Hubble TCT, Docker BB, Rutherfurd ID (2010) The role of riparian trees in maintaining riverbank stability: a review of Australian experience and practice. Ecol Eng 36(3):292–304. https://doi.org/10.1016/j.ecoleng.2009.12.006

Kumar R, Ambasht RS, Srivastava AK, Srivastava NK (1996) Role of some riparian wetland plants in reducing erosion of organic carbon and selected cations. Ecol Eng 6(4):227–239. https://doi.org/10.1016/0925-8574(95)00059-3

Li M-H, Eddleman K (2002) Biotechnical engineering as an alternative to traditional engineering methods: a biotechnical streambank stabilization design approach. Landsc Urban Plan 60(4):225–242. https://doi.org/10.1016/S0169-2046(02)00057-9

Munjal M, Behal D (2011) Urban ecosystem changes around natural drains: case of Becharji Nālā of Vishwamitri River in Vadodara City. Int J Soc Sci 9:313–322. https://doi.org/10.30954/2249-6637.04.2020.14

Oelschlaeger M (2007) Ecological restoration, Aldo Leopold, and beauty. Environ Philos 4(1 & 2):149–162

Poulin M, Andersen R, Rochefort L (2013) A new approach for tracking vegetation change after restoration: a case study with peatlands. Restor Ecol 21:363–371. https://doi.org/10.1111/j.1526-100X.2012.00889.x
Raval SR (2023) Ecological Restoration. Journal of Landscape Architecture (New Delhi) 75: 46–47
Society for Ecological Restoration (2021) Ten principles underpin good ecosystem restoration throughout the United Nations Decade 2021–2030. A Joint Press Release dated Tuesday, 21 Sept 2021. Accessed from: https://www.ser.org/news/579457/Ten-Principles-Underpin-Good-Ecosystem-Restoration-throughout-the-United-Nations-Decade-2021-2030.htm
Srivastava NK, Lal R, Regnald M (2009) Role of selected riparian herbs in reducing soil erosion and nutrient loss under simulated rainfall. Environmental Earth Sciences 61(2):405–417. https://doi.org/10.1007/s12665-009-0353-5
United Nations Environment Programme (2021) Becoming #GenerationRestoration: ecosystem restoration for people, nature and climate. UNEP/FAO, Nairobi
USDA–NRCS United States Department of Agriculture Natural Resources Conservation Service (2007) Part 654 National Engineering Handbook: Stream Restoration Design. Washington DC
Weinstein MP, Litvin SY, Krebs JM (2014) Restoration ecology: ecological fidelity, restoration metrics, and a systems perspective. Ecol Eng 65:71–87
Zhang T et al (2020) Probable pangolin origin of SARS-CoV-2 associated with the COVID-19 outbreak. Curr Biol 30:1346–1351

Correction to: An Experiment in Quasi-Eco-Restoration of a Creek Slope: A Collaborative Approach on a University Campus in India

Padamnabhi S. Nagar, Shishir R. Raval, Haribhai R. Kataria, Ankur Rajwadi, Snehal Chavda, Jaydeep J. Sharma, Gulzar Malek, Roshan D. Virola, and Tanmay Rohit

Correction to:
Chapter 13 in: D. Mittal, D. K. Gupta (eds.),
Ecological Restoration in India: Perspective and Practices,
https://doi.org/10.1007/978-981-96-8943-9_13

The original version of this chapter was inadvertently published with errors. The following corrections have been made after original publication.

1. The surname of the last author has been updated to read as Tanmay Rohit.
2. The acknowledgement section has been updated as below:

In this collaborative restoration work, individuals and large groups both played critical roles. We are grateful to Nurul, Kamartaha Saiyed, Agardeep, Atul, the Botanical Garden staff, the Green Army (made up of the Faculty of Science, mainly Botany Department, students), University officials, the Botany Head, the Dean of the Faculty of Science, and the collaborators from a professional landscape architecture firm, saa..synergies: Annagrace Babu, Sharan Vyas, Bhoomi Shah, Purva Mistry, Tatsat Shah, and Shivangi Patel.

The updated version of this chapter can be found at
https://doi.org/10.1007/978-981-96-8943-9_13

D. Mittal, D. K. Gupta (eds.), *Ecological Restoration in India: Perspective and Practices*, https://doi.org/10.1007/978-981-96-8943-9_14

MIX
Papier aus verantwortungsvollen Quellen
Paper from responsible sources
FSC® C105338

If you have any concerns about our products,
you can contact us on
ProductSafety@springernature.com

In case Publisher is established outside the EU,
the EU authorized representative is:
Springer Nature Customer Service Center GmbH
Europaplatz 3, 69115 Heidelberg, Germany

Printed by Libri Plureos GmbH
in Hamburg, Germany